The 'Ecosystem Approach' in International Environmental Law

The ecosystem approach, broadly understood as a legal and governance strategy for integrated environmental and biodiversity management, has been adopted within a wide variety of international environmental legal regimes and provides a narrative, a policy approach and in some cases legally binding obligations for States to implement what has been called a 'new paradigm' of environmental management. In this last respect, the ecosystem approach is also often considered to offer an opportunity to move beyond the outdated anthropocentric framework underpinning much of international environmental law, thus helping re-think law in the Anthropocene.

Against this background, this book addresses the question of whether the ecosystem approach represents a paradigm shift in international environmental law and governance, or whether it is in conceptual and operative continuity with legal modernity. This central question is explored through a combined genealogical and biopolitical framework, which reveals how the ecosystem approach is the result of multiple contingencies and contestations, and of the interplay of divergent and sometimes irreconcilable ideological projects. The ecosystem approach, this book shows, does not have a univocal identity, and must be understood as both signalling the potential for a decisive shift in the philosophical orientation of law and the operationalisation of a biopolitical framework of control that is in continuity with, and even intensifies, the eco-destructive tendencies of legal modernity. It is, however, in revealing this disjunction that the book opens up the possibility of moving beyond the already tired assessment of environmental law through the binary of anthropocentrism and ecocentrism.

Vito De Lucia is a PostDoc Fellow at the K. G. Jebsen Centre for the Law of the Sea, Faculty of Law, UiT The Arctic University of Norway.

Part of the Law, Justice and Ecology series

Series Editor: Anna Grear
Law School, Cardiff University, UK

for information about the series and details of previous and forthcoming titles, see https://www.routledge.com/Law-Justice-and-Ecology/book-series/LAWJUSTECO

A GlassHouse Book

The 'Ecosystem Approach' in International Environmental Law

Genealogy and Biopolitics

Vito De Lucia

Routledge
Taylor & Francis Group
a GlassHouse book

First published 2019
by Routledge
2 Park Square, Milton Park, Abingdon, Oxon OX14 4RN

and by Routledge
52 Vanderbilt Avenue, New York, NY 10017

a GlassHouse book

Routledge is an imprint of the Taylor & Francis Group, an informa business

British Library Cataloguing-in-Publication Data
A catalogue record for this book is available from the British Library

Library of Congress Cataloging-in-Publication Data
Names: De Lucia, Vito, author.
Title: The 'ecosystem approach' in international environmental law: genealogy and biopolitics/Vito De Lucia.
Description: New York: Routledge, 2019. |
Series: Law justice and ecology | Includes index.
Identifiers: LCCN 2018052716 (print) | LCCN 2018056110 (ebook) | ISBN 9781315150772 (ebk) | ISBN 9781138557260 (hbk)
Subjects: LCSH: Environmental law, International. | Ecosystem management.
Classification: LCC K3585 (ebook) | LCC K3585. D38 2019 (print) | DDC 344.04/6–dc23
LC record available at https://lccn.loc.gov/2018052716

ISBN: 978-1-138-55726-0 (hbk)
ISBN: 978-1-315-15077-2 (ebk)

Typeset in Galliard
by Integra Software Services Pvt. Ltd.

Contents

Preface

This book is based on my PhD dissertation, which I defended in April 2016. While writing a PhD dissertation is to a significant extent a lonely endeavour, the work that led to this book has benefited from the crucial support, formal and informal, material and moral, of many people. I wish to extend my gratitude firstly to my two supervisors: Anna Grear, for showing unwavering faith throughout, and for being a mentor and a friend; and Tore Henriksen, for helping me stay down to earth and on course throughout. I am also very grateful to the assessment committee members, Christina Allard, Inger Johanne Sand and Karen Morrow.

The PhD would not have been possible without the financial support of the Norwegian Research Council, and the project "Bærekraftig utvikling av havområder i nord. Muligheter og trusler". The K.G. Jebsen Center for the Law of the Sea and the Faculty of Law, UiT The Arctic University of Norway, made it all possible, by offering excellent material conditions for doing research, and a stimulating academic and social environment. I wish to thank them in the persons of Tore Henriksen, the Center Director, and Hege Brækhus and Trude Haugli, successive Deans of the Faculty during my doctoral fellowship. A big thank you also to Christin Skjervold (for all the patient support, administrative and otherwise).

I also need to thank all my past and present colleagues at the K. G. Jebsen Center of the Law of the Sea and at the Faculty of Law, UiT The Arctic University of Norway. While all in different ways have offered significant contributions, I wish to thank some in particular, though in no particular order: Anna Nylund, Christina Allard, Signe Bush, Maria Madalena das Neves, Jussi Pedersen, Kristoffer Svendsen, Vegard Helland and Roger Stelander Magnussen. I wish also to thank Raul Primicerio, Svein Anders Noer Lie, Riccardo Baldissone and Andreas Kotsakis for endless conversations on ecology, philosophy, law, Italian theory, Foucault and genealogy. A special thank goes to Elise Johansen, for a crucial reminder at a critical juncture: it is possible to write a PhD and still have time for one's family.

The transformation of the PhD dissertation into this book (which is a little over half the original length and has an entirely new chapter on genealogy) has also benefited from the support and input of several people. I wish to thank in this respect especially the anonymous reviewers who helped shape the book in its current form with many insightful comments and suggestions. I am also grateful to my research assistant, Daniela Toma, whose patient work was invaluable for the final preparation of the manuscript.

Finally, I wish to thank my family: Karoline, my wife, and my children, Gylve, Tellef and Elvira (Dad is finally done with "the book"!). It is to them, and to the memory of my parents Giuseppe and Gaetana, that I dedicate this work.

Kvaløysletta, 7 October 2018
Vito De Lucia

Part I

Setting the stage

1 Setting the stage

Introduction: time of crisis. Ecology, law and the Anthropocene

We live in a time of ecological emergency. Indeed, unfolding ecological crises intersect and overlap at multiple scales and across multiple domains. The Millennium Ecosystem Assessment, a particularly authoritative source, speaks of 'substantial and largely irreversible loss in the diversity of life on Earth', and finds that 'approximately 60% ... of the ecosystem services examined ... are being degraded or used unsustainably, including fresh water, capture fisheries, air and water purification, and the regulation of regional and local climate, natural hazards, and pests'.[1] Even more dramatically, UNEP's 5th Geo Report emphatically articulates its main message thus: '[t]he scale, spread and rate of change of global drivers are without precedent. Burgeoning populations and growing economies are pushing environmental systems to destabilizing limits'.[2]

The world's fisheries are in a state of decline, with over 80 per cent[3] (and as much as 90 per cent)[4] of the fish stocks for which assessment information is available reported as being fully exploited or overexploited, and almost 30 per cent of capture fisheries considered to be overexploited, depleted or recovering from depletion.[5] However, even these dismal figures may be optimistic, according to some critics of the methodology that the FAO used to arrive at the preceding figures.[6] Other models provide far gloomier figures, putting the

1 Millennium Ecosystem Assessment, *Ecosystems and Human Well-being: Synthesis*, Washington, DC: Island Press, 2005, p. 1. Similar messages are reported in S. Butchart et al., 2010, 'Global Biodiversity: Indicators of Recent Declines', 328:5982, *Science* 1164; and Food and Agricultural Organization of the United Nations (FAO), *State of the World's Forests*, 2011.

2 UNEP, Global Environment Outlook 5: Environment for the Future We Want, UN Doc. DEW/1417/NA, 2012a, p. 4.

3 Ibid.

4 O. Hoegh-Guldberg et al., 2015, *Reviving the Ocean Economy: The case for action – 2015*, Geneva, WWF International, p. 26.

5 Food and Agriculture Organization of the United Nations (FAO), *The State of World Fisheries and Aquaculture. Opportunities and Challenges*, FAO, 2014, p. 37.

6 See e.g. World Ocean Review, *The Future of Fish – The Fisheries of the Future*, Maribus, 2013, pp. 43–5.

percentage of depleted or overexploited fish stocks at over 56 per cent.[7] Regardless of the static image of how things are, however, there is broad agreement on the dynamic trend, which is towards a global worsening of the situation of most fish stocks.[8] Perhaps, then, it is not far-fetched to entitle an editorial as dramatically as 'the end of fish'.[9]

In addition to this already dramatic picture, global warming and the ensuing climatic changes – both present and future – exacerbate every single one of the environmental problems mentioned, and create new ones at every imaginable temporal and spatial scale. Global atmospheric concentrations of carbon dioxide, methane and other greenhouse gases (GHGs) have increased markedly because of human activities since the beginning of the industrial revolution, and now significantly exceed pre-industrial levels.[10] The global increases in the concentration of GHGs 'are due primarily to fossil fuel use and land use change, while those of methane and nitrous oxide are primarily due to agriculture', as the Fourth Assessment Report of the Intergovernmental Panel on Climate Change (IPCC) reported.[11]

The Fifth Assessment report of the IPCC sharpened its message, stating that '[w]arming of the climate system is unequivocal, and since the 1950s, many of the observed changes are unprecedented over decades to millennia. The atmosphere and ocean have warmed, the amounts of snow and ice have diminished, sea level has risen, and the concentrations of greenhouse gases have increased'.[12] Moreover, '[e]ach of the last three decades has been successively warmer ... than any preceding decade since 1850', and the period 1983–2012 'was likely the warmest 30-year period of the last 1400 years' in the Northern

7 Also, FAO uses (still) maximum sustainable yield (MSY) as the measure of what it calls 'biologically sustainable' levels of fishing. MSY is in itself a contested measure and may hide significantly the level of overexploitation of current stocks. Indeed, FAO also recognizes how its 'results are based on single-species assessments and it is ecologically impossible to harvest all species at the MSY level simultaneously. Therefore, some stocks may need to have their abundance maintained above the MSY level to avoid ecosystem overfishing.' FAO, *State of World Fisheries*, p. 41.

8 See e.g. Hoegh-Guldberg et al., *Reviving the Ocean Economy*, and World Ocean Review, *Future of Fish*; FAO, *State of World Fisheries.* considers fishstock productivity to be stagnating (or even 'stable'), but this has been heavily criticized by e.g. D. Pauly and R. Froese, 'Comments on FAO's State of Fisheries and Aquaculture, or SOFIA 2010', 36:3 *Marine Policy* 2012, 746–7, which speak of an actual 'strong biomass decline'.

9 A. Novogratz and M. Velings, 'The end of fish', *Washington Post*, 3 June 2014, http://www.washingtonpost.com/posteverything/wp/2014/06/03/the-end-of-fish/ (Accessed 26 September 2018).

10 IPCC, 'Summary for Policymakers' in S. Solomon et al. (eds), *Climate Change 2007: The Physical Science Basis, Contribution of Working Group I to the Fourth Assessment Report of the Intergovernmental Panel on Climate Change*, Cambridge University Press, 2007, p. 2.

11 Ibid.

12 IPCC, 'Summary for Policymakers', in T. Stocker et al. (eds), *Climate Change 2013: The Physical Science Basis. Contribution of Working Group I to the Fifth Assessment Report of the Intergovernmental Panel on Climate Change*, Cambridge University Press, 2013, p. 4.

Hemisphere.[13] Working Group II, furthermore, issued a specialized report dedicated to impacts, adaptations and vulnerability.[14] This report described the wide range of impacts already observed, whose severity is set to increase due to a momentum already built into the climate system (due to past emissions whose effects have not yet materialized). Such observed effects include an increase in the severity and frequency of extreme weather events; sea level rise (which is already threatening low-lying Pacific islands States); alteration of hydrological systems, affecting water resources in terms of quantity and quality and negative effects on crop yields.[15] Moreover, each of these effects is poised to be further exacerbated in the future, and to compound the effects on the world's ecosystems due to non-climatic stressors. It may be useful to recall that, in a recent address, climatologist James Hansen (Director of the NASA Goddard Institute for Space Studies) reiterated the warning that we find ourselves in a 'planetary emergency'.[16] UN Secretary General, Ban-ki Moon, addressing the High-level Plenary meeting at the recent High-level Segment of the 18th Conference of the Parties to the UN Framework Convention on Climate Change, was equally clear, as he tried to urge the Parties to move forward: '[l]et us be under no illusion. This is a crisis.'[17]

The inadequacy of environmental law in the Anthropocene

The primary response of law to contemporary ecological crises takes the form of environmental law. Environmental law, as a separate branch or field of law, emerged in the early 1970s[18] and has enjoyed sweeping success, at least if judged by the growth of environmental legislation at all levels of legal articulation. Environmental law however, legal theory increasingly suggests,

13 Ibid., p. 5.

14 IPCC, 'Summary for policymakers', in C. Field et al. (eds), *Climate Change 2014: Impacts, Adaptation and Vulnerability. Part A: Global and Sectoral Aspects. Contribution of Working Group II to the Fifth Assessment Report of the Intergovernmental Panel on Climate Change*, Cambridge University Press, 2014, pp. 1–32.

15 Ibid., p. 4

16 Hansen has offered such a warning consistently in the last years. The latest occasion (at the time of writing) was a lecture given at the 'Energy Emergency, Energy Transition' global trade union roundtable, convened by Cornell University's Global Labor Institute (GLI), a program of the Worker Institute at Cornell, and the Rosa Luxemburg Foundation, and held on 10 October 2012. The lecture is available at: http://www.youtube.com/watch?v=-dbmqhon5TY (Accessed 26 September 2018).

17 UN News Center, 'Doha meeting must take decisive action to tackle growing crisis of climate change', 4 December 2012, http://www.un.org/apps/news/story.asp?NewsID=43671&C#.UMDNsoN2PwA (Accessed 26 September 2018).

18 Though in some accounts environmental law has gone through two development phases: the first 'classic' phase spans the period 1850s to, roughly, the 1960s; the second phase begins in the early 1970s and is the 'sustainable development' phase. Thus J. Holder, 2000, 'New Age: Rediscovering Natural Law', in M. D. A. Freeman, ed., *Current Legal Problems*, 53:1, Oxford University Press, pp. 165–7.

suffers from decisive structural inadequacies that prevent it from meaningfully addressing the contemporary ecological emergencies outlined in the previous section.[19] Anna Grear suggests in this respect, that law (law in general, as a practice and as a theory that encompasses, but exceeds environmental law) has 'failed thus far to respond in any way that really counts'.[20] Klaus Bosselmann similarly observes that *environmental* law merely mitigates the ecological damages deriving from industrial activities.[21] While certainly able to save some trees, observes further Bosselmann, environmental law is losing the forest, an observation echoed by other scholars in similar, if not equivalent terms, and with even more force.[22] It is in this sense then, that environmental law, while responding to a number of immediate problems, has not done so in any way that really counts.

Indeed, environmental law has arguably predominantly functioned as a mitigating instrument whose strategic aim is that of *containment*: the ecological disruption of economic and industrial activities is to be mitigated, contained or otherwise reduced *to the extent possible*, and in a cost-effective manner, in line with liberal environmentalism.[23] The questions raised within the context of environmental legal and policy discourse are not so much about *whether* certain activities should be carried out, but rather *how, where* and *under which conditions.*[24] Incorporating best available technologies (BATs), best environmental practices (BEPs), and emphasizing the internalization of environmental costs, environmental law maintains for the most part a 'business as usual' tone[25] and a largely procedural character. This can perhaps be usefully understood as a *merely descriptive* process of legal regulation, where the threshold between legal and illegal is demarcated on the basis of feasibility, a feasibility routinely captured in legal instruments and legal texts

19 See, among a growing literature, M. M'Gonigle and L. Takeda, 2013, 'The Liberal Limits of Environmental Law: A Green Legal Critique', 30:3 *Pace Environmental Law Review*, 1005; K. Bosselmann, 'Losing the Forest for the Trees: Environmental Reductionism in the Law', 2:8 *Sustainability* 2010, 2424; Holder, 'New Age'.

20 A. Grear, 'Towards a New Horizon: in Search of a Renewing Socio-Juridical Imaginary', 3:5 *Oñati Socio-Legal Series* 2013a, 966,.970.

21 Bosselmann, 'Losing the Forest'.

22 See in particular M'Gonigle and Takeda, 'Liberal Limits', especially section IV; see also M. Wood, 'Advancing the Sovereign Trust of Government to Safeguard the Environment for Present and Future Generations (Part I): Ecological Realism and the Need for a Paradigm Shift', 39:1 *Environmental Law Review* 2009, 43.

23 S. Bernstein, *The Compromise of Liberal Environmentalism*, New York: Columbia University Press, 2001.

24 M. Hasley, 'Majesty and Monstrosity: Delueze and the Defence of Nature' in A. Philippopoulos-Mihalopoulos (ed.), *Law and Ecology. New Environmental Foundations*, London: Routledge (a Glasshouse Book) 2011, p. 224, where he refers in particular to the debate about uranium mining in relation to a particular World Heritage site.

25 Holder, 'New Age', p. 167.

with formulations such as, e.g., 'as far as possible' and 'as appropriate'[26] and 'cost effective',[27] or through references to prevailing and generally accepted standards.[28] And while these open-ended references allow the adaptation of legal principles and regimes to changing technical practices and technological innovations, environmental regulation remains a mirror of the practices of the regulated actors and environmental law acts as recipient of available norms established in other functional realms, such as technology, economics and science, repackaged in a legal vocabulary.[29]

Environmental law, legal modernity and anthropocentrism

Environmental law's key problematics are a reflection or a consequence of the broader legal ideology within which it emerged. In this sense, environmental law inevitably reproduces the key thresholds, concepts and categories of legal modernity, and its underlying worldview.[30] There is abundant literature outlining, describing and exploring the ways in which legal modernity and its categories affect environmental law in multiple ways.[31] However, a very brief outline of the ways in which environmental law inherits key thresholds, concepts and categories will prove useful for the purposes of this book. The expression legal modernity refers, in the context of this book, to the particular legal inflection or articulation of modernity as a socio-cultural formation. While modernity itself is

26 See e.g. Articles 5, 6, 7, 8, 9, 10, 11 and 14 of the UN Convention on Biological Diversity (adopted 5 June 1992, entered into force 29 December 1993) 1760 UNTS 79.

27 See e.g. United Nations Framework Convention on Climate Change (adopted on 9 May 1992, entered into force on 21 March 1994) 1771 UNTS 107, which at Article 3(3) states that 'policies and measures to deal with climate change should be cost-effective'.

28 See e.g. United Nations Convention on the Law of the Sea (adopted 10 December 1982, entered into force 16 November 1994), 1833 UNTS 3; Articles 61 and 119, or Part XII, section 5.

29 B. Lange, 'Focuauldian-inspired Discourse Analysis: A Contribution to Critical Environmental Scholarship? ' in A. Philippopoulos-Mihalopoulos (ed.), *Law and Ecology: New Environmental Foundations*, London: Routledge Glasshouse Books, 2011. To illustrate this in practice, Lange uses the example of the administrative procedure of authorization of transgenic agricultural products in the EU. More broadly on the relation between law and the norm from a socio-legal perspective, see F. Ewald, 'Norms, Discipline, and the Law', 30, *Representations Special Issue: Law and the Order of Culture*, 1990, 138.

30 For a more detailed discussion of such thresholds, concepts and categories see, e.g. V. De Lucia, 'Re-embodying Law: Transversal Ecology and the Commons' in R. Thomas-Pellicer, V. De Lucia and S. Sullivan S. (eds), *Law, Philosophy, Ecology: Exploring Re-embodiments*, Routledge, Glasshouse Books, 2016.

31 See e.g. De Lucia, 'Re-embodying Law'; A. Grear, 'The Vulnerable Living Order: Human Rights and the Environment in a Critical and Philosophical Perspective', 2:1 *Journal of Human Rights* 2011, 23; Grear, 'New Horizon'; L. Godden, Nature as Other: The Legal Ordering of the Natural World: PhD Thesis, Queensland: Faculty of Law, Griffith University, 2000; M'Gonigle and Takeda, 2013, 'Liberal Limits'; P. Fitzpatrick, *Mythology of Modern Law*, London, New York: Routledge, 1992.

a very complex and contested concept,[32] with temporal, cultural and theoretical framing traversing and stretching its semantic space and conceptual articulations,[33] I take modernity to indicate a particular cultural formation characterized by a set of specific ontological, epistemological and axiological commitments. These commitments, in turn, have a fundamental role in shaping law, as we shall briefly see in the remainder of this section.

A central element of modernity is arguably the Cartesian separation between mind and matter.[34] This binary idea has simultaneously ontological, epistemological and axiological implications, and underpins a series of modulations and permutations that are translated into a series of binaries. Thus, modernity arrays the world in subjects and objects, cultures and natures, wilful actions and mechanistic motions, rational minds and material bodies. In turn, these binaries characterize all the central categories of *legal* modernity. Law, moreover, has historically operationalized these binaries (and it has often legitimized them, given the particular authoritative role law has within the cultural context of modernity). Generally, law and culture are arguably entangled,[35] and are mutually constitutive.[36] Law itself is a particular 'cultural artefact',[37] that can only be understood by way of reference to the broader socio-cultural formation within which law, or a particular idea of law, emerges.[38] Conversely, to understand a socio-cultural formation 'one cannot fail to consider law'.[39] As such, '[l]aw is a major articulation of a culture's self-concept, representing the theory of society and environment within that culture'.[40] Moreover, as part of

32 B. De Sousa Santos, *Toward a New Common Sense. Law, Science and Politics in a Paradigmatic Transition*, New York, London: Routledge, 1995.

33 See e.g. P. Osborne, 'Modernity is a Qualitative, Not a Chronological, Category', in F. Barker, P. Hulme and M. Iverson (eds), *Postmodernism and the Rereading of Modernity*, Manchester: Manchester University Press, 1992; M. Foucault, 'What is Enlightenment?' in P. Rabinow (ed.), *Ethics: Subjectivity and Truth (Essential Works of Foucault 1954–1984, volume 1)*, City of Westminster: Penguin Books, 2000a, p. 309; A. Hornborg, *The Power of the Machine, Global Inequalities of Economy, Technology, and Environment*, Walnut Creek, CA: Altamira Press, 2001, p. 161; G. De Anna, 'Modernità e Immanenza: l'Azione Umana in Tommaso D'Aquino e Thomas Hobbes' in L. Parisoli (ed.), *Il Soggetto e la Sua Identità. Mente e Norma, Medioevo e Modernità*, Palermo: Officina di Studi Medievali, 2010.

34 Grear, 'New Horizon'.

35 L. Rosen, 2008, *Law as Culture: An Invitation*, Princeton, NJ: Princeton University Press; J. C. Smith and D. N. Weisstub, *The Western Idea of Law*, London: Butterworth-Heinemann, 1984.

36 Law and culture are mutually constitutive. As Clifford Geertz argues, Law produces a specific vision of a community, and not just an echo of it; C. Geertz, *Local Knowledge: Further Essays in Interpretive Anthropology*, London: Fontana Press, 1983, p. 218. See also R. Brooks, R. Jones and R. Virginia, *Law and Ecology: The Rise of the Ecosystem Regime*, Aldershot: Ashgate, 2002.

37 H. Petersen, *Gender and Nature in Comparative Legal Cultures*, in D. Nelken (ed.), *Comparing Legal Cultures*, Aldershot: Dartmouth, 1997, p. 142.

38 Rosen, *Law as Culture*, pp. 6–7. See also, on this esp. P. Grossi, *L'Ordine Giuridico Medievale*, Laterza, 2006.

39 Rosen, *Law as Culture*, pp. 6–7.

40 Smith and Weisstub, *Western Idea*, London: Butterworth-Heinemann, 1984, p. vii.

a 'distinctive way of imagining the real',[41] law produces and shapes the world in particular ways. Law is thus not merely descriptive, but also operational and *performative*.[42] The reference to performativity highlights, importantly, the 'constitutive power'[43] of law; that is, 'its capacity to create the meaning by which people understand the social environment in which they live, and their place in it'.[44] As law *categorizes* and *orders*, it defines the boundaries and the forms of the real.[45] Yet, somewhat circularly, these operations are fully enmeshed in a particular cultural formation, they *always already* imply a commitment to a prior ontological belief: as law participates in the production of the world (a *certain* world), it also simultaneously re-produces *certain* ontological premises.[46]

In the environmental legal field, the operationalization of legal modernity and its binary worldview has traditionally taken policy shape through tools such as protected areas. Such tools have enacted the central split between nature and human culture, but also another permutation of that binary split, one between chaos and order. The protection that environmental law affords the 'majestic' aspects of nature rely upon acts of partition and classification in which 'islands of wildness ... are conceivable only on the basis of an ongoing and generalized ecological violence'.[47] This, Hasley suggests, is a dialectic between monstrous and majestic that traverses the entirety of environmental law.[48] Thus environmental law has arguably functioned as a legitimating mechanism of the ontological thresholds underpinning legal modernity, operating as it does as a 'force

41 Geertz, *Local Knowledge*, p. 184.

42 Thus, e.g. C. Grzegorczyk, 'Le concept de bien juridique: l'impossible définition?', 24 *Archives de philosophie du droit*, 1979, 259. Others have characterized the autonomous force of law as discursive or productive (see e.g. G. Frankenberg, 'Critical Comparison: Re-thinking Comparative Law', 26 *Harvard International Law Journal* 1985, 411), but the attribute 'performative' seems most adequate to emphasize not only the discursive force of law, or its capacity to 'produce' (ideas, practices of inclusion/exclusion, worlds) but that of a particular willful enactment (in the artistic sense) of a scenography and/or choreography. The performance of law thus is in close proximity to the capitalist spectacle (see in this respect G. Debord,*The Society of the Spectacle*, London: Black and Red, 1997).

43 R. Cotterrel, 'Law in Culture', 17:1 *Ratio Juris* 2004, 3.

44 Ibid. As Cotterrel observe the relationship between law and culture can be – and is – articulated in multiple ways in legal scholarship. He identifies six such ways: (1) law's dependence on culture; (2) law's recognition of culture; (3) law's domination of culture; (4) law as an object of cultural struggle; (5) law as a cultural projection; (6) law's stewardship of culture.

45 M. Davies, *Asking the Law Question*, 3rd edn, Lawbooks, 2008, esp. chapter 1. See also more specifically D. Delaney, 'Making Nature/Marking Humans: Law as a Site of (Cultural) Production', 9:3 *Annals of the Association of American Geographers* 2004, 487 and Godden, *Nature as Other*.

46 See C. Grzegorczyk, 'Le concept de bien'. See also K. Tuori, *Critical Legal Positivism*, Aldershot: Ashgate, Applied Legal Philosophy series, 2002.

47 Hasley, 'Majesty and Monstrosity', p. 219; L. Kotzé and T. Marauhn (eds), *Transboundary Governance of Biodiversity*, The Hague: Martinus Nijhoff, 2014, p. 124.

48 Hasley, 'Majesty and Monstrosity'.

or line designed to keep chaos ... at arm's length',[49] separating 'the sacred [from] the abject'.[50]

Another important aspect that environmental law inherits from legal modernity is the centrality of the subject, and the peripheral role that nature occupies, as the subject's environment. Indeed, the very name *environmental* law signals the reproduction of this cultural 'malaise'.[51] The idea of the environment as that which surrounds us, contains in this sense two crucial assumptions that environmental law internalizes and then operationalizes: a separation between man and nature (that is, its environment) and, relatedly, between subject and object; and the modern obsession with the centre, and in particular with a *human* centre.[52] The environment, observes Philippopoulos-Mihalopoulos, is little more than a 'dervish-like outside that whirls like a frilly skirt around a stable pivot', a pivot that is 'not only stable, fixed and unyielding but significantly "central"'.[53] This means, Philippopoulos-Mihalopoulos continues, but one thing: 'one can do away with the frills and stick to the ... centre'.[54]

The question of the centre is indeed a decisive perspective for legal modernity. The entire edifice of modernity as a socio-cultural formation, and of legal modernity as its legal articulation, revolves arguably around a centre: the subject that knows, wills, owns, regulates. This fundamental philosophical commitment is usually referred to as anthropocentrism, and has ontological, epistemological and axiological implications that traverse environmental law – in its intimately intertwined domestic, supranational and international articulations.

Anthropocentrism is a central element of discussion in environmental philosophy and environmental ethics.[55] In fact, some understand the very core of environmental ethics to be, in ultimate analysis, 'a set of critiques ... of the anthropocentric worldview'.[56] Anthropocentrism is also widely identified as the

49 Ibid., pp. 218–19.

50 Ibid., p. 219. In a similar fashion, Lee Godden emphasizes how modernity constructs nature as other, and in doing so allows only one of two alternative views: either as an object of control – through property rights – or as 'wilderness to be preserved apart from human society', Godden, *Nature as Other*, p. 2. See also S. Chaplin, 'Fictions of Origin: Law, Abjection, Difference', 16:2 *Law and Critique* 2005, 161, 165–6, which describes law as a 'dividing line that serves to exclude filth', that is, to separate the filth of industrial modernity from the purity that protected areas are supposed to guarantee and protect, through law.

51 A. Philippopoulos-Mihalopoulos, 'Towards a Critical Environmental Law' in A. Philippopoulos-Mihalopoulos (ed.), *Law and Ecology. New Environmental Foundations*, New York: Routledge, 2011a, p. 159.

52 See, *ex pluribus*, Holder, 'New Age' and Philippopoulos-Mihalopoulos, 'Towards a Critical'..

53 A. Philippopoulos-Mihalopoulos, 'Looking for the Space between Law and Ecology' in A. Philippopoulos-Mihalopoulos (ed.) *Law and Ecology. New Environmental Foundations*, New York: Routledge, 2011b, p. 22.

54 Ibid., p. 22.

55 See, *ex pluribus*, P. Curry, *Ecological Ethics: An Introduction*, 2nd edn, London and Malden: Polity, 2011, pp. 54–6; D. R. Keller (ed.) *Environmental Ethics: The Big Questions*, Malden, MA: Wiley-Blackwell, 2010; H. Rolston III, *A New Environmental Ethics: The Next Millennium for Life on Earth*, London: Routledge, 2012.

56 Keller, *Environmental Ethics*, p. 62.

crucial problem that affects legal modernity and particularly environmental law (and more broadly the modern legal ordering of nature, regardless of the branch of law under discussion).[57] Indeed, some theorists identify the principle of anthropocentrism as the very organizing principle of environmental law.[58] Yet the term anthropocentrism is considered 'loaded and ambiguous',[59] and has been described as 'one of the most contentious concepts of ecological ethics',[60] despite the fact that 'the critique of anthropocentrism [is] so familiar nowadays so as to be taken almost for granted without argument'.[61] Indeed other terms are sometimes used, with the same conceptual intention, such as homocentrism, human speciesism, species chauvinism etc.,[62] but as Curry underlines, the term anthropocentrism remains probably the best choice since it is solidly established in the literature.[63] So, given the centrality it has within the context of legal modernity, and its centrality in most ecological critiques of law,[64] it is important to consider it carefully.

57 Thus, among many others, Bosselmann, 'Losing the Forest'; D. Wilkinson, 'Using Environmental Ethics to Create Ecological Law' in J. Holder and D. McGillivray (eds), *Locality and Identity: Environmental Issues in Law and Society*, Aldershot: Dartmouth, 1999; C. Cullinan, *Wild Law: A Manifesto for Earth Justice*, South Africa: Siber Ink, 2002; S. Emmenegger and A. Tschentscher, 'Taking Nature's Rights Seriously: The Long Way to Biocentrism in Environmental Law', 6:3 *Georgetown International Environmental Law Review* 1994, 545; A. Gillespie, *International Environmental Law, Policy and Ethics*, Oxford: Clarendon Press, 1997.

58 Thus e.g. M. Cecchetti, *La Disciplina Giuridica Della Tutela Ambientale Come 'Diritto Dell'Ambiente', federalismi.it*, 25, Rivista di Diritto Pubblico Italiano, Comunitario e Comparato, 2006, in relation to the principle of anthropocentrism in the Italian constitution; S. Grassi, 'Tutela dell'Ambiente (diritto amministrativo)', in A. Falzea, P. Grossi and E. Cheli (eds), *Enciclopedia del Diritto: Annali I [Accertamento – Tutela], Giuffrè*, 2007, p. 1114, in relation to EU law; Gillespie, *International Environmental Law* and A. Gillespie, *The Illusion of Progress: Unsustainable Development in International Law and Policy*, London: Earthscan, 2001, particularly at pp. 15–21, in relation to international law.

59 S. Vogel, *Against Nature: The Concept of Nature in Critical Theory*, New York: SUNY Press, 1996, p. 160.

60 Curry, *Ecological Ethics*, p. 54.

61 Vogel, *Against Nature*, p. 160.

62 See, *ex pluribus*, R. Routley and V. Routley, 'Against the Inevitability of Human Chauvinism' in K. E. Goodpaster and K. M. Sayre (eds), *Ethics and Problems of the 21st Century*, Notre Dame, IN: University of Notre Dame Press, 1979; T. Hayward, 'Anthropocentrism: A Misunderstood Problem', 6:1, *Environmental Values* 1997, 49, 52–4. Hayward in particular distinguishes speciesism ('the arbitrary refusal to extend moral consideration to relevantly similar cases') from human chauvinism (which implies that the 'interests of humans must always take precedence') and from anthropocentrism (whose 'ineliminable element … is marked by the impossibility of giving meaningful moral consideration to cases which bear no similarity to any aspect of human cases'). In other words, anthropocentrism simply recognizes that 'as long as the valuer is a human, the very selection of criteria of value will be limited by this fact'. All quotations from Hayward, 'Anthropocentrism', p. 56.

63 Curry, *Ecological Ethics*, p. 55.

64 See, *ex pluribus*, Bosselmann, 'Losing the Forest'; Cullinan, *Wild Law*; M. Tallacchini, *Diritto per la Natura. Ecologia e Filosofia del Diritto*, Turin: Giappichelli Editore, 1996.; Gillespie, *International Environmental Law*.

By simple reference to the literal meaning of the word, anthropocentric means human-centred. In this primary sense, a worldview based on such an outlook implies the centrality of human beings in the world. This centrality is premised on a number of factors, such as the capacity for agency, the ability to reason, to communicate symbolically and through speech, all elements that lend support to the human exceptionalism thesis.[65]

A crucial element necessary to frame the concept in a broader sense has to do with the location of value. Anthropocentric outlooks locate value *intrinsically* – that is independently of external attributions or valuations – *only* in the case of human beings. All else, in this view, has value only *instrumentally*, that is, in light of the benefits that such entities may provide to humanity. Restating this in a Kantian vocabulary, in this view only humans are ends, by virtue of their reason, while other entities are only means. As far as definitions go, Curry offers a very succinct one that captures the key aspects of anthropocentrism: the 'unjustified privileging of human beings ... at the expense of other forms of life'.[66] Anthropocentrism however, is not a monolithic concept, rather, it exists in a variety of forms that are arrayed along a gradient, including outright 'resourcism' on one end,[67] and stewardship on the other.[68]

The 'deep contradiction' of environmental law

However, environmental law is located at an epistemological crossroads. If, in fact, environmental law's conceptual and theoretical apparatus upholds and reproduces the thresholds and ideo-ontological trajectories of modernity, it is simultaneously a vehicle for the postmodern demands of the ecological crises to find a way into law. Besides an increasing number of individual contributions,[69] several novel strands of environmental legal theory have begun focusing on the epistemological fractures traversing environmental law. Through the insights of ecology, as both a science and an ethical framework, theoretical endeavours such as Earth Jurisprudence,[70] Law for Nature,[71] Green Legal Theory,[72]

65 See ibid., pp. 3–18.

66 Curry, *Ecological Ethics*, p. 55.

67 P. Curry, 'Re-Thinking Nature: Towards an Eco-Pluralism', 12:3 *Environmental Values* 2003, 337, p. 338.

68 C. Palmer, 'Stewardship: A Case Study in Environmental Ethics' in R Berry (ed.), *Environmental Stewardship: Critical Perspectives: Past and Present*, London: T&T Clark, 2006.

69 These will be scholars from whom I will repeatedly draw in the course of the book, so I will refrain from presenting a list here.

70 See, e.g. Cullinan, *Wild Law* and P. Burdon, *Earth Jurisprudence: Private Property and the Environment*, London and New York: Glasshouse/Routledge, 2014.

71 See Tallacchini, *Diritto*. For a comparison of these latter two theories I refer the reader to V. De Lucia, 'Towards an Ecological Philosophy of Law: A Comparative Discussion', 4:2, *Journal of Human Rights and the Environment* 2013a, 167.

72 M'Gonigle and Takeda, 'Liberal Limits'.

Ecological Law[73] and Critical Environmental Law[74] are trying to re-frame, re-organize or fundamentally problematize the very idea of environmental law, or its central concepts and categories.

Environmental law, in this respect, exhibits a significant degree of self-reflexivity. However, while self-reflexively aware of how the 'environmental problematic' can only be addressed by exploring and addressing its root causes, through a change in paradigm, '*environmental law itself does not address this problematic; it operates within it*'.[75] This is a problem that affects in different ways many of the attempts at re-articulating environment law in an ecocentric direction, including most of the theoretical strands just mentioned[76] (and this is, importantly, a key consideration also in relation to the 'ecosystem approach', also caught in this contradiction). Environmental law then is arguably affected by a 'deep contradiction',[77] a contradiction that actually 'extends, rather than resolves' the pervasiveness and intensity of environmental problems.[78] In other words, and here lies the tragedy of environmental law, it understands where the problems lie, while simultaneously recognizing that the solutions are forever outside of its grasp; worse, every attempt at a solution entrenches the problems further and more deeply.

Ecology

While environmental law is thoroughly caught in its deep contradiction, things nonetheless appear to be changing, as many legal theorists note.[79] Ecology is contributing to a destabilization of the very premises of legal modernity. Environmental law, to be sure, has had a structural dependency on ecology as a science since its very beginning, as the identification of the problems, thresholds and objectives have always required expert ecological knowledge.

Ecology, however, in its double epistemic role of science and worldview, has also prompted deep ontological, epistemological and axiological reflections.[80] As

73 K. Bosselmann and M. Montini, 'The Oslo Manifesto: From Environmental Law to Ecological Law: A Call for Re-Framing Law and Governance', IUCN WCEL Ethics Specialist Group Workshop, IUCN Academy of Environmental Law Colloquium, University of Oslo, 21 June 2016.

74 Philippopoulos-Mihalopoulos, 'Towards a Critical'; V. De Lucia, 'Beyond Anthropocentrism and Ecocentrism: A Biopolitical Reading of Environmental Law', 8:2 *Journal of Human Rights and Environment* 2017a, 181.

75 M'Gonigle and Takeda, 'Liberal Limits', pp. 1019–20.

76 See e.g. De Lucia, 'Towards an Ecological', p. 167.

77 M'Gonigle and Takeda, 'Liberal Limits', p. 1005.

78 Ibid.

79 Philippopoulos-Mihalopoulos, 'Towards a Critical' and 'Looking for the Space', p. 23; Brooks et al., *Law and Ecology*; Emmenegger and Tschentscher, 'Taking Nature's Rights Seriously'; Tallacchini, *Diritto*.

80 See on this esp. Tallacchini, *Diritto*; M. Tallacchini, 'A Legal Framework from Ecology', 9:8 *Biodiversity and Conservation* 2000, 1085; see also De Lucia, 'Re-embodying Law'.

explained by philosopher of ecology Kevin deLaplante, ecology 'has been viewed … as a foundational source for the development of metaphysical, epistemological and normative views'[81] on the relationship between humanity and nature. In this sense then, ecology is not simply a field of scientific knowledge, but offers inspiration for a new worldview.[82] The ecological worldview *both* represents (epistemologically) *and* demands (culturally and, crucially, legally) a paradigm shift.[83] The centre around which the environment 'skip[s] and turn[s]'[84] seems to be vanishing: ecology dissolves the subject as the pivot of the system, and instead operates without any stable centre, or, paradoxically with a multiplicity of centres each contingently and provisionally activated as subjectivities by a particular question and through a particular perspective. Moreover, ecology *integrates*: it demands the integration of 'laws relating to living things and those that relate to the inanimate physical media that support them';[85] it demands the integration of politically and jurisdictionally fragmented legal institutions, or rather their continuous re-alignment along shifting ecological boundaries;[86] it demands the integration of humans and/in nature.[87]

The 'ecosystem approach' in international environmental law: a paradigm shift?

The 'ecosystem approach' in international environmental law: a paradigm shift?

It is at this pivotal intersection, where law and ecology meet, that we can find the theme of this book: the 'ecosystem approach'. The 'ecosystem approach', broadly understood as a legal and governance 'strategy for the integrated management of land, water and living resources',[88] is increasingly being adopted within a wide variety of international environmental legal regimes, from international watercourses[89] to

81 K. deLaplante, 'Environmental Alchemy: How to Turn Ecological Science into Ecological Philosophy', 26:4 *Environmental Ethics*, 2004, p. 361.

82 Ibid.

83 See e. g. Wood, 'Sovereign Trust'.

84 Philippopoulos-Mihalopoulos, 'Towards a Critical', p. 22.

85 W. Howarth, 'The Progression towards Ecological Quality Standards', 18:3 *Journal of Environmental Law* 2006, 4.

86 See e.g. J. Brunnée, and S. Toope, 'Environmental Security and Freshwater Resources: A Case for International Ecosystem Law', 5:1 *Yearbook of International Environmental Law* 1994, 41, among many others.

87 Thus e.g. R. Grumbine, 'What is Ecosystem Management?' 8:1 *Conservation Biology* 1994, 27.

88 Decision V/6 'Ecosystem Approach', COP5, Nairobi, 15–26 May 2000.

89 For a current review of the ecosystem approach in international watercourses see O. McIntyre, 'The Emergence of an "Ecosystem Approach" to the Protection of

oceans,[90] from biodiversity[91] to fisheries,[92] from Antarctica[93] to climate adaptation.[94]

The rapid and widespread success of the 'ecosystem approach' derives arguably from what it promises, and to whom it promises it. Responding to hopes of arresting and reversing the increasingly negative trends of resource depletion and ecological degradation affecting most ecosystems in the world,[95] the 'ecosystem approach' in fact, 'at least according to its advocates', promises to 'protect the environment, maintain healthy ecosystems, preserve biological diversity, and achieve sustainable development',[96] all at once. In this respect, the 'ecosystem approach' translates a number of key ontological and epistemological insights of ecology into law, and its promise rests on four interrelated elements: integration, integrity, information and iteration. First, the 'ecosystem approach' promises integration, challenging the traditionally fragmentary approach of environmental law. It promises to integrate laws that regulate living resources with laws that regulate pollution and degradation of the physical environment; it aims at integrating, within a transversal ecosystem

International Watercourses Under International Law', 13:1 *Review of European, Comparative and International Environmental Law* 2004, 1.

90 M. Belsky, 'Using Legal Principles to Promote the "Health" of an Ecosystem', 3 *Tulsa Journal of Comparative and International Law* 1995, 183, p. 196. See also Report on the work of the United Nations Open-ended Informal Consultative Process on Oceans and the Law of the Sea at its Seventh Meeting (hereinafter ICP-7 Report), 17 July 2006; Convention for the Protection of the Marine Environment of the North-East Atlantic (OSPAR Convention), adopted 22 September 1992, entered into force 25 March 1998, 2354 UNTS 67; OSPAR Commission, *The North-East Atlantic Environment Strategy of the OSPAR Commission for the Protection of the Marine Environment of the North-East Atlantic 2010–2020*, OSPAR Agreement 2010 – 2013; Statement on the Ecosystem Approach to the Management of Human Activities, 'Towards An Ecosystem Approach To The Management Of Human Activities', JMM1, Bremen, 25–26 June 2003, Agenda item 6. R. Long, Legal Aspects of Ecosystem-Based Marine Management in Europe' in A. Chircop., S. Coffen-Smou and M. McConnell (eds), *Ocean Yearbook*, The Hague: Martinus Nijhoff, 2012.

91 See in particular Decision II/8, 'Preliminary Consideration of Components of Biological Diversity Particularly Under Threat and Action Which Could Be Taken Under the Convention', COP2, Jakarta, 6–17 November 1995, and Decision V/6, 2000, op. cit.

92 See in general Food and Agriculture Organisation (FAO), *Fisheries Management: The Ecosystem Approach to Fisheries, FAO Technical Guidelines for Responsible Fisheries* (No. 4, Suppl. 2. Food and Agriculture Organizations of the United Nations 2003).

93 Convention on the Conservation of Antarctic Marine Living Resources (CCAMLR), (adopted 20 May 1980, entered into force 7 April 1982) 1329 UNTS 48.

94 The ecosystem-based adaptation programme to climate change is being promoted by the United Nations Environment Programme (UNEP), the United Nations Development Programme (UNDP) and the International Union for Conservation of Nature (IUCN), see e.g. http://web.unep.org/coastal-eba/what-is-eba (Accessed 26 September 2018).

95 The Millennium Ecosystem Assessment speaks for example of 'substantial and largely irreversible loss in the diversity of life on Earth', and finds that 'approximately 60% … of the ecosystem services examined … are being degraded or used unsustainably', Millennium Ecosystem Assessment, *Ecosystems and Human Well-being.*, p. 1.

96 R. Lackey, 1998a, 'Seven Pillars of Ecosystem Management', 40 *Landscape and Urban Planning* 1998a, 21.

perspective, fragmented jurisdictional and political boundaries; it promises, moreover, to integrate the social and the ecological dimensions into a single conceptual and operative framework.[97] The 'ecosystem approach', additionally, promises epistemic integration, by incorporating a number of central ecological principles in law. Transcending the immediate, singular management objective, the 'ecosystem approach' shifts focus, and recognizes how each individual component is linked to a wide array of inter-temporal and cross-scale ecosystem processes.[98] Thus, the 'ecosystem approach' promises to expand the scope of legal concern as regards space, time and complexity. Moreover, the 'ecosystem approach' focuses on cumulative effects and impacts. Cumulative effects are changes to ecosystems determined by a combination of past, present and future actions or events.[99] In this respect, the 'ecosystem approach' is 'inherently precautionary in nature',[100] as recognized by the International Law Commission (ILC).[101]

The 'ecosystem approach' then arguably moves past both 'classic' and 'modern' articulations of environmental law[102] and towards an 'ecological' inflection of law that, incorporating scientific ecology and ecocentric ethics, signals a shift to a new phase of its development, a phase by some characterized as 'little short of a paradigm shift'.[103] The 'ecosystem approach' may then offer an answer to calls for an urgent and complete re-think of law in the Anthropocene. It may promise *radical change.* It is in relation to this promise that this book will interrogate the 'ecosystem approach'. The starting question in this book is thus *whether and to what extent the 'ecosystem approach' represents a paradigm shift in international environmental law.*

97 At least in certain interpretations, perhaps most in international environmental law, as we shall see in the course of this book.

98 Secretariat of the Convention of Biological Diversity, *The Ecosystem Approach*, CBD Guidelines, 2004a.

99 From this perspective, any action or event interacts with the cumulative context within which it occurs, and precautionary thresholds may be lowered as a consequence. For one example of how cumulative effects and the precautionary principle may interact see B. Sage, 'Precautionary Coastal States' Jurisdiction', 37 *Ocean Development and International Law*, 2006, 359, 370.

100 O. McIntyre, 'The Ramsar Convention on Wetlands and General International Water Law: Mutually Supportive Regimes', Report prepared for the Ramsar Convention Secretariat, 2014a, p. 17.

101 '[T]he obligation to protect the ecosystems of international watercourses', maintains the ILC, is 'a general application of the principle of precautionary action', ILC, *Report of the International Law Commission on the Work of its Forty-Ninth Session* 195, p. 280, quoted in McIntyre, 2014a, 'Ramsar Convention', p. 17, fn 99; McIntyre interprets ILC's statement as 'categorical', ibid.

102 This distinction in 'phases' of environmental law is Jane Holder's; J. Holder, 'New Age: Rediscovering Natural Law', 53:1 *Current Legal Problems* 2000, 151. Similarly, in relation to the promise of a paradigm shift, Brooks et al., *Law and Ecology*. and M. Wallace et al., 'Moving toward Ecosystem Management: Examining a Change in Philosophy for Resource Management', 3 *Journal of Political Ecology* 1996, 1.

103 Ibid., p. 167. See also, for an early discussion of the relation between ecology (as a science) and law, D. Tarlock, 'The Nonequilibrium Paradigm in Ecology and the Partial Unraveling of Environmental Law', 27 *Loyola L.A. Law Review* 1994, 1009.

The concept of paradigm

Since the driving question in this book is whether the 'ecosystem approach' represents a paradigm shift, a few words to clarify the concept of paradigm are warranted. The word paradigm, in its original Greek etymology, means a pattern, reference or example. Plato used the idea of paradigm to describe the model necessary for the creation of the Cosmos.[104] Today there are two main ways to understand the concept of paradigm. One is linked to the work of philosopher of science Thomas Kuhn. Paradigm in the Kuhnian sense refers to a set of 'universally recognized scientific achievements that for a time provide model problems and solutions to a community of practitioners'.[105] In this sense, a paradigm is 'a framework containing the basic assumptions, ways of thinking, and methodology that are commonly accepted by members of a scientific community'.[106] This is a widely accepted notion of paradigm. There is however another, broader, understanding of the concept of paradigm that is particularly relevant for the purposes of this book. From this broader perspective, a paradigm is both a 'cognitive framework'[107] *and* a cultural framework. A paradigm is in other words a broad set of cultural mentalities and patterns of cognition capable of providing a 'network of coordinates through which the bearers of [a given] culture perceive reality and construct their mental image of the world'.[108] In other words, a paradigm is a particular horizon of sense. Both concepts of paradigm are relevant for our purposes, and both are invoked, as we shall see in the next section, in the literature on the 'ecosystem approach'. However, it is the second and broader understanding of paradigm as a cognitive and cultural framework that more centrally reflects the promises of the 'ecosystem approach' in relation to a radical re-thinking of law, and thus the claims that it represents a paradigm shift.

Two narratives of paradigm shift

The central role of ecology in relation to paradigmatic shifts in environmental law was already recognized, particularly in the US, in the 1990s, where some scholars and commentators, most notably Dan Tarlock, began speaking of a 'partial unravelling' of environmental law due to the emergence of a new paradigm in ecological sciences.[109] There are arguably two primary ways in which the narrative of the paradigm shift is invoked in the literature. While they are usually intertwined with one another, they can be distinguished for analytical

104 R. Archer-Hind (ed.), *The Timaeus of Plato*, 'Genesis', 27c – 34a, London: Macmillan, 1888.

105 T. Kuhn, *The Strucutre of Scientific Revolutions*, Chicago: University of Chicago Press, 2nd edn, 1970, enlarged, reprinted in *International Encyclopedia of Unified Science*, Volume 2, Number 2, p. viii.

106 http://dictionary.reference.com/browse/paradigm (Accessed 26 September 2018).

107 Ibid.

108 A. Gurevich, *Categories of Medieval Culture*, London: Routledge & Kegan Paul, 1985, p. 13. Gurevich is referring more precisely to 'worldview'. There is clearly an overlap between the concept of worldview, and the concept of paradigm deployed in this book.

109 Tarlock, 'Nonequilibrium Paradigm'.

purposes. Moreover, as we shall see, this distinction will prove useful throughout the book. In fact, if on the one hand this distinction marks the limit of environmental law in relation to its deep contradiction, it simultaneously charts the potential trajectory that a radical articulation of the 'ecosystem approach' may follow. In this respect, the distinction can be formulated in the following terms. A first way in which the narrative of the paradigm shift is deployed is primarily related to a policy or management perspective. This deployment can be called the 'weak' narrative of paradigm shift.

Hartie, Klaphake and Schliep for example, discuss in general the science and policy debates around the 'ecosystem approach' and its characterization as a paradigm shift, and underline how the concept of the 'ecosystem approach' 'can be interpreted as an attempt to codify basic elements of a new paradigm in resource management'.[110] Similarly, Kidd, Plater and Frid suggest that 'we are in the midst of a paradigm shift in planning and management of the natural environment and resources'.[111] Galindo-Leal and Bunnell[112] as well as Tudela and Short[113] make in turn explicit reference to a Kuhnian type of paradigm shift (albeit the latter two with particular regards to the context of fisheries management). Uy and Shaw similarly suggest that the 'ecosystem approach' is a paradigm shift. However, they argue that this shift is taking place slowly. They emphasize how the 'ecosystem approach' 'is not a revolutionary concept … but a *slow revolution* … built on decades of experience of thousands of individuals in natural resource management … leading to a change from existing, well-established paradigms'.[114] Both Pyhälä[115] and Hopkins,[116] furthermore, speak of the 'ecosystem approach' in terms of a paradigm shift towards sustainability,

110 V. Hartje, A. Klaphake and R. Schliep, 'The International Debate on the Ecosystem Approach Critical Review International Actors Obstacles and Challenges', *BfN – Skripten*, 2003, 80, 12.

111 S. Kidd, A. Plater and C. Frid, *The Ecosystem Approach to Marine Planning and Management*, New York: Routledge, 2011, p. 1. They identify three main reasons for considering that the ecosystem approach is a paradigm shift: the recognition of the dependence of human well-being on ecosystem health, the recognition of the humans in nature paradigm and a holistic and integrated, rather than sectoral, view of management, ibid., p. 29.

112 C. Galindo-Leal and F. Bunnell, 'Ecosystem Management: Implications and Opportunities of a New Paradigm', 71 *The Forestry Chronicle* 1995, 601.

113 S. Tudela and K. Short, 2005, 'Paradigm Shifts, Gaps, Inertia, and Political Agendas in Ecosystem-Based Fisheries Management', 300 *Marine Ecology Progress Series* 2005, 282, 283–4.

114 N. Uy and R. Shaw (eds), *Ecosystem-based Management*, Bradford, UK: Emerald Group Publishing Limited, 2012, p. 6. (my emphasis)

115 M. Pihälä, 'HELCOM Baltic Sea Action Plan: An Ecosystem Approach to the Management of Human Activities' in M. Reckermann, K. Brander, B. MacKenzie and A. Omstedt (eds), *Climate Impacts on the Baltic Sea: From Science to Policy*, Heidelberg: Springer Verlag, 2012, pp. 20–1.

116 C. Hopkins, 'The Concept of Ecosystem Health and Association with the Ecosystem Approach to Management and Related Initiatives', *ICES BSRP/HELCOM/UNEP Regional Sea Workshop on Baltic Sea Ecosystem Health Indicators*, 30 March–1 April 2005, Sopot, Poland.

linked directly to the frameworks of ecosystem services and natural capital (a crucial linkage, as we will see in Chapter 9). Hatcher and Bradbury, finally, speak of 'a paradigm of ecosystem management', and characterize thusly ecosystem management as a 'paradigm shift'.[117] They further emphasize however, how, like all paradigm shifts, the 'ecosystem approach' is caught between the failure of existing paradigms and the risk of uncritical acceptance in light of its promises.[118] Furthermore, others speak of paradigm shift in relation to international water law,[119] to international fisheries law,[120] or more broadly to marine management[121] or ocean resources management.[122] What is common in this narrative, and what prompted me to characterize it as 'weak', is that the paradigm shift that the 'ecosystem approach' embodies is considered internal to the overall dynamic of sustainable development.

A second way, which can be called the 'strong' narrative of paradigm shift, is more radically linked to a challenge to fundamental categories that ecology pose to legal modernity. This narrative has both a descriptive and a normative dimension (i.e. the 'ecosystem approach' *is* a paradigm shift, and the 'ecosystem approach' *demands* a paradigm shift). Thus the 'ecosystem approach is characterized as 'fundamental shift',[123] as a 'quite radical' conceptual framework that 'calls for a change in the entire field within which opportunities and problems are examined – a change from a view of environment in a political or people-oriented context to a view of politics in an "ecosystem context"'.[124] Wallace et al. point to an ongoing shift investing 'attitudes about natural resources and the environment' in what is most often characterized as an 'ecocentric' direction. They present a table of 'shifts' (they call it 'changes in science and politics') from 'Enlightenment thought' (that is, from modern rationality) to

117 B. Hatcher and R. Bradbury, 'Marine Ecosystem Management. Is the Whole Greater than the Sum of the Parts?' in D. Rothwell and D. L. VanderZwaag (eds), *Towards Principled Ocean Governance: Australian and Canadian Approaches and Challenges*, New York: Routledge, 2013, p. 205.

118 Ibid., p. 206.

119 A. Kibaroglu, A. Kirschner, S. Mehring and R. Wolfrum, *Water Law and Cooperation in the Euphrates-Tigris Region*, Leiden: Martinus Nijhoff, 2013, p. 86.

120 M. West, 'The Future of the International Law of Capture Fisheries', 23 *Journal of Northwestern Atlantic Fisheries Sciences* 1998, 19, esp. p. 23ff.

121 Kidd, Plater and Frid, *Ecosystem Approach*.

122 R. Boyle, 'The Implications of a Paradigm Shift in Ocean Resource Management for Fisheries Stock Assessment' in R. Beamish and B. Rothschild (eds), *The Future of Fisheries Science in North America*, Heidelberg: Springer, 2009.

123 Wallace et al., 'Moving toward', p. 1; see also M. Goldman, P. Nadasdy and M. Turner (eds), *Knowing Nature: Conversations at the Intersection of Political Ecology and Science Studies*, Chicago: University of Chicago Press, 2011, p. 140, which observe how the notion of 'paradigm shift' is increasingly invoked in wildlife management literature, where the ecosystem approach has an increasingly important role.

124 J. Vallentyne and A. Beeton, 1988, 'The 'Ecosystem' Approach to Managing Human Uses and Abuses of Natural Resources in the Great Lakes Basin', 15:1 *Environmental Conservation* 1988, 58.

'ecosystem management', thus implicating a 'paradigm shift', even though the expression 'paradigm' does not appear explicitly in their texts.[125] Grumbine also spoke implicitly in terms of paradigm shift, indicating how the 'ecosystem approach' could contribute to dismantling the 'fence' that has segregated humans from nature.[126] From a normative perspective, finally, Berkes suggests, within the specific context of fisheries, that an 'ecosystem approach' to fisheries management *has to be* and *needs* to be 'revolutionary'.[127]

Some preliminary clarifications

The 'levels of articulation' of environmental law

It is important at this time to offer some preliminary clarifications. The first relates to the idea of environmental law. The main concern of this book is with international environmental law. However, environmental law as a discourse is not fragmented in a multiplicity of jurisdiction-specific discourses, especially from the perspective of an international legal scholar. Indeed, there is arguably an increasing porosity between international, supranational and domestic articulations of environmental law.[128] In many ways then, international environmental law is *one* element in a global discourse of environmental law and governance articulated through a set of nested hierarchies and circular relationships.[129] This porosity derives, on the one hand, from the fact that many key principles of environmental law – now common to international, supranational and domestic legal systems – develop at first within the context of domestic legal experiences, from whence they travel upwards to the international level;[130] or, conversely, find their initial articulation in the international arena, and are only subsequently absorbed into domestic legal systems.[131] In this respect, it has been noted how

125 Wallace et al., 'Moving Toward'.

126 Grumbine, 'What is Ecosystem Management?', p. 35.

127 F. Berkes, 'Implementing ecosystem-based management: evolution or revolution?', 13:4 *Fish and Fisheries*, 2011, 465, p. 466 and p. 473 respectively.

128 Indeed, the notion of 'transnational' law has gained prominence in the last decades, and there is at least one academic journal dedicated to transnational environmental law.

129 For an understanding of environmental law as 'global law' see e.g. R. Percival, 'Global Law and the Environment', 86 *Washington Law Review* 2001, 579.

130 Typical in this respect is the precautionary principle, which is said to have first emerged within the legal context of Germany as the so-called *Vorsorgeprinzip*, see S. Boehmer Christiansen, 'The Precautionary Principle in Germany: Enabling Government' in T. O'Riordan and J. Cameron (eds), *Interpreting the Precautionary Principle*, London: Earthscan Publications Ltd., 2001; see also J. Wiener, 'Something Borrowed for Something Blue: Legal Transplants and the Evolution of Global Environmental Law', 27 *Ecology Law Quarterly* 2001, 1295.

131 Notable examples are the Polluter Pays Principle, developed within the context of OECD in the early 1970s; or the principle of Sustainable Development, that has percolated from the international level to the European and then to the domestic jurisdictions of member States. Just to trace the absorption of the principle in the Italian jurisdiction, this occurred in 2008,

'interbreeding between national and international law occurs more frequently than has been acknowledged'.[132]

This porosity does not necessarily translate into a conceptual and legal equivalence at the level of doctrinal or jurisprudential interpretation, given the complexities of integrating a rule or a principle of law into different legal cultures and systems.[133] However, it does show conceptual proximity, and points also, importantly, to a common discursive and conceptual space.[134]

In this book, I will thus refer to and develop arguments drawing from all three levels, sometimes seamlessly and sometimes after more explicit explanations of how such transversality is useful and warranted.

The 'ecosystem approach': a note on terminology

Given the complex and often confusing deployment of terminology in relation to the 'ecosystem approach', a preliminary explanatory note on terminological matters is necessary at the outset. While the 'ecosystem approach', in determinate form, will be used to refer to a conglomerate of articulations, I do not intend to suggest a monolithic concept by such a formulation. Rather, I intend to refer to an 'emergent trajectory',[135] which contains the multiplicity of articulations of the concept, but that is also traversed by all the tensions that I will highlight in the course of the book. Thus, in order to avoid any confusion, this referent is marked with single quotation marks: the 'ecosystem approach'. The reader is advised that in every such instance I am referring to an open container that reflects the elusive and multiple nature of a concept that is otherwise, for the most part, referred to in the literature through a determinate form. This despite a multiplicity of formulations, inflections, articulations. I have chosen this particular formulation (i.e. 'ecosystem approach'), due to its increasing presence in the literature and in formal legal documents. However, in certain contexts, the expression 'ecosystem approach' may refer to *both* this broad concept,

when law 4/2008 amended the 2006 'environmental code' (Law 152/2006) by inserting article 3-quarter, entitled 'the principle of sustainable development', thus bringing Italian environmental legislation in line with EU environmental law; see also D. Tarlock, 'The Influence of International Environmental Law on U.S. Pollution Control Law', 21 *Vermont Law Review* 1997, 759 and Wiener, 'Something Borrowed'.

132 Ibid., p. 1307

133 There is a large comparative legal literature dedicated to the complexities involved in these processes of rule migration; see, in particular G. Teubner, 'Legal Irritants: Good Faith in British Law or How Unifying Law Ends Up in New Differences', 61 *Modern Law Review* 1998, 11.

134 It is however useful to note how this common discursive and conceptual space also offers crucial resources for doctrinal interpretations of key principles of environmental law, so the proximity of these levels of articulation of environmental law obtains, arguably across legal methodologies (doctrinal, socio-legal etc.).

135 I am grateful to Anna Grear for this formulation.

and a more concrete, particular articulation, such as within the context of the Convention on Biological Diversity (CBD).

Relatedly, I will subsume under the expression 'ecosystem approach' the entire spectrum of ecosystem-oriented frameworks existing within scientific and international environmental law and policy discourse. I will nonetheless still occasionally use other expressions such as ecosystem management or ecosystem-based management, in direct quotations, or if warranted by the context of the discussion (for example if the study I am discussing uses the expression ecosystem management). I am aware, however, that the different terminologies may point to significant substantive differences; or by converse that different expressions are sometimes used to refer to the *same* framework. I will indeed discuss explicitly these questions, and the problematics they raise, in Chapter 4.

Finally, at other junctures, when I particularly want to emphasize plurality, I may use the expression 'ecosystem approaches', and, unless I am quoting directly from certain documents which deploy the plural, I will be referring to the same emergent trajectory.

Situating the book: theoretical and methodological approach

It is at this point also useful to situate the book from a theoretical and methodological perspective. It is thus important to note at the outset that I will not carry out a doctrinal analysis aimed at establishing the current legal status of the 'ecosystem approach' in international environmental law. By contrast, my theoretical and methodological perspective can be broadly located within critical legal scholarship.[136] What distinguishes critical legal scholarship from scholarship that possesses what Becket – speaking of international law – calls '*esprit critique*'[137] is that critique in the former sense starts from the premise that the problems of law are not merely 'marginal or exceptional', but 'endemic, consistent and

136 Critical Legal Studies has had two 'waves' or 'generations'. The first emerged as a more critical declension of American legal realism in the 1960s and 1970s, and deployed a largely Marxists critical apparatus. Typical of this wave of critical legal studies are the works of Roberto Unger (such as, e.g. *Law in Modern Society: Toward a Criticism of Social Theory*, New York: Free Press, 1976) and Duncan Kennedy (e.g. 'Form and Substance in Private Law Adjudication', 89:8 *Harvard Law Review* 1979a, 1685 and 'The Structure of Blackstone's Commentaries', 1979b, 28 *Buffalo Law Review* 1979b, 205). This wave eventually waned. A second wave resuscitated the critical legal studies movement in Europe, and more specifically in England, but deployed a different set of theoretical and methodological tools, most especially poststructuralism. For more details, see, e.g. M. Tushnet, 'Critical Legal Studies: A Political History', 100:5 *The Yale Law Journal* 1991, 1515; T. Murphy, 'BritCrits: Subversion and Submission, Past, Present and Future', 10 *Law and Critique* 1999, 237; J. Schlegel, 'Critical Legal Studies' in A. Hadden and A. Brophy (eds), *A Companion to American Legal History*, Chichester: Wiley-Blackwell, 2013.

137 J. Beckett, 'Critical International Legal Theory', Oxford Bilbiographies, http://www.oxfordbibliographies.com/view/document/obo-9780199796953/obo-9780199796953-0007.xml (Accessed 26 September 2018).

structural'.[138] The main research question – whether and to what extent the 'ecosystem approach' can be considered a paradigm shift – must necessarily pass through the critical reconstruction of the complex genealogies and of the competing, and sometimes irreconcilable, narratives and ideological projects within which it is situated.[139] In order to do that, and mindful of Elizabeth Fisher's observation that environmental legal scholarship is open to methodological pluralism,[140] I will deploy two interrelated theoretical and methodological frameworks that will underpin respectively Part II and Part III of this book: genealogy and biopolitics. Each will be discussed in detail in individual chapters (respectively Chapter 2 and Chapter 8), however, they are, together, two complementary elements of an 'analytics of biopolitics'.

Originally outlined by Thomas Lemke, the methodological approach called 'analytics of biopolitics'[141] aptly combines the concept of biopolitics and the method of genealogy. Deploying an 'analytics of biopolitics' as the main framework of analysis[142] allows critical environmental legal analysis to *open* rather than close.[143] It is in this respect that this book is located in that critical space that has been tentatively called 'critical environmental law'.[144] Unlike analyses carried out under the guidance of a binary conceptual framework, an analytics of biopolitics does not aim at finding 'an ultimate and objective representation of reality'. Its goal is rather that of problematizing and situating the object of critical analysis within its larger genealogical context. The crucial goal of an analytics of biopolitics, then, is to 'generate problems', and to 'ask questions that have not yet been asked' in order to 'destabilize' what appears as natural or self-evident.[145]

Similarly, critical environmental law aims to outline the slippages that intervene at the margins of the intersection between law and ecology. It does not offer a simplified, linear narrative of increasing interpenetration between law and ecology (a narrative in which law becomes increasingly ecocentric as the relationship between law and ecology goes through various stages: 'quarrel', 'initial

138 Ibid.

139 This argument has been already formulated in condensed form in V. De Lucia, 'Competing Narratives and Complex Genealogies. The Ecosystem Approach in International Environmental Law', 27:1 *Journal of Environmental Law* 2015a, 91.

140 Fisher observed that mature environmental legal scholarship requires '*some thought-out*' methodological position, where the term "some" stands to indicate the methodological pluralism of environmental legal scholarship, E. Fisher et al., 'Maturity and Methodology: Starting a Debate about Environmental Law Scholarship', 21:2 *Journal of Environmental Law* 2009, 213, emphasis in the original.

141 T. Lemke, *Biopolitics. An Advanced Introduction*, New York: New York University Press, 2011, Chapter 9, p. 177ff.

142 Ibid.

143 Opening, rather than producing closures, is an explicit goal of the theoretical framework called critical environmental law, see Philippopoulos-Mihalopoulos, 'Towards a Critical'.

144 Ibid., but see all contributions in the collection.

145 Lemke, *Biopolitics*, p. 123.

embrace', 'courtship' and 'marriage').[146] An analytics of biopolitics by contrast, transposed to the specific (critical environmental legal) terrain of the 'ecosystem approach', radically problematizes analyses that understand ecological or ecosystem regimes as those regimes where 'the science of ecology is applied through environmental laws'.[147] The emphasis falls, rather on the *double* genealogy of concepts that are *simultaneously* scientific and legal, and on their biopolitical problematization.

The structure of the book

As mentioned, the question driving the research that led to this book is *whether and to what extent the 'ecosystem approach' represents a paradigm shift in international environmental law*. Some preliminary questions however, must be tackled at the outset. Indeed, before being able to assess whether or not the 'ecosystem approach' represents an important shift in paradigm, it is crucial to understand what the 'ecosystem approach' is. Is there a uniform, universal concept of the 'ecosystem approach', or are there multiple articulations? Are these multiple articulations simply a variation on a theme, perhaps due to the novelty of the concept, which inevitably leads to a series of local instabilities and context-specific differences? Or are they rather different concepts that latch onto the same semantic referents? In order to answer both the preliminary questions and the central question, the book is structured in two parts and eleven chapters, including this introduction. Part II is dedicated to a genealogical study of the 'ecosystem approach'. The main argument of Part II is that, despite claims to the contrary, there are still irreducible problems in identifying the 'ecosystem approach' with any degree of stability. While some argue that these problems are contingent and that there is a 'real' or 'better' understanding of the ecosystem approach that can be arrived at, Part II will show how the 'ecosystem approach' is characterized by a *constitutive* and *irreducible* complexity. The most that can be done, I argue in Part II, is to speak of a *plurality* of articulations of the 'ecosystem approach', and to show what the prevailing articulations are within particular regimes and what the tensions are that emerge due to underlying complex genealogies and the competing narratives at work therein.

In Chapter 2 I introduce in some detail the genealogical method of inquiry, its aims, and I also outline my deployment of it against the broader context of genealogy as a critical method. I will also explain the benefits of its application to the study of the 'ecosystem approach'.

Chapter 3 provides an initial and broad location of the 'ecosystem approach'. After a brief historical account aimed at showing the early history of the concept, the 'ecosystem approach' is located through (a) presenting a number of scientific, policy and legal definitions of the concept and (b) discussing some of

146 See e.g. Brooks et al., *Law and Ecology*, p. 369.
147 Ibid., p. 2.

the elements that are considered, by scholarship or by institutional actors, to be common to all or most articulations of the concept. It is exactly this attempt at locating the 'ecosystem approach', however, that offers an initial idea, to be further cemented in subsequent chapters, of how the concept is ultimately inserted in a complex discursive field where a plurality of perspectives interact, compete and conflict.

Chapter 4 addresses three preliminary questions in order to help further locating the 'ecosystem approach' within the context of international environmental law. The first question hinges on the different institutional genealogies of the 'ecosystem approach', which I discuss in terms of 'clusters'. This distinction offers analytical clarity in relation to the particular genealogies – in relation to the prevailing narratives, value orientations and overall legal and policy objectives – that affect the concrete articulations of the 'ecosystem approach'. This preliminary framing is necessary before being able to move on to a more detailed deconstruction of the 'ecosystem approach'. The second question relates to the way that the 'ecosystem approach' is framed and understood. In this respect, I articulate a distinction between a wide and a narrow concept of the 'ecosystem approach'. This distinction is relevant insofar as it allows mapping the field, and it further allows deciphering the ways in which the 'ecosystem approach' is 'read' or articulated into particular regimes. A third question relates to the ways in which the 'ecosystem approach' is labelled. It is argued that there is a 'labelling knot' that needs to be disentangled to be able to navigate in a field affected by a disparate set of nomenclatures – sometimes the same thing is called by different names, and sometimes the same name refers to different things.

Chapter 5 explores the underlying origin of some of the complexities, contestations and ambiguities affecting the 'ecosystem approach', and the reason they are largely inevitable and irreducible. In particular, this chapter focuses on the problematics that arise from recognizing that ecology is a framework of ambiguity, both at the most general level of ecology as a field of knowledge, and at the more particular level of the concept of ecosystem, the central naturalistic underpinning of the 'ecosystem approach'. Such recognition moreover, rather than prompting us to seek the most reliable or true articulation of ecology, alerts us to the irreducible genealogical nature of this field of knowledge, in a way that further corroborates the usefulness of the genealogical approach taken in this book, particularly in relation to the 'ecosystem approach'.

The competing narratives and the complex genealogies underlying both ecology in general and the concept of ecosystem more specifically, are also central to the 'ecosystem approach'. Situated within a space of conflicting values, the 'ecosystem approach' is *one* of the terrains where irreducible contestations over the values underlying and informing ecological science and environmental law take place. These contestations are usually framed within a narrative space where anthropocentrism and ecocentrism represent opposite poles along a gradient of intermediate positions (such as biocentric or weak anthropocentric positions). Chapter 6 explores precisely the conflicting values and the competing narratives underlying the 'ecosystem approach', in order to further demonstrate its complex genealogical character. The chapter first outlines, in broad terms, the conflicting values and the competing narratives involved, in order to describe the field of tensions within which the 'ecosystem approach' finds articulation(s); then it illustrates how each narrative may find expression in a particular articulation of the 'ecosystem approach'. With this chapter Part II ends.

Part III shifts the critical register in order to displace the modern obsession with the '-centrisms' still upheld in most ecological critiques of law (and reflected in the anthropocentrism/ecocentrism binary). To this end, I introduce a biopolitical framework of analysis. Part III first builds its theoretical framework, and then applies it to a real world case, the Convention on Biological Diversity, in order to show how such a framework offers significantly more critical purchase than the usual binary framework premised on the concepts of anthropocentrism and ecocentrism.

Chapter 7 begins the shift in critical register. The competing narratives identified in the previous chapters, and their key conceptual referents (anthropocentrism and ecocentrism), will be problematized in order to prepare the terrain for a full shift from the binary anthropocentric/ecocentric framework, to a biopolitical framework of analysis.

Chapter 8 effectively shifts the critical register and presents the concepts of biopower and biopolitics. Biopolitics in particular, technically one of the two modes of biopower, offers to be, this chapter argues, the best interpretive register available to a critical legal engagement with the 'ecosystem approach'. After presenting the concepts of biopower and biopolitics, and their relationship, in the theory of Michel Foucault, the *locus classicus* of biopolitical thinking, the chapter discusses some key elaborations of biopolitics produced by three Italian political philosophers: Agamben, Negri and Esposito. What emerges is a dual consideration. On the one hand, biopolitics is a novel way to frame power and its operations, one that has become particularly decisive in what Esposito calls a second articulation of modernity, where its mode or tonality becomes prevalent vis-à-vis sovereign power. On the other hand, there is a deep, intimate continuity between sovereignty and biopolitics, so that, with Esposito, it is possible to argue that modernity is always biopolitical. As the literature on biopolitics tends to address the biopolitical government of human populations, the chapter concludes with the articulation of an ecological declension of biopolitics, that is, with a transposition of the biopolitical framework to nature, and to environmental law.

Chapter 9 offers a case study to test the critical purchase of the biopolitical framework. This test case is the biodiversity regime, and more specifically its central convention, the 1992 Convention on Biological Diversity (CBD). The relevance of the CBD as a test case is twofold. First, it is arguably the CBD that has carried out the most comprehensive effort to develop the 'ecosystem approach' as a conservation strategy. Second, within the CBD is perhaps the legal regime where the tensions between the competing narratives discussed in this book are best visible.

Chapter 10 is the last substantive chapter. It aims to show that the 'ecosystem approach', while susceptible of a number of productive readings enabled by its 'productive ambiguities',[148] even as the tensions and the competing narratives that traverse the concept, and the genealogical 'play of forces' within which it is inserted, tend to be subjected to a decisive biopolitical resolution. Yet it is precisely at the aporetic core of the biopolitical resolution that one may find some open, unstable, ambiguous terrain where the critical legal scholar can operate counter- or, better, non-hegemonically. While apparently without remainder in terms of its conceptual and material reach, in fact, the biopolitical closure can never actually entirely fill the genealogical field of discursivity where the 'ecosystem approach' is situated. It can only provisionally *hegemonize* it. Biopolitics then, as a prevalent framework of control and mode of operation of power, cannot but leave *some* space for the articulation of counter-trajectories, for the utterance of counter-readings, for the drawing out of antagonist interpretive registers that are available through the productive ambiguities that inhabit the conceptual field of the 'ecosystem approach'. The final part of the chapter finally answers the original question: is the 'ecosystem approach' a paradigm shift?

148 I am grateful to Anna Grear for this formulation, which she uttered during one of our conversations, and which has stuck, insofar as it captures well a core concern of this book. See also, A. Grear, 2012a, 'Human Rights, Property and the Search for "Worlds Other"', 3:2 *Journal of Human Rights and Environment* 2012a, 173, where she speaks of 'productive ambivalence'.

Part II

A genealogical reading of the 'ecosystem approach'

2 Genealogy as a legal method

Introduction

Genealogy, as Foucault illustrates, is a history that does not search for a single point of origin; it does not attempt to find the true essence, the original identity, of concepts and phenomena; it rejects a linear understanding of history[1] where 'values unravel inexorably towards their perfection'.[2] Genealogy, rather, implies the theorization 'of the opacity of the origin',[3] an origin which does not, and cannot, contain any 'uncontaminated essence',[4] but only 'laceration' and 'multiplicity'.[5] The concept of genealogy in this respect, means

> to identify the accidents, the minute deviations – or conversely, the complete reversals – the errors, the false appraisals, and the faulty calculations that gave birth to those things that continue to exist and have value for us; it is to discover that truth or being does not lie at the root of what we know and what we are, but the exteriority of accidents.[6]

However, genealogy also, crucially, focuses on 'the various systems of subjection' and on the 'hazardous play of domination'.[7] History, from this perspective, 'is always produced through a particular stage of forces',[8] As Foucault argues:

1 'History moves one way; it is the linear progress of gradual disclosure of essences or values: the becoming-realized of the values of freedom, equality or rights', C. Douzinas and A. Gearey, *Critical Jurisprudence. The Political Philosophy of Justice*, Oxford and Portland, OR: Hart Publishing, 2005, p. 54.

2 Ibid.

3 R. Esposito, *Bíos. Biopolitics and Philosophy*, Minnesota: University of Minnesota Press, 2008, p. 79.

4 Ibid., p. 80.

5 Ibid.

6 M. Foucault, 'Nietzsche, Genealogy, History', in D. F. Bouchard (ed.), *Language, Counter-Memory, Practice: Selected Essays and Interviews*, Ithaca, NY: Cornell University Press, 1977, p. 146.

7 Ibid., p. 148.

8 Ibid., pp. 148–9.

> humanity does not gradually progress from combat to combat until it arrives at universal reciprocity, where the rule of law finally replaces warfare; humanity installs each of its violences in a system of rules and thus proceeds from domination to domination.[9]

Indeed, in a genealogical approach, '[e]very thing and process is the site of a - struggle',[10] so that history can never exist 'beyond the play of forces'.[11] It follows that '[e]very phenomenon … has many antagonistic senses that depend on the perspective of the powers that try to dominate it'.[12] Every concept, understood genealogically, is the contingent result of an agonistic conflict of forces. Law, importantly, is arguably a privileged point of emergence of such conflicts and contestations.[13] All legal concepts, therefore, can be approached genealogically, and are arguably the result of an ensemble of competing forces, each trying to assert hegemony.[14]

Genealogy, in summary, is aimed at the reconstruction of the contingencies and contestations (what Foucault, following Nietzsche, calls respectively descent[15] and emergence[16]) constituting the complex history of phenomena, concepts and, importantly, law. A genealogical approach, moreover, and importantly, leads to a methodology that is inevitably pluralist and perspectival,[17] insofar as every concept is capable of having 'many antagonistic senses that depend on the perspective of the powers that try to dominate it'.[18]

9 Ibid., p. 151.
10 Douzinas and Geary, *Critical Jurisprudence*, p. 49.
11 Ibid. The concept of 'play of forces' is crucially internal to the methodology adopted in this book. The formulation is duly credited to its proper source, but future occurrences, while surrounded by single quotation marks, will not be accompanied by a footnote with the same point reference, as it will be transposed into the natural language of the genealogical method, and transformed into a term of art, notwithstanding the acknowledgement of the original formulation.
12 Ibid., p. 50.
13 G. Bascherini, 'Italian Theories. Spunti attorno all'esperienza giuridica a partire da un recente saggio di Roberto Esposito', *Costituzionalismo*, 2013, http://www.costituzionalismo.it/articoli/438/ (Accessed 26 September 2018).
14 P. Monateri, 'Legal Formants and Competitive Models: Understanding Comparative Law from Legal Process to Critique in Cross-System Legal Analysis', 2008, http://ssrn.com/abstract=1317302, (Accessed 26 September 2018). Similarly, Tigar and Levy, who speak of law as "twice fragile": law is in fact a privileged site of struggles and contestations, which take different forms according to the particular legal terrain within which they occur. Law (legal ideology) thus is twice fragile in the sense that, on the one side, it may be forced to accommodate new content in its existing forms through reformist social struggles, and, on the other, change may begin from within which, suddenly, may present an altogether new form of law and legal ideology, a discontinuity as it were; M. Tigar and M. Levy, *Law and the Rise of Capitalism*, Delhi: Aakar Books, 2005, p. 293.
15 Foucault, 'Nietzsche', pp. 80–3.
16 Ibid., pp. 83–6.
17 Douzinas and Gearey, *Critical Jurisprudence*, esp. p. 49ff.
18 Ibid., p. 49.

Genealogy and genealogies

Albeit genealogy is most immediately associated with Foucault, Foucault is but one element in the 'shared tradition' of genealogy.[19] Other notable genealogists are Fredrik Nietzsche (genealogy of morals), Charles Darwin (genealogy of humans), Quentin Skinner (genealogies of liberty and of the modern State), Bernard Williams (genealogy of truth and truthfulness), Edward Craig (genealogy of knowledge) and earlier still, proto-genealogists David Hume and Thomas Hobbes (genealogy of sovereign power).[20] It is thus perhaps best to speak of genealogies, in the plural, also considering that not only may the object of genealogies differ (e.g. morals, liberty, truth etc.), but also the particular methodological inflection.

In this respect, Craig has proposed a taxonomy that distinguishes between three main typologies: vindicatory genealogies, subversive genealogies and explanatory genealogies. Ian Koopman, who has written an influential book on genealogy as critique,[21] adds also proto-genealogies, that is embryonic, preliminary or incomplete articulations. I shall follow such taxonomy for the purposes of this chapter, which is at least pedagogically useful. Now a few words on each.

Proto-genealogies such as Hume's, which he himself calls 'experimental reasoning', entail 'formulating hypotheses about what causes could have led to given effects'.[22] This experimental reasoning however remains only an embryonic genealogical orientation, as it remains unidimensional. The question 'how did things unfold?' does not open up into why have they unfolded this way, and how else could have they unfolded? No specific inquiry is made into what relations were forged between power and fields of knowledge in the processes of formation of certain practices, an inquiry which is a key element of the genealogical method. Vindicatory genealogies, such as Williams', on the other hand, use genealogy in order to vindicate, as it were, a certain state of affairs by anchoring its legitimacy to its pedigree. Williams' genealogy in this respect is emphatically *not* a critical methodology. Another set of genealogies are subversive. Nietzsche's genealogy is the key example.[23] Nietzsche's genealogy is a form of critical interpretation, and as such 'there are counterfactual elements in the interpretation', considering that '[t]o understand how something could have come about implies that a different outcome was also possible'.[24]

19 I. Koopman, *Genealogy as Critique. Foucault and the Problems of Modernity*, Indiana University Press, 2013, p. 8ff. discusses also the two strands of reception of Foucault's work in Anglo-Saxon/American context, literary studies and philosophy.

20 Ibid., p.18ff.; D. Hoy, 'Nietzsche, Hume and the Genealogical Method', in Y. Yovel (ed.), *Nietzsche as an Affirmative Thinker*, Martinus Nijhoff, 1986, pp. 20–38 in particular suggests Hume is a proto-genealogist.

21 Koopman, *Genealogy as Critique*.

22 Ibid., p. 21.

23 E. Craig, *Knowledge and the State of Nature. An Essay in Conceptual Synthesis*, OUP, 1991; Koopman, *Genealogy as Critique*.

24 Koopman, *Genealogy as Critique*, p. 31.

Subversive genealogy aims at subverting the current state of affairs, aims at denaturalizing 'reality', as brilliantly shown in Nietzsche's genealogy of morals.

Subversive and vindicatory genealogies, however, suffer according to Ian Koopman from the so-called 'genetic fallacy'.[25] This fallacy implies that the justification or de-legitimation of present practices rests on the circumstances of their emergence.[26] As Foucault himself recognized, 'no matter how base and despicable the emergence of certain of our practices has been, we can neither reject nor legitimate these practices on the basis of the terms of their emergence alone'.[27]

In yet another sense and still using Craig's 'map', genealogy can be explanatory and/or neutral. Before we give in to disbelief over the very possibility of a neutral stance, let me clarify that neutral here is a contrasting word, to be understood against the positively aimed genealogies of vindication and subversion. 'The mode of neutrality', Koopman for example suggests, 'is the mode of doubt, indeterminacy, and vagueness. It is in this sense that genealogical problematization [an important notion I will return to in the next section] is neutral – it leaves us in doubt, with questions, and unprepared to pronounce a verdict'.[28] We will return to this particular point, to, as it were, problematize it through asserting a different theoretical kinship of genealogy than Koopman's. However, we can anticipate that, even where a genealogy is not prepared to pronounce a verdict, neutrality cannot endure. Indeed, genealogy is 'one crucial aspect of a broader apparatus of critical inquiry whose primary functions are that of genealogical problematization and pragmatist reconstruction',[29] but that is couched, or can be couched within a broader strategy, parts of which may be legitimately, and shall we say, inevitably, normative. Genealogy then is 'one methodological part of a fuller ensemble of critical methodologies'.[30] Indeed, in this book a first genealogical part will be complemented by a second part where an additional critical theoretical framework will be articulated and then applied to a particular case.

We have thus far, briefly seen that there are a variety of genealogies, but what is *my* genealogy? What is the genealogical method that *I* intend to articulate as an important method for finding a critical register through which to produce *critical* environmental legal scholarship, and critically deconstruct and reconstruct the 'ecosystem approach'? The key is the genealogical notion of problematization.

25 Ibid., p. 60ff. Koopman however diminishes the importance and role of the genetic fallacy to a weak version where the normative conclusions are not invalidated, but only weakened by reliance on genetic reasoning.

26 Ibid., p. 61.

27 Ibid., p. 87.

28 Ibid., p. 60.

29 Ibid., p. 147.

30 Ibid.

Genealogy as problematization

Problematization is a key element of Foucault's genealogical method, especially in Ian Koopman's interpretation of Foucault. But what is problematization and how does it relate to genealogy? How does it enact genealogy? For Foucault, problematization is a dual concept harbouring a set of tensions that impress it with a radical dynamic and, as it were, a sharp genealogical character.[31] Foucault in fact understood problematization first as an activity, as a critical inquiry underpinned by a verbal form: to problematize. This is perhaps a partially familiar sense. However, Foucault understood problematization also in nominal form, as the *object* of the critical inquiry. A problematization, in this respect, is the starting point of the inquiry, and it refers to two different though entangled notions. In a first sense, nominal problematizations refer to problems that elicit certain responses. A practice, concept, principle or rule, for example, may no longer offer sufficient purchase on (physical or social) reality; it may no longer facilitate the achievement of its underlying objective. This 'weakness' is precisely what elicits the emergence of a novel practice aimed at countering the weakness. Yet nominal problematizations refer also, in a second and somewhat reciprocal sense, to the emergent practice that through its very emergence renders problematic, displaces, denaturalizes, existing practices, concepts, principles or rules. An existing practice may become inadequate, in other words, either because it does not achieve its intended purpose, or because a new and emerging practice appears more promising.

To bring it closer to the environmental legal field, a nominal problematization may refer both to certain environmental legal practices and principles that no longer seem to offer useful results, such as the fragmented and sectoral approach to environmental protection or the legal consideration of nature as an object susceptible of proprietary regulation. It may also refer however, to novel practices, such as the 'ecosystem approach', that problematize, as it were, existing practices through their very emergence, and contribute to rendering visible their inadequacy to achieve their intended purpose, and thus contribute to their displacement. These two types of nominal problematization are obviously deeply entangled, yet they offer two reciprocal *perspectives* that may be usefully maintained distinct for analytical and methodological purposes, particularly in light of the active aspect of problematization: the critical inquiry.

As a critical inquiry, problematization *intensifies* the nominal problematization. Indeed, it takes it seriously. Yet it also tries to penetrate its complexity and to illustrate, expose, explore how each new practice, principle, or concept is contingently articulated and contested from various perspectives of power, each trying to hegemonize and dominate it. A critical environmental law then must take seriously the problematizations linked, in our case, to the emergence of the 'ecosystem approach' – the legal and governance weaknesses that prompted its

31 Ibid, p. 99ff.

emergence and the displacing effect that it has had on existing concepts and practices. Simultaneously however, it means problematizing its emergence along multiple vectors of inquiry, tracing the contingencies and contestations that are either actively operative or archived in the trajectories of its descent and emergence.

Problematization, in short, is critique focusing on contingency and complexity,[32] and genealogy is a multi-vector practice of problematization, focusing on the intersections and entanglements between knowledge and power, and their effects on conceptual or material trajectories.

Legal narratives: towards a genealogy of the 'ecosystem approach'

Having offered an outline of genealogy as a method for a critical environmental law, it is important now to discuss albeit briefly, the question of the 'sources' of law, and their usage. In order to explore the 'ecosystem approach' genealogically, it is important to capture the entire set of legally relevant patterns and discourses underlying the contingencies involved in its descent, and the contestations involved in its emergence. The relevant material exceeds thus the scope of the doctrinal theory of sources even further than the already flexible scenario of sources facing international environmental lawyers.[33] I use the term 'legal narratives' to capture these patterns and discourses, while generally referring to the idea, central to critical legal scholarship, to read law *outside* of the formal legal text.

Following Robert Cover, I further suggest that the processes of formation of legal meaning must be located well beyond the confines of formal legal processes. Cover argues that '[n]o set of legal institutions or prescriptions exists apart from the narratives that locate it and give it meaning'.[34] Law then, is 'not merely a system of rules to be observed, but a world in which we live',[35] and can be fully understood only through narratives. Indeed, as Cover emphasizes – a point which will be explored in relation to the 'ecosystem approach' – each legal decision, text or concept is situated between competing narratives.[36] Thus law 'cannot escape [its] location in a normative universe', it cannot escape that

32 Ibid., p. 48.

33 I am referring in particular to the increasing porosity between law and political commitments, represented by the proliferation of soft law in international environmental law, famously described by Dupuy as 'either not yet, or not only law', P. Dupuy, 'Soft Law and the International Law of the Environment', 12:2 *Michigan Journal of International Law*, 1991, 420. For a more general account of material sources of law, including, importantly, legal history, see e.g. R. Kolb, 'Legal History as a Source of International Law' in S. Besson and J. D'Aspremont (eds), *The Oxford Handbook of the Sources of International Law*, OUP, 2017.

34 R. Cover, 1983, 'The Supreme Court, 1982 Term – Foreword: Nomos and Narrative', 97:4 *Harvard Law Review* 1983, 4.

35 Ibid., p. 5.

36 R. Cover, 'Violence and the Word', 95 *Yale Law Journal* 1986, 1601. Similarly, Tigar and Levy, *Rise of Capitalism.*

it originates and ends 'in the narratives that are the trajectories plotted upon material reality by our imaginations'.[37]

The concept of legal narrative finds further inspiration in the insights of two conceptual frameworks: Sacco's idea of 'legal formants';[38] and Lowe's idea of 'interstitial norms'.[39] Legal formants are *all* those elements that contribute to the formation of legal meaning. They include the dogmatic theory of sources of law;[40] the reasons and argumentations used by jurists to reach a particular conclusion;[41] and what Sacco calls 'cryptotypes',[42] that is, the *habitus* of a jurist, as influenced and conditioned by the broader historical, cultural, social and economic context within which a particular jurist is situated. Legal formants however – and this is what is of most interest for my purposes here – include what Sacco calls 'declamatory statements'. These declamatory statements, suggests Sacco, 'may not be strictly legal',[43] but rather 'propositions about philosophy, politics, ideology or religion'.[44] And while they may be in contradiction to what Sacco calls, by contrast, the 'operational rules' of law,[45] they crucially contribute to understanding law within its ideological context.[46] As Sacco observes, it would be 'difficult' [47] as well as 'inadequate and unfair' [48] to explain law without reference to the underlying organizing ideas of a culture or of an ideology.[49] Moreover, the more ambiguous and indeterminate a legal

37 Cover, 'The Supreme Court', 5.

38 R. Sacco, 'Legal Formants: A Dynamic Approach to Comparative Law (Installment I of II) ', 39:1 *The American Journal of Comparative Law* 1991a, 1.

39 V. Lowe, 'Sustainable Development, Unsustainable Argument' in A. Boyle and C. Freestone (eds), *International Law and Sustainable Development. Past Achievements and Future Challenges*, Oxford: Oxford University Press, 1991, and V. Lowe, 'The Politics of Law-Making: Are the Method and Character of Norm Creation Changing?' in M. Byers (ed.), *The Role of Law in International Politics. Essays in International Relations and International Law*, Oxford: Oxford University Press, 2000.

40 Standard, obviously, to each particular legal domain or legal order.

41 Sacco, 'Legal Formants'.

42 R. Sacco, 'Legal Formants: A Dynamic Approach to Comparative Law (Installment II of II) ', 39:2 *The American Journal of Comparative Law* 1991b, 343, 384–385, but also throughout.

43 Sacco, 'Legal Formants', p. 32.

44 Ibid.

45 E.g. ibid., p. 30, but Sacco uses the expression throughout

46 Ibid., pp. 30–34. Indeed, they may make visible the slippage that exists between the ideology inscribed in, and articulated through, declamatory statements (relating, for example, to universal suffrage), and the underlying operational rule which, in point of fact, denies that ideological declamation, see ibid., p. 33

47 Ibid., p. 32.

48 Ibid.

49 Sacco refers for example to the crucial role such declamatory statements had in the Soviet legal system, especially in the form of a particular ideological origin (i.e. the ideas of Marx and Engels) underlying the production of many soviet legal categories, Sacco, 'Legal Formants', p. 32; likewise, Sacco further suggests, it would be difficult to explain 'canon law without the notion of God', ibid.

formulation or principle is, the wider the set of legal formants that may come into play, a key consideration in relation to the object and focus of this book.

Legal formants are specifically oriented towards accounting for the complex genealogy of legal norms, and the set of practices and actors involved. The idea of interstitial norms, developed by Lowe in specific relation to international law, offers a similar approach, yet one more closely linked to the production of legal meaning in relation to broad legal concepts like the 'ecosystem approach'.[50] Interstitial (or secondary) norms, argues Lowe, operate at the interstices that exist between primary norms. They are norms or concepts, which, while lacking the precise normativity of primary rules, articulate 'another species of normativity',[51] an 'interstitial normativity'[52] that pushes and pulls the 'boundaries of true primary norms when they threaten to overlap or conflict with each other'.[53] Indeed, especially once adopted by a tribunal in relation to a case to be decided, it is 'the *very* idea' contained in these interstitial concepts (and Lowe uses the concept of sustainable development as deployed in the *Gabcikovo-Nagymaros* case[54] as a paradigmatic case) that 'point … towards a coherent approach to [the] decision' in cases where primary norms are in conflict.[55] An important characteristic of interstitial norms is that, by 'occupying the field', they exercise 'an immense gravitational pull' on the entire field of law.[56]

What is of decisive importance for my purposes here however, is not what interstitial norms are, but the way in which they arise and the modalities of their development. Interstitial norms, suggests Lowe, have no author, but rather 'emerge'.[57] Interstitial norms are characterized by the diffusion of participation in their development, a development which 'may involve a very wide range of contributors',[58] and are shaped by a 'much wider range of concepts and social pressures … than is ordinarily the case'.[59] And while interstitial norms are 'drawn out' by lawyers, there is a large and open set of actors able to contribute to their formation and development.[60] Indeed, Lowe suggests that there 'is no limit, in principle, to the category of persons who may contribute to the

50 Lowe, 'Sustainable Development' and Lowe, 'Politics of Law-Making'.
51 Lowe, 'Sustainable Development', p. 21.
52 Ibid., p. 31.
53 Ibid.
54 ICJ, *Gabcikovo-Nagymaros Project* (Hungary/Slovakia), Judgment, I. C. J. Reports 1997, p. 7.
55 Lowe, 'Politics of Law-Making', p. 217.
56 Ibid.
57 Ibid., p. 219.
58 Ibid., p. 220.
59 Ibid.
60 Lowe uses the example of the role an NGO such as Greenpeace may have in relation to the determination of the content of a principle such as sustainable development, ibid, p. 219. It must be noted that Lowe develops the concept of interstitial norms specifically in the context of international law

development of interstitial norms',[61] in such a way that these norms are the site 'where general culture obtrudes most clearly into the processes of legal reasoning'.[62]

The concept of legal narratives shifts thus attention from the formal sources of law, to the processes and the discourses from which legal principles and concepts emerge. This approach is especially useful in relation to concepts such as the 'ecosystem approach', understood, as it is in this book, as an elusive, ambiguous and contested concept. Indeed, the more ambiguous and indeterminate a legal formulation or principle is, the wider the set of legal formants, social pressures and legal narratives that may arguably come into play. This in turn resonates with the view that international environmental law is a 'propitious breeding ground' for broad 'indefinite' (and contested) principles and concepts,[63] so that it has been characterized as 'hot'[64] or 'postmodern'.[65] Approaching the analysis of the 'ecosystem approach' from this conceptual and methodological angle is thus arguably very fruitful and allows for a rich critical analysis.

From this methodological perspective then, the documents, definitions and descriptions included in the analysis will not be selected based on their formal legal status, but on their substantive relevance in terms of their contribution to the articulation of the 'ecosystem approach' within both scientific contexts and international normative, institutional, policy and political contexts.

61 Ibid., p. 219.

62 Ibid., p. 221.

63 N. De Sadeleer, *Environmental Principles: From Political Slogans to Legal Rules*, Oxford: Oxford University Press, 2002, p. 1.

64 E. Fisher, 'Environmental Law as Hot Law', 25:3 *Journal of Environmental Law* 2013, 347.

65 De Sadeleer, *Environmental Principles.*

3 Locating the 'ecosystem approach'

Introduction

In the previous chapter I have outlined the genealogical method, its aims and strengths, and its relationship with the 'ecosystem approach'. I have also shown how it is important to look at narratives, in order to capture the full set of relevant 'sources' for the purposes of a genealogical analysis. For the rest of Part II, and starting with this chapter, I will deploy the genealogical method in order to problematize, in the terms discussed in the previous chapter, the 'ecosystem approach'.

The aim of this chapter is to provide an initial and broad location of the 'ecosystem approach' through a series of interlinked steps. First, the chapter will offer a brief historical account aimed at showing the early history of the concept so as to highlight its genealogical character. This will show how the development of the 'ecosystem approach' has proceeded with hiccups and is linked to a series of contingencies that reflect its genealogical emergence, rather than a linear progress. Subsequently, the chapter reviews a number of scientific, policy and legal definitions of the concept. This serves the double aim of delineating the conceptual space where the 'ecosystem approach' is located, and further highlighting its genealogical character, evident in the series of perspectives from which the various, and at times conflicting, articulations emerge. Finally, the chapter discusses some of the elements that are considered, by scholarship or by institutional actors, to be common to all or most articulations of the concept. The goal is emphatically not that of recovering an 'origin', but rather identifying the contingencies implicated in the historical development of the concept along the entire multiplicity of vectors, and through some of the key junctures where interpretations may have clashed, certain orientations may have prevailed etc. It is exactly this attempt at locating the 'ecosystem approach', I suggest, that offers an initial idea, to be further cemented in subsequent chapters, of how the concept is ultimately inserted in a complex discursive field where a plurality of perspectives interact, compete and conflict, and hence are best understood and disentangled by way of the genealogical method.

Historical antecedents

The first modern attempts at developing an approach to the management of natural resources cognizant of the importance of ecological processes, and employing what can be described as 'ecosystem-oriented research', according to legal historian Harry Scheiber, date back over a century ago.[1] This account – Scheiber admits - does not represent 'a standard view ... accepted by all historians of ecological science',[2] and links the emergence of an 'ecosystem research vision'[3] to the development of fisheries science in Northern Europe. Towards the end of the nineteenth century European fisheries scientists, faced with the novel understanding that fisheries were not inexhaustible, began shifting attention towards the 'habitat of fish species and the relationship of fishery population dynamics to the physical environment'.[4] This effort was pioneered by Norwegian zoologist and fisheries scientist Johan Hjort[5] and other key figures from Scotland, Denmark and Germany, whose central focus was that of understanding 'fish population dynamics in terms of ecosystem relationships within fishery habitats'.[6]

While this was a short-lived effort, as World War I brought a virtual end to both commercial fisheries and deep-water scientific research,[7] a significant institutional innovation remains as a legacy of this period, namely the foundation in 1902 of the International Council for the Exploration of the Sea (ICES). ICES, incidentally, would eventually play a significant role in the elaboration of the 'ecosystem approach' within the ocean and fisheries cluster in contemporary times.

As Scheiber's account shows, the subsequent development of marine living resources management (which he primarily locates in North America) focused mostly on single-species management and led to the development of practical harvesting methodologies which eventually converged into what is today known as the 'Maximum Sustainable Yield' (MSY) method, still a central concept in fisheries management.[8] An ecosystem orientation however 'continued to be pursued vigorously in limnology'[9] and, by the 1930s, ecosystem analyses began to appear in the context of animal and other land-based ecological studies.[10]

1 Thus H. Scheiber, 'From Science to Law to Politics: An Historical View of the Ecosystem Idea and Its Effect on Resource Management', 24 *Ecology Law Quarterly* 1997, 631.

2 Ibid., p. 635. Further, the author admits to depart 'significantly' or even 'radically' from other more orthodox views such as that of D. Worster, *Nature's Economy: The Roots of Ecology*, San Francisco: Sierra Club Books, 1977, see Scheiber, 'From Science to Law', fn 12.

3 Ibid., p. 636.

4 Ibid.

5 Hjort 'laid the intellectual foundations of modern fisheries "recruitment" research, as well as advancing the larger ecological approach to fishery habitats and ecosystem relationships', ibid., p. 637.

6 Ibid., p. 638.

7 Ibid., p. 637.

8 See E. Hey, 'The Persistence of a Concept: Maximum Sustainable Yield', 27:4 *The International Journal of Marine and Coastal Law* 2012, 763.

9 Ibid., p. 639.

10 Ibid.

Later on, in the context of environmental policy, a series of workshops on wildlife management held in the early 1970s sponsored by the World Wide Fund for Nature (WWF), and attended by, among others, significantly, the International Union for Conservation of Nature (IUCN), began reviewing the scientific basis of conservation and began using language that would eventually organize around an ecosystem-based approach to management.[11] In the opinion of some scholars, '[t]he statement on living resources produced by this program is recognized as the first formalized set of principles addressing ecosystem management'.[12]

One of the central targets of these workshops, and of the statement captured in a set of principles published in 1978,[13] was the concept of MSY that would become central in fisheries management. MSY was (and is) considered to be a fundamental hindrance to the sound ecological management of wildlife for two reasons: its lack of attention to the wider ecosystems within which species exist and the incentives it created for commercial industries to build overcapacity into their operations.[14] These principles, which included the need to maintain ecosystems at a desirable state in order to ensure sustainable yields and thus avoid irreversible changes and satisfy the needs of present and future generations,[15] were also to feed into the then ongoing process of negotiation of the Law of the Sea Convention (UNCLOS).[16] The principles were subsequently 'updated' by Mangel and colleagues, who distilled seven principles further encouraging a wider ecosystem perspective and emphasizing the linkages between social and ecological systems.[17]

With particular reference to international environmental law, it is the 1972 Stockholm Declaration on the Human Environment, which formally introduced the concept of ecosystem to international law and policy narratives. The 1974 Polar Bear Agreement[18] utilized the ecosystem as a legal concept in its provision requiring parties to take what today might be called an 'ecosystem approach' – and is certainly an ecosystem-oriented one – to the conservation of polar bears.[19] The agreement built on the increasing recognition that a 'bear should … preferably not

11 M. Forst, 'The Convergence of Integrated Coastal Zone Management and the Ecosystems Approach', 52:6 *Ocean and Coastal Management* 2009, 294.

12 Ibid., p. 295.

13 S. Holt and L. Talbot, 'New Principles for the Conservation of Wild Living Resources', 43:2 *Wildlife Monographs, Supplement to The Journal of Wildlife Management* 1978, 59.

14 Forst, 'The Convergence', 294, p. 295.

15 Holt and Talbot, 'New Principles'.

16 These principles were finally captured in Holt and Talbot, 'New Principles', Annex 4 'Application to the Law of the Sea'.

17 M. Mangel et al., 'Principles for the Conservation of Wild Living Resources', 6:2 *Ecological Applications*, 1996.

18 Agreement on the Conservation of Polar Bears of 1973, 13 ILM 13 (1974).

19 For a brief historical overview of the events that led to the adoption of the Polar Bear Agreement see, e.g. T. Larsen, 1975, 'Progress in Polar Bear Research and Conservation in the Arctic Nations', *Boston College Environmental Affairs Law Review*, 4:2, 295.

be considered only as a unique and isolated object worth protection', but should rather be seen 'as a part of the Arctic ecosystems'.[20] Article II of the Polar Bear Agreement reflects this 'ecosystem' view, which came to prevail during the preparatory and drafting work of the IUCN Polar Bear Specialist Group, providing that:

> Each Contracting Party shall take appropriate action to protect the ecosystems of which polar bears are a part, with special attention to habitat components such as denning and feeding sites and migration patterns, and shall manage polar bear populations in accordance with sound conservation practices based on the best available scientific data.[21]

The 'ecosystem approach' was also discussed in relation to international freshwater law during the early 1970s, and more specifically in the context of the Great Lakes Water Quality Agreement, where the concept was introduced in 1978.[22]

Albeit succinct, this account of the early history of the 'ecosystem approach' is useful in one specific respect. It shows how the genealogical descent of the 'ecosystem approach' is punctuated by starts and stops, by *faux pas* and re-calibrations, all linked as much to the contingencies of history as to the contexts within which its development can be located. It also shows how there is no single point of origin from which the 'ecosystem approach' progressed towards its current form, but rather a constellation of junctures and contexts. Already in this first, most basic respect then (and we will return to this point in chapter 4) the 'ecosystem approach' can be understood as the result of a complex set of genealogical strands located in different domains (oceans, fisheries, wildlife, marine mammals) and involving a different set of actors (conservation scientists, managers, lawyers, NGOs, States).

The next step will be to try and locate the 'ecosystem approach', and the conceptual space it occupies.

Locating the 'ecosystem approach'

Responding to the question, what is the 'ecosystem approach', is arguably a difficult task from both a scientific and a legal perspective.[23] The concept is

20 IUCN, *Polar Bears, Proceedings of the Third Working Meeting of the Polar Bear Specialist Group*, Survival Service Commission of IUCN, Morges, 7–10 February 1972, IUCN publications new series, Supplementary Paper No. 35, p. 71.

21 Agreement on the Conservation of Polar Bears of 1973, op. cit., Article II.

22 Great Lakes Water Quality Agreement of 1978, 30 UST 1383 as amended in 35 UST 2370 (1983) and 2185 UNTS 504 (1987). A new amendment was adopted in 2012 through the Protocol Amending the Agreement between Canada and the United States of America on Great Lakes Water Quality Agreement of 1978, as amended on 16 October 1983, and on 18 November 1987 (signed 7 September 2012, entered into force 12 February 2013).

23 Thus for example Long, 'Legal Aspects'.

'interpreted differently in different contexts',[24] and 'has proven difficult to define in a simple manner'.[25] Indeed, the concept is often defined by resorting to a set 'of common words used in uncommon ways, of common words used with multiple meanings, of common words used with mutually exclusive definitions, of new words used in old ways'[26] (a consideration that I will return to in Chapter 4).

Things have undoubtedly changed since the time when biologist Robert Grumbine lamented that the 'ecosystem approach' (or ecosystem management, as he calls it) was still being 'perceived by many as a buzzword'.[27] However, if Grumbine, in 1994, could discount the great variety of interpretations of the 'ecosystem approach' in light of its novelty,[28] in 1999 discussions still frequently revealed that experts 'often [held] widely divergent views of what [the 'ecosystem approach'] entails',[29] despite the 'general agreement that the "ecosystem approach" is of fundamental importance'.[30] In 2006, during the United Nations Open-ended Informal Consultative Process on Oceans and the Law of the Sea at its seventh meeting, it was still generally recognized that 'there is no universally agreed definition of an ecosystem approach'.[31] Importantly, recent scholarship still recognizes how there isn't any uniform understanding or definition of the concept.[32]

However, notwithstanding the difficulties of providing a stable definition – particularly across regimes and domains – of the 'ecosystem approach', some scholarship suggests that there is 'substantial agreement'[33] on at least *some* core elements of the concept.[34] In fact, recent discussions of the 'ecosystem approach', after initial acknowledgements of the difficulties associated with stabilizing a definition,

24 United Nations Open-ended Informal Consultative Process on Oceans and the Law of the Sea, 7th Report (hereinafter ICP-7), 17 July 2006, Part A, Section 1, para 6.
25 Secretariat of the CBD, 2004, p. 3.
26 R. Lackey, 'Ecosystem Management: In Search of the Elusive Paradigm', 4:2 *Human Ecology Review* 1998b, 107.
27 R. Grumbine, 'Reflections on "What Is Ecosystem Management?"', 11:2 *Conservation Biology* 1997, 41. Grumbine first discussed the ecosystem approach in Grumbine, 'What is Ecosystem Management?'
28 Grumbine, 'What is Ecosystem Management?', p. 28.
29 E. Maltby, 'Some European Perspectives on the Ecosystem Approach' in H. Korn et al. (eds), *Report on the Scientific Workshop on the Ecosystem Approach – What Does It Mean for European Ecosystems?* Bonn: Bundesamt für Naturschutz, 1999 p. 27.
30 Ibid.
31 ICP-7, 2006, Part A, Section 1, para 6.
32 E. Kirk, 'The Ecosystem Approach and the Search for an Objective and Content for the Concept of Holistic Ocean Governance', 46:1 *Ocean Development and International Law* 2015, 33, p. 37.
33 A. Trouwborst, 'The Precautionary Principle and the Ecosystem Approach in International Law: Differences, Similarities and Linkages', 18:1 *Review of Comparative, European and International Environmental Law* 2009, 26, p. 28.
34 Trouwborst identifies three core elements, ibid. p. 28; within the context of fisheries other authors list up to seven elements, A. Fabra and V. Gascón, 2008, 'The Convention on the

tend to focus on those general features or elements that characterize most if not all understandings of the concept.[35] Lack of a clear and precise definition is in this respect often not considered to constitute an important hindrance in relation to the ability to operationalize the concept. McIntyre, for example, suggests that both 'the constitutive elements' and 'the practical means for the implementation' of the 'ecosystem approach ... have become increasingly well understood in recent years'.[36] Long[37] as well as Fabra and Gascón[38] similarly argue that conceptual and definitional difficulties do not detract from the ability to implement the 'ecosystem approach' in practice. Contrary views exist, emphasizing by contrast how the 'ecosystem approach', while successful as a label, suffers from operationalization failure.[39] However, they are arguably a minority view.

In an institutional setting, relevant examples of this attitude are offered by the United Nations Open-ended Informal Consultative Process on Oceans and the Law of the Sea (UNICPOLOS),[40] the CBD, and the Arctic Council, as we shall see presently. While UNICPOLOS recognizes that the concept is 'interpreted differently in different contexts',[41] its report provides quite a long and complex list of 'agreed consensual elements'.[42] Similarly, within the context of the CBD, it is recognized that it 'has proven difficult to define [the ecosystem approach] in a simple manner'.[43] Yet the CBD endorses the so-called Malawi Principles, which define the 'ecosystem approach' in terms of a set of 12 principles and a supporting set of operational guidelines, both of which have been revised following a first phase of practical implementation.[44] The Arctic Council, finally, suggests that the time

Conservation of Antarctic Marine Living Resources (CCAMLR) and the Ecosystem Approach', 23 *International Journal of Marine and Coastal Life* 2008, 567, p. 569.

35 Thus for example Long, 'Legal Aspects'; Trouwborst, 'The Convention'; Fabra and Gascón, 'The Convention'; ICP-7 Report, op. cit.; within the context of the Arctic Coucil see The Protection of the Arctic Marine Environment Working Group (PAME), *The Ecosystem Approach to Management*, Concept Paper, 2014.

36 O. McIntyre, 'The Protection of Freshwater Ecosystems Revisited: Towards a Common Understanding of the "Ecosystems Approach" to the Protection of Transboundary Water Resources', 23:1, *Review of European Comparative, and International Environmental Law (RECIEL)*, 2014b, 88.

37 Long, 'Legal Aspects'.

38 Fabra and Gascón, 'The Convention', p. 567.

39 S. Tudela and K. Short, 'Paradigm Shifts, Gaps, Inertia, and Political Agendas in Ecosystem-based Fisheries Management', 300:1 *Marine Ecology Progress Series* 2005, 241.

40 An ongoing consultative process organized under the aegis of the United Nations Division for Ocean Affairs and the Law of the Sea (DOALOS).

41 ICP-7, 2006, op. cit., part a, Section 1, para 6.

42 Ibid.

43 Secretariat of the CBD, *The Ecosystem Approach*, 2004b, p. 3. https://www.cbd.int/doc/publications/ea-text-en.pdf (Accessed 26 September 2018)

44 See e.g. Expert Meeting on The Ecosystem Approach, *Review of the Principles of the Ecosystem Approach and Suggestions for Refinement: A Framework for Discussion*, Montreal 7–11 July 2003.

for discussion is over (a discussion summarized in many ways in the report of the Expert Group on Ecosystem-Based Management),[45] and that the 'ecosystem approach', while an 'evolving concept', is 'ready for implementation'.[46]

If discussion is over, however, it should be easy to locate the 'ecosystem approach' through a sufficiently stable definition and a sufficiently clear and narrow set of elements that characterize it. In order to move closer to a more precise understanding of the 'ecosystem approach' then, the next two subsections will, in turn, review a number of key definitions, and then a number of attempts at distilling some common elements from the significant variety of articulations of the 'ecosystem approach' across a range of legal regimes and policy documents.

Many definitions

Discussions of the 'ecosystem approach' start, naturally, from definitions. This section will also offer definitions as a starting point. However, it must be noted at the outset that there is an (over-)abundance of – scientific, policy and legal – definitions, each offering a particular semantic, substantive or conceptual perspective. Meffe and colleagues, for example, list 14 different 'selected' definitions of the 'ecosystem approach' (the selection implying that there are significantly more definitions available in the literature), along with their own, each emphasizing different key elements.[47] Schaepfer reports six definitions out of a larger set from which he draws.[48] Brooks and colleagues identify 11 definitions.[49] Grumbine in turn reviews 33 articulations of the 'ecosystem approach' from which he extrapolates some core themes and his own definition.[50] This section will review only *some* of the existing definitions. It will draw on scientific, policy and legal definitions for a number of reasons. First, many definitions occupy multiple dimensions simultaneously (the science/policy demarcation is particularly porous in relation to the 'ecosystem approach') so that their interplay is an important element in the identification of the 'ecosystem approach'. Second, this approach affords a fully stereographic view of how the 'ecosystem approach' is framed across different domains, as well as across

45 Arctic Council, Ecosystem-Based Management in the Arctic Report submitted to Senior Arctic Officials by the Expert Group on Ecosystem-Based Management, (J05-ARCTIC-007, 2013) 2013.

46 The Protection of the Arctic Marine Environment Working Group (PAME), *The Ecosystem Approach to Management*, Concept Paper, 2014, p. 2.

47 G. Meffe et al., *Ecosystem Management: Adaptive, Community-Based Conservation*, 1st edn, Washington, DC: Island Press, 2002.

48 R. Schlaepfer, *Ecosystem-Based Management of Natural Resources: A Step towards Sustainable Development*, IUFRO Occasional Paper No. 6, International Union of Forest Research Organizations, Washington, DC: IUFRO Secreteriat, 1997, pp. 8–9.

49 Brooks et al., *Law and Ecology.*

50 Grumbine, 'What is Ecosystem Management?'

different legal regimes. In this second respect, all definitions are threads in the legal narratives that shape the 'ecosystem approach'. Such a stereographic view, moreover, provides useful empirical referents in relation to some analytic distinctions that I will outline in Chapter 4 (as regards a wide and a narrow concept of the 'ecosystem approach' and as regards the different institutional clusters where the 'ecosystem approach' has emerged and developed).

Scientific and scholarly definitions

As already noted, there exists an abundance of definitions of the 'ecosystem approach', whose focus and completeness vary greatly.[51] Some definitions emphasize the ecological dimension, placing socio-economic 'deliverables' (such as 'feet of timber, total catch, or visitor days')[52] outside of the 'ecosystem approach'.[53] An important example in this respect is offered by a definition crafted in 1996 by a team of scientists affiliated to the Ecological Society of America, which understands the 'ecosystem approach' primarily in relation to ecological sustainability. The 'ecosystem approach' is defined by this group as a management framework that is 'adaptable' (through monitoring and research) and is 'based on our best understanding of the ecological interactions and processes necessary to sustain ecosystem structure and function'.[54] Similarly, and perhaps even more clearly ecologically oriented, Clark and Zaunbrecher define the 'ecosystem approach' as a type of management that uses 'system-wide concepts to ensure that all plants and animals in ecosystems are maintained at viable levels in native habitats and natural ecosystems processes are perpetuated indefinitely'.[55]

This orientation is also visible in a scholarly definition often cited in the legal literature. Brunné and Toope, in a seminal article on the 'ecosystem approach' in the context of international freshwater law, suggest that '[s]imply put', the ecosystem approach

51 Uy and Shaw, *Ecosystem-based Management*, p. 6; see also Meffe et al., *Ecosystem Management.*

52 N. Christensen et al., 'The Report of the Ecological Society of America Committee on the Scientific Basis for Ecosystem Management', 6:3 *Ecological Applications* 1996, 665, p. 669.

53 Ibid.

54 Christensen et al., 'The Report', pp. 668; the full definition reads as follows: 'Ecosystem management is management driven by explicit goals, executed by policies, protocols, and practices, and made adaptable by monitoring and research based on our best understanding of the ecological interactions and processes necessary to sustain ecosystem composition, structure, and function. Ecosystem management includes sustainability, goals, sound ecological models and understanding, complexity and connectedness, the dynamic character of ecosystems, context and scale, humans as ecosystem components, adaptability, and accountability'; ibid., p. 668–9.

55 T. Clark and D. Zaunbrecher, 'The Greater Yellowstone Ecosystem: The Ecosystem Concept in Natural Resource Policy and Management', 5:3 *Renewable Resources Journal* 1987, 8, pp. 11–12.

> requires consideration of the whole system rather than individual components. Living species and their physical environments must be recognized as interconnected, and the focus must be on the interaction between different subsystems and their responses to stresses resulting from human activity. Not only does interconnectedness imply management approaches that are broad-based in a spatial sense; it requires as well that human interaction with and use of the environment respect the need for maintaining 'ecosystem integrity', in other words, the system's capacity for self-organization.[56]

Others however focus more squarely on the socio-ecological linkages that the 'ecosystem approach' aims to capture. Grumbine offers a definition based on the inclusion of what an extensive literature review revealed to be ten dominant themes in ecosystem management.[57] The 'ecosystem approach', he suggests, is a management framework that integrates 'scientific knowledge of ecological relationships within a complex sociopolitical and values framework toward the general goal of protecting native ecosystem integrity over the long term'.[58] Indeed, for Grumbine (who, as we will see, is a central supporter of ecocentric articulations of the 'ecosystem approach', a consideration that hints at the complexities involved), 'human values play a dominant role in ecosystem management goals'.[59]

These definitions align with the perception, perhaps most readily shared by international water law scholars, that the emergence of the 'ecosystem approach' can be linked to an increasing focus on the protection of 'purely ecological interests'.[60] Others however, through the socio-ecological linkage, arrive at a shift in perspective whereby the structural link becomes a goal orientation that functionalizes ecological protection towards human needs. Some in this respect define the 'ecosystem approach' as '[t]he careful and skillful use of ecological, economic, social, and managerial principles in managing ecosystems to produce, restore, or sustain ecosystem integrity and desired conditions, uses, products, values, and services over the long run'.[61]

Inclusive definitions and concepts of the 'ecosystem approach' in which the ecological and the socio-economic are fully intertwined, though often with the emphasis tilted towards the latter, are increasingly gaining favour. A good

56 Brunnée and Toope, 'Environmental Security', p. 55.

57 Grumbine's definition and summary of the central elements of the ecosystem approach are still used by the literature (especially in the United States) as the starting point for analysis Thus, e.g. J. Ruhl, 'Ecosystem Services and Ecosystem Management – How Good a Fit? in K. Robbins (ed.), 2013, *The Laws of Nature: Reflections on the Evolution of Ecosystem Management Law and Policy*, Akron: University of Akron Press.

58 Grumbine, 'What is Ecosystem Management?', p. 31.

59 Ibid.

60 McIntyre, 'The Protection', p. 16.

61 Overbay 1992, as quoted in Meffe et al., *Ecosystem Management*, at Box 2.2. (there is no page number).

example is a 2005 scientific consensus statement that frames the 'ecosystem approach' (referred to as 'ecosystem-based management'), as:

> an integrated approach to management that considers the entire ecosystem, including humans. The goal of ecosystem-based management is to maintain an ecosystem in a healthy, productive and resilient condition so that it can provide the services humans want and need. Ecosystem-based management differs from current approaches that usually focus on a single species, activity or concern; it considers the cumulative impacts of different sectors.[62]

In this definition the increasingly significant intersection of the 'ecosystem approach' with the conceptual framework of ecosystem services is quite visible. This is an important point, but for now, it is sufficient to merely make a note of it, and I will return at length on this intersection in Chapter 9.[63]

Policy and legal-institutional definitions

Having offered a representative, though not exhaustive, review of definitions from the scientific and scholarly literature, we can now turn to policy and institutional contexts. In the following I will review *some* of the many existing policy and legal definitions, drawing especially from the context of legal regimes representing the three clusters which will be discussed in the next chapter. A natural starting point is the CBD. As noted above, the definition given to the 'ecosystem approach' in the report of the Malawi Workshop, subsequently endorsed by the Conference of the Parties to the CBD in decision V/6, describes the 'ecosystem approach' as 'a strategy for the integrated management of land, water and living resources that promotes conservation and sustainable use in an equitable way'.[64] This is a very concise definition focusing on the aspect of integration. Importantly, however, the 'ecosystem approach' is also to be 'based on the application of appropriate scientific methodologies focused on levels of biological organization, which encompass the essential structure, processes,

62 K. McLeod et al., *Scientific Consensus Statement on Marine Ecosystem-Based Management*, 2005. The statement was signed by 219 scientists and policy experts, and published by the Communication Partnership for Science and the Sea, p. 1, http://marineplanning.org/wp-content/uploads/2015/07/Consensusstatement.pdf (Accessed 26 September 2018).

63 For an in-depth and critical analysis of this interaction, see V. De Lucia, 'A Critical Interrogation of the Relation between the Ecosystem Approach and Ecosystem Services', 27:2 *RECIEL* 2018a, 104. See also S. Brels, D. Coates and F. Loures, *Transboundary Water Resources Management: The Role of International Watercourse Agreements in Implementation of the CBD*, CBD Technical Series 40, Secretariat of the Convention on Biological Diversity, Montreal, 2008; A. Rieu-Clarke and C. Spray, 'Ecosystem Services and International Water Law: Towards a More Effective Determination and Implementation of Equity?', 16:2 *Potchefstroom Electronic Law Journal* 2013, 12; McIntyre, 'The Protection'; McIntyre, 'Ramsar Convention'.

64 Decision V/6, 2000, op. cit., Annex A para 1.

functions and interactions among organisms and their environment'.[65] The CBD has developed a comprehensive framework in relation to the 'ecosystem approach', with a set of principles and operational guidance. This will be discussed in detail in Chapter 9. The definition just reported is sufficient at this point, for the purposes of this section.

The FAO, which has also devoted significant resources to the development of policy and technical guidelines in relation to the 'ecosystem approach' (albeit in the sectoral context of fisheries), has defined the 'ecosystem approach' as a management framework striving 'to balance diverse societal objectives, by taking into account the knowledge and uncertainties about biotic, abiotic and human components of ecosystems and their interactions and applying an integrated approach to fisheries within ecologically meaningful boundaries'.[66]

The Commission of the Convention for the Protection of the Marine Environment of the North-East Atlantic (OSPAR Commission), which has carried out significant and important work in relation to the 'ecosystem approach' to the prevention, reduction and control of marine pollution, albeit in a regional context, defines it as

> a comprehensive integrated management of human activities based on the best available scientific knowledge about the ecosystem and its dynamics, in order to identify and take action on influences which are critical to the health of the marine ecosystems, thereby achieving sustainable use of ecosystem goods and services and maintenance of ecosystem integrity.[67]

This definition was subsequently adopted by the International Council for the Exploration of the Sea (ICES), a key scientific player in the development of the 'ecosystem approach' in a marine context,[68] and further reiterated in a joint statement on the 'ecosystem approach' of the OSPAR Commission with the Baltic Marine Environment Protection Commission – Helsinki Commission (HELCOM).[69] UNEP, finally, describes the 'ecosystem approach' in terms of its 'emphasis on integrating human needs with conservation practice' while recognizing 'the inter-connectivity between ecological, social-cultural, economic

65 Report of the Workshop on the Ecosystem Approach, Lilongwe, Malawi, 26–28 January 1998, para 8; (hereinafter Malawi Report).

66 FAO, 2003, op. cit., p. 14.

67 OSPAR Convention, BDC, Dublin, 20–24 January 2003, Summary Record, Annex 13, para 6, emphasis mine. See also Statement on the Ecosystem Approach to the Management of Human Activities, 2003, op.cit., (n.4), ANNEX 5 (Ref. §6.1), para 5.

68 International Council for the Exploration of the Sea (ICES), 2005, 'Guidance on the Application of the Ecosystem Approach to Management of Human Activities in the European Marine Environment', ICES Cooperative Research Report no. 273, 4, para 4(1).

69 Statement on the Ecosystem Approach to the Management of Human Activities, 2003, op. cit., Agenda item 6, ANNEX 5 (Ref. §6.1).

and institutional structures'.[70] However, it also more concisely defines the 'ecosystem approach' as 'an integrated process to conserve and improve ecosystem health that sustains ecosystem services for human well-being'.[71]

Conclusions

The preceding sections have illustrated how there exist a significant number of definitions of the 'ecosystem approach', some of which have been reviewed. As anticipated by early commentators,[72] many of the definitions deploy a combination of concepts and terms that rather than clarifying the meaning of the 'ecosystem approach', render it more ambiguous by using 'common words ... in uncommon ways, ... common words ... with multiple meanings, ... common words ... with mutually exclusive definitions' and finally 'new words used in old ways'.[73] An important consideration that emerged from the review is that different *orientations* traverse the definitions of the 'ecosystem approach'. Some definitions, for example (such as the 2005 consensus statement), while focusing on ecosystem health and resilience, are clearly located within a horizon of utility, in the sense that ecosystem functioning is to be preserved 'so that it can provide the services humans want and need'.[74] Others focus more squarely on the integrity of ecosystems and on 'purely ecological interests',[75] without any immediate functionalization to human utility. I will return to these differences at length in subsequent chapters, so for the time being I will ask the reader to simply note these remarks.

This situation is however not surprising, to the extent that, as often remarked in the literature, it reflects the fact that particular epistemic communities, interest groups or institutional contexts infuse the 'ecosystem approach' with disciplinary, institutional and sometimes parochial goals or perspectives.[76] Hence ecologists, government agencies, multi-stakeholder groups and natural resource users' groups all propose significantly different definitions, with emphases consistent with the main stakes for each group.[77] Similarly, it is perhaps not surprising that the different goals pursued by legal regimes or institutional actors translate into different emphases given to different aspects of the 'ecosystem

70 UNEP, *Securing a Green Economy through Ecosystem Management*, International Ecosystem Management Partnership, UNEP-IEMP, 2011a, p. 6.

71 Ibid.

72 Lackey, 'Ecosystem Management'.

73 Ibid., 107.

74 McLeod et al., *Scientific Consensus*, p. 1.

75 McIntyre, 'Ramsar Convention', p. 16.

76 Thus Christensen et al., 'The Report', as well as Brooks et al., *Law and Ecology*; Christensen et al. for example considers that the 'diversity of definitions brings to mind the fable of the six blind men and the elephant, with each definition reflecting the specific perspective, agenda, or interest of the author', each however believing that they may have 'grasped the entire elephant', Christensen et al., 'The Report', p. 668.

77 Brooks et al., *Law and Ecology*, pp. 268–9.

approach'. However, this differentiation runs counter to those narratives that portray the 'ecosystem approach' as a paradigm shift, at least to the extent that multiplicity does not sit well alongside a well defined articulation of a paradigm shift narrative if the very boundaries of the involved paradigms remain elusive. This is another signal of the genealogical complexity of the 'ecosystem approach', and of how genealogy offers a suitable method for an inquiry into the multiple lines of emergence and descent that underpins the 'ecosystem approach'.

Common elements of the 'ecosystem approach'

Another useful way to gain a better understanding of the 'ecosystem approach', and thus to locate the conceptual space it occupies, is to review existing attempts (both doctrinal and institutional) to distil a set of core elements that are shared by the various formulations of the concept, either as a *de minimis* set, or as a set of stipulated elements, the presence or absence of which operate as an indicator, rather than as a precise threshold or minimum requirement.

A useful starting point is, again, Grumbine's work, which very early on, by way of an extensive literature review, distils a number of key elements from ten primary themes that he extrapolates from the review.[78] Grumbine then identifies five central 'goals' of the 'ecosystem approach' (which he calls ecosystem management), namely the maintenance of viable populations of all native species *in situ*; the maintenance of 'representation' of all native ecosystem types within protected areas; the maintenance of ecological integrity; management over long ('evolutionary') temporal scales and the objective to 'accommodate human use and occupancy within these constraints'.[79]

Much more recently, and with particular reference to ocean management, Kimball identifies three 'primary principles' that underlie what she calls an ecosystem-based approach: the first emphasizes that ecosystems must be exploited within the limits of their functioning; the second emphasizes the need for a holistic appreciation of all stressors when taking management decisions; and the third suggests that 'the appropriate geographic scale for assessment and response actions' should be determined by 'ecosystems and their linkages'.[80]

Based on existing practice and with specific reference to living marine resources management, Fabra and Gascón suggest that 'implementation of an "ecosystem approach" in relation to marine policies would comprise at least' seven elements.[81] These include setting ecosystem objectives 'in relation to the

78 The ten themes are: hierarchical context; ecological boundaries; ecological integrity; data collection; monitoring; adaptive management; interagency cooperation; organizational change; humans embedded in nature; values. Grumbine, 'What is ecosystem Management?', p. 31.

79 Grumbine, 'What is Ecosystem Management?', p. 31.

80 L. Kimball, *International Ocean Governance. Using International Law and Organizations to Manage Marine Resources Sustainably*, Gland, Switzerland and Cambridge, UK: The World Conservation Union, 2001, p. 46.

81 Fabra and Gascón, 'The Convention', p. 569

state, services and functioning of the ecosystems';[82] management based on the best scientific knowledge; adaptive management; preservation of biodiversity; participatory governance; policy integration; and the consideration of cumulative effects and long temporal scales.[83] The authors add an eighth element, which they seem to consider not yet stabilized: the need to establish marine protected areas.[84] It may be useful to note how most of these elements are taken from the work done within the context of the CBD on the 'ecosystem approach'.[85]

Trouwborst, finally, extrapolating from a legal perspective, also tries to arrive at a set of *de minimis* elements characterizing the concept. He identifies, more concisely, only three core elements upon which there is, in his opinion, substantial agreement: (1) the holistic management of human activities (2) the requirement to base management on the best available knowledge and (3) the aim of satisfying human needs in ways which do not compromise the integrity and health of ecosystems.[86]

A most recent attempt to arrive at a unified set of elements of the 'ecosystem approach', with the aim of overcoming the fact that 'lack of consensus on its definition has precluded the use of a universal implementation framework',[87] uses a frequency analysis of principles embedded in a number of definitions across domains and regimes in order to arrive at fifteen key principles, including, among others and in order of descending frequency of appearance: ecosystem connections; spatial and temporal scales; adaptive management; use of scientific knowledge; stakeholder involvement; ecological integrity; the coupling of social and ecological systems; and the role of uncertainty.[88]

Attempts at delimiting the 'ecosystem approach' around a (variable) number of core elements can be found also within an institutional context. An important example is offered by the FAO, given the considerable work devoted by FAO to the conceptual and technical development of 'ecosystem approach'. The FAO has indeed identified a number of elements that, it suggests, can be considered as common elements of the 'ecosystem approach', and more precisely of the ecosystem approach to fisheries management (EAF). According to the FAO, most of the 'principles and conceptual elements [of EAF] are already contained in a number of binding or voluntary arrangements, agreements, conventions (global or regional), codes, etc., of direct or indirect relevance to fisheries',[89] and

82 Ibid.
83 Ibid.
84 Ibid.
85 For a detailed analysis of the ecosystem approach in the CBD I refer the reader to Chapter 9.
86 Trouwborst, 'Precautionary Principle', pp. 26–8.
87 R. Long, A. Charles and R. Stephenson, 2015, 'Key Principles of Marine Ecosystem-based Management', 57 *Marine Policy* 2015, p. 53.
88 Ibid., pp. 53–60.
89 FAO, 2003, op. cit., p. 2.

particularly in the Code of Conduct on Responsible Fisheries.[90] The 'ecosystem approach' to fisheries, in other words, is already part and parcel of international conservation practices, and has already found specific legal expression. The main challenge resides then in ensuring coherence of this fragmented picture, and in its implementation.[91] In a 2003 FAO Technical Paper, Garcia et al. identify 17 'principles' of the 'ecosystem approach' to fisheries.[92] These principles are derived from 'various forms of ecosystem management or ecosystem-based management described in the literature or adopted formally by states',[93] and include, *inter alia*: precaution; ecosystem integrity; human and ecosystem well-being; impact minimization; institutional integration; participatory governance; maximum biological productivity; and species interdependence.

An especially important institutional example, for reasons that will be noted presently, is offered by the United Nations Open-ended Informal Consultative Process on Oceans and the Law of the Sea (UNICPOLOS). At its seventh meeting, UNICPOLOS 'agreed by consensus to a number of elements relating to ecosystem approaches and oceans'.[94] The list is complex, and very long and divided into two sets. The first set comprises those elements that should normally be present for a management framework to be following an 'ecosystem approach'. There are 14 elements in this set, including, *inter alia*: 'conservation of ecosystem structures and their functioning and key processes in order to maintain ecosystem goods and services;[95] the balancing of diverse social objectives;[96] the use of best available knowledge;[97] participatory governance;[98] precaution;[99] the appropriate balance between, and integration of, conservation and sustainable use of marine biological diversity'.[100] A second list comprises those elements required 'for the improved application of an ecosystem approach',[101] such as, *inter alia*: identification of ecologically based

90 Ibid., p. 2. Thus also Fabra and Gascón, 'The Convention', p. 571, and W. Turrell, *The Policy Basis of the "Ecosystem Approach" to Fisheries Management*, Aberdeen, UK, 2004: EuroGOOS Publication, 21, p. 20.

91 Thus also Long, who maintains that 'the absence of a universally accepted definition of the "ecosystem approach" or "ecosystem-based management" in international or EU law does not appear to have led to any intractable problems regarding the implementation of the concept in practice', Long, 'Legal Aspects', p. 421.

92 S. Garcia, et al., 'The Ecosystem Approach to Fisheries: Issues, Terminologies, Principles, Institutional Foundations, Implementation and Outlook', FAO Fisheries Technical Paper 443, Food and Agriculture Organization of the United Nations Rome, 2003, pp. 21–6.

93 Ibid., p. 22.

94 ICP-7, 2006, p. 1.

95 ICP-7, 2006, para 6, letter (a).

96 Ibid., letter (e).

97 Ibid., letter (g).

98 Ibid., letter (f).

99 Ibid., letter (h).

100 Ibid., letter (m).

101 ICP-7, 2006, para 8.

management areas;[102] assessment of ecosystem health and indicators;[103] adaptive management;[104] ecosystem monitoring;[105] and addressing the 'root causes' of ecosystem degradation.[106] Noteworthy is the method of selection – consensus – reflecting a political process rather than a scientific methodology which on the other hand informed not only the attempts available in the scientific literature, but also other institutional attempts such as FAO's.

Finally, perhaps the most crucial attempt (considering its influence) at arriving at a delineation of the concept of the 'ecosystem approach' by way of indicating its key elements is offered by the CBD. While details will be discussed at length in Chapter 9, it is useful to note already here how the CBD endorsed the aforementioned Malawi Principles, a set of 12 principles of which the 'ecosystem approach' is comprised, along with a set of four operational guidelines.

From the preceding review, a number of conclusions can be drawn. First, the level of generality of some of these elements (or of entire sets) is such that they provide arguably very little guidance as to how and where to identify an 'ecosystem approach' as distinguished from other resources governance and management frameworks. To say, for example, that one of the key elements of the 'ecosystem approach' is that it is based on the best available science does not really add anything: neither in and of itself, nor vis-à-vis other resources management approaches (who would deliberately not use the best available science?). Second, the particular elements that are listed are sometimes not only vague, but also, and importantly, ambiguous and contested concepts in and of themselves. This important problem shows how even when some precision can be achieved in relation to the elements that characterize a management approach or framework as an 'ecosystem approach', this precision is achieved at the expense of moving the ambiguity from the concept itself to some of its underlying and constitutive elements. This applies to concepts like integrity or health, which are central to almost all descriptions (and definitions) of the 'ecosystem approach' reviewed, implicitly or explicitly, and are often the underlying reference in the construction of legal obligations to protect ecosystems.[107] Moreover, some sets of elements represent primarily a specialized application of the 'ecosystem approach', and hence may not provide sufficient guidance in

102 Ibid., letter (b).
103 Ibid.
104 Ibid., letter (e).
105 Ibid., letter (c).
106 Ibid., letter (f).
107 Such is the case especially within the field of international water law. See e.g. the Great Lakes Water Quality Agreement, where ecological integrity stands central to its legal framework, and, for a doctrinal analysis (albeit dated), A. D. Tarlock, 'International Water Law and the Protection of River System Ecosystem Integrity', 10:2 *Brigham Young University Journal of Public Law* 1996, 181. Recent legal scholarship has indeed begun articulating the case for ecological integrity to fill the role of an environmental *gründnorm*, R. Kim and K. Bosselmann, 'International Environmental Law in the Anthropocene: Towards a Purposive System of Multilateral Environmental Agreements', 2:2 *Transnational Environmental Law* 2013, 285.

relation to a broader or universal framework.[108] Finally, even if some common elements may be, and indeed are, extrapolated, their commonality may still be decisively counteracted by the conflicting orientations that traverse the 'ecosystem approach' as well as some of its constitutive elements, as will become clearer in Chapter 5.

Conclusions

This chapter has endeavoured to broadly locate the 'ecosystem approach'. First, a brief historical outline has shown the multiplicity of historical antecedents that synchronically and diachronically have in different ways tried to integrate an ecosystem perspective into various domains of resource management. This brief historical account has shown the genealogical complexity underpinning the 'ecosystem approach' as it has made visible how the development (or better, the 'descent') of the 'ecosystem approach' is punctuated by starts and stops, by *faux pas* and re-calibrations, all linked as much to the contingencies of history as to the contexts within which its development can be located.

Second, starting from the recognition that responding to the question 'what is the ecosystem approach' is difficult from both a scientific and a legal perspective, the third, fourth and fifth sections have tried more concretely to locate the 'ecosystem approach' by way of reviewing existing definitions and existing attempts to delineate common elements constituting the 'ecosystem approach'. What has emerged thus far is that, on the one hand, there are significant variations in emphases and orientations (particularly in relation to the distinction between what can be tentatively described as a utilitarian and an ecological orientation); and on the other, that attempts at extrapolating core or common elements that may help delineate an 'ecosystem approach' are many and varied, may include very few or very many core elements, sometimes offer effectively no practical guidance, or are otherwise peculiar to a specific context, which limits their utility across regimes.

What emerges ultimately is that what can be located is a broad outline of the *conceptual space* the 'ecosystem approach' occupies, rather than a well-defined concept.

108 This is particularly true in the case of FAO's ecosystem approach to fisheries, which is explicitly considered as an evolution of older fisheries management framework, rather than a paradigm shift, Thus FAO, 2003, op. cit.

4 The multiplicity of the 'ecosystem approach'

Histories, concepts, names

Introduction

As I have argued elsewhere,[1] the 'ecosystem approach' is located within a discursive field of competing narratives and is the result of complex genealogies, situated within a 'play of forces'. My contention is, therefore, that there are ecosystem *approaches*, in the plural, traversing different ideological affiliations and responding differently to the different institutional and regulatory contexts into which they are inserted.

Emerging on the international legal scene in the 1970s by way of the penetration of the language of ecosystems into international law,[2] the 'ecosystem approach' has, from the outset, developed in parallel in a multiplicity of institutional contexts. It is therefore probably more appropriate to speak of a plurality of ecosystem approaches, or of a set of strands whose articulations are potentially quite different, even though they all refer to the same underlying conceptual framework (i.e. ecosystem ecology). Moreover, some of these articulations are widely construed, so that in order to speak of an 'ecosystem approach', it is sufficient that the concept of ecosystem be a significant element of a legal regime. This is the case of the 'ecosystem approach' to international freshwaters, for example. But even more widely, the 'ecosystem approach' can be read and interpreted even into regimes that in substance, if not in language and form, seem to embrace *some* elements of the ecosystem approach. Other articulations refer more narrowly to the 'ecosystem approach' only as a management framework with a set of governing principles and operational guidelines which are far more detailed than the principles that may be implied by the more generic deployment of the concept of ecosystem. The latter is, in many respects, a policy framework and a management methodology, while the former offers law a new set of principles based on some form of normativity immanent in nature.

In this chapter I will address three questions in order to try and further locate the 'ecosystem approach' within the context of international environmental law. The first question relates to the way that the 'ecosystem approach' is framed

1 De Lucia, 'Competing Narratives'.
2 And in particular in the Stockholm Declaration.

and understood. The second section will articulate a distinction between a wide and a narrow concept of the ecosystem approach. This distinction is relevant insofar as it enables the creation of a map of the field, and it further enables deciphering the ways in which the 'ecosystem approach' is read into particular regimes. The second question hinges on the different institutional genealogies of the 'ecosystem approach', which I discuss in terms of clusters. This distinction offers analytical clarity in relation to the particular genealogies – in relation to the prevailing narratives, value orientations and overall legal objectives – that affect the concrete articulations of the 'ecosystem approach'. The final question relates to the ways in which the 'ecosystem approach' is labelled. It is argued that there is a 'labelling knot' that needs to be disentangled in order to be able to navigate in a field affected by a disparate set of nomenclatures where the same thing is sometimes called by different names, and where sometimes the same name is used in reference to different things.

Wide and narrow concept of the ecosystem approach

This section will discuss two distinct concepts of 'ecosystem approach', namely a wide concept and a narrow concept. The two concepts, or understandings, are not necessarily demarcated by sharp boundary lines. Indeed, reading the literature, the sense is that this distinction, which is arguably operative, remains undetected on the part of most, if not all, commentators, even as they implicitly operationalize this analytical distinction.[3] However, the present author considers that, at least as a preliminary question, this distinction must be explicitly discussed as it will help to clarify some of the ambiguities affecting the 'ecosystem approach' which will be considered later in this book. For this purpose, the next two sections will discuss, respectively, the wide and the narrow concept of the 'ecosystem approach'.

Wide concept

Some scholars understand the 'ecosystem approach' in a very wide sense. There are two routes to this wide interpretation. One goes through the concept of ecosystem, the other through what we can call the notion of 'effective' or 'essential' equivalence.

The first route, which can be called the formal route, takes as its starting point the fact that the 'ecosystem approach' is fundamentally linked to the concept of ecosystem. A wide understanding of the 'ecosystem approach' thus reads

3 See e.g. Belsky, 'Using Legal Principles to Promote the "Health" of an Ecosystem', 3 *Tulsa Journal of Comparative and International Law*, 1995, 3, 183,; and in general the discussion of the 'ecosystem approach' in Chapter 13 as regards in particular United Nations Convention on the Law of the Sea (UNCLOS), (adopted 10 December 1982, entered into force 16 November 1994) 1833 UNTS 3.

the concept, in general, into any environmental regime that employs and deploys the concept of ecosystem, from which specific legal consequences can be drawn. In this sense the concept of ecosystem is a crucial way for understanding as well as for framing environmental law and regulation. This perspective leads some theorists to 'see' the 'ecosystem approach' in a wide variety of regimes understood as implementing an 'ecosystem approach' not only directly or explicitly, but also implicitly and indirectly. From this wide perspective then, the 'ecosystem approach' is at work in the Convention on the Conservation of Antarctic Marine Living Resources (CCAMLR);[4] in the UN Fish Stocks Agreement;[5] in the UN Convention on the Law of the Non-Navigational Uses of International Watercourses (Watercourses Convention);[6] and in the Agreement on the International Dolphin Conservation Program.[7] In this view moreover, some scholars[8] consider that even UNCLOS has incorporated an 'ecosystem approach', to the extent that it imposes on States duties to protect the marine environment, including 'rare and fragile *ecosystems* as well as habitat of depleted, threatened or endangered species and other forms of marine life'.[9]

There is, however, a second, yet in many ways overlapping, route. This can be called the substantive route, since it reads the 'ecosystem approach' into legal regimes based on whether a particular regime *essentially* or *effectively* incorporates an 'ecosystem approach', even if there is no formal deployment of the concept or language of ecosystems. Moreover, the implicit inclusion of an 'ecosystem approach' can be inferred from 'broad consideration of biodiversity and the importance of the natural environment and its related functions and services'.[10] For example, the concept of wise use, central to the Ramsar

4 Fabra and Gascón, 'The Convention'.

5 See e. g. E. Metzer, *The Quest for Sustainable International Fisheries: Regional Efforts to Implement the 1995 United Nations Fish Stock Agreement*, Ottawa: NRC Research Press, 2009.

6 See e.g. McIntyre, 'The Emergence'.

7 Thus Y. Tanaka, *A Dual Approach to Ocean Management. The Cases of Zonal and Integrated Management in International Law*, Aldershot: Ashgate, 2008, p. 78.

8 Thus e.g. Belsky, 'Using Legal Principles'; J. Morishita, 'What is the ecosystem approach for fisheries management?', 32:1 *Marine Policy* 2008, 19; H. Wang, 'Ecosystem Management and Its Application to Large Marine Ecosystems: Science, Law, and Politics', 35:1 *Ocean Development & International Law* 2004, 41.

9 UNCLOS, Article 194(5), emphasis mine. Birnie, Boyle and Redgewell maintain that this obligation goes 'beyond the old customary rule based on the *Trail Smelter* arbitration' and extends the duty of States to 'the marine environment as a whole', including areas beyond national jurisdiction, or global commons, in accordance with Principle 21 of the Rio Declaration': Birnie et al., *International Law*, p. 387.

10 E. Metzer, *The Quest for Sustainable International Fisheries. Regional Efforts to Implement the 1995 United Nations Fish Stock Agreement*, Ottawa: NRC Research Press, 2009, p. 144. Metzer lists in this respect the Convention on Wetlands of International Importance especially as Waterfowl Habitat of 1971, the Convention on International Trade in Endangered Species of Wild Flora and Fauna of 1973 and the Bonn Convention on the Conservation of Migratory Species of Wild Animals of 1979.

Convention, while not articulated in the language of the 'ecosystem approach', is understood to be substantively congruent with it.[11] Wise use, it is claimed, can be understood as 'the longest-established example among intergovernmental processes of the implementation of what have become more recently to be known as ecosystem approaches for the conservation and sustainable development of natural resources'.[12] Even the Convention on International Trade in Endangered Species of Wild Fauna and Flora (CITES)[13] has been considered to embody, albeit implicitly, an 'ecosystem approach', to the extent that listing criteria set out under CITES[14] 'allow risks posed by diminished or degraded habitat, introduced species and pollution to be taken into account, in addition to overharvesting'.[15] In this respect, the requirement in article IV(3) of CITES that 'exports of CITES-listed species must be monitored so as to ensure the role of the species in its ecosystem' reflects, some argue, 'an aspect of the ecosystem approach'.[16]

Finally, the FAO also follows a 'substantive' line of reasoning. The FAO takes the view that while the specific language of the 'ecosystem approach … may not yet be common in international instruments, regional conventions or arrangements and national legislation, the underlying principles and conceptual objectives … appear in many of them'.[17]

Narrow concept

The narrow concept of the 'ecosystem approach', on the other hand, focuses on a more technical, or narrowly methodological, meaning and refers to the 'ecosystem approach' as defined, for example, within the CBD, that is, as a 'strategy for the integrated management of land, water and living resources'.[18] Generally, *this* 'ecosystem approach' is understood as a management framework,

11 C. Finlayson et al., 'The Ramsar Convention and Ecosystem-Based Approaches to the Wise Use and Sustainable Development of Wetlands', 14:3/4 *Journal of International Wildlife Law & Policy* 2011, 176, p. 191.

12 Ibid., p. 177.

13 Convention on International Trade in Endangered Species of Wild Fauna and Flora (adopted 3 March 1973, entered into force 1 July 1975) 993 UNTS 243.

14 Adopted with CITES Resolution Conf. 9.24 (Rev. CoP16), 'Criteria for amendment of Appendices I and II', Ninth meeting of the Conference of the Parties Fort Lauderdale (United States of America), 7–18 November 1994

15 Kimball, *International Ocean Governance*, p. 33.

16 D. Currie, *Ecosystem-Based Management in Multilateral Environmental Agreements: Progress towards Adopting the Ecosystem Approach in the International Management of Living Marine Resources*, WWF, 2007; http://assets.panda.org/downloads/wwf_ecosystem_paper_final_wlogo.pdf (Accessed 26 September 2018), p. 39.

17 Garcia et al., 'Ecosystem Approach', p. 15. Thus also T. Aqorau, 'Obligations to protect marine ecosystems under international conventions and other instruments' in M. Sinclair and G. Valdimarsson (eds), *Responsible Fisheries in the Marine Ecosystem*, Wallingford, UK and Cambridge, MA: FAO and CAB International, 2003.

18 Decision V/6, 2000, op. cit.

or methodology, and is considered a policy principle, rather than a legal concept.[19] Understood as a methodology, the 'ecosystem approach', rather than indicating a particular obligation to protect natural ecosystems and the conceptual and material breadth of such obligations, is more narrowly a technical and management instrument (or at best a best practice) intended to facilitate the fulfilment of primary substantive obligations. For example, The Conference of the Parties to the CBD has explicitly acknowledged how the 'ecosystem approach' does not possess a legally binding character, but it is rather intended to offer a framework of 'flexibility and experimentation' in the implementation of the substantive obligations under the CBD.[20]

In many respects, this concept of the 'ecosystem approach' can be understood as the application of a style of research characterized by the qualifier 'an ecosystem approach to....'.[21] As deLaplante observes, this style or method can be equally applied to environmental management, to public education, to human well-being, to fisheries, or to urban development.[22] Its main feature is that it entails situating the focus of the investigation within its broader environmental context, on the assumption that at least some of its properties depend on interactions and relations with the environment within which the focus is situated.[23]

A second meaning of a narrow approach refers to its sectoral focus. The 'ecosystem approach' to fisheries is a prime example of such narrow sectoral understanding. Within the context of fisheries management in fact, the 'ecosystem approach' is utilized precisely as a style of research, with a double purpose. On the one hand, it has been recognized how, compared with traditional fisheries management practices, it is today necessary to integrate into management decisions a large ensemble of elements affecting the abundance of fish stocks. On the other, the traditional goal of achieving the standard goals of fisheries management, that is, that of ensuring the rational, optimal or sustainable utilization of the fisheries resources, remains.

19 Thus Birnie et al., *International Law*, p. 386. They based their position largely on the fact that the ICJ has 'consistently', as they put it, refused 'to redraw maritime boundaries in accordance with environmental or ecosystem considerations', p. 386, fn 33. For a contrary opinion however see Wang, which argues that 'the obligation to adhere to the ecosystem-based management', with particular reference to the marine environment and to marine living resources, 'has been established in international law', H. Wang, 'Ecosystem Management and Its Application to Large Marine Ecosystems: Science, Law, and Politics', 35:1 *Ocean Development & International Law*, 2004, 41, p. 54. See also B. Sage, 'Precautionary Coastal States' Jurisdiction', 37 *Ocean Development and International Law* 2012, 359, and Long, 'Legal Aspects'.

20 Decision VII/11, 'Ecosystem approach', COP7, Kuala Lumpur, 9–20 and 27 February 2004, Annex II, Section A (1)(2).

21 K. deLaplante and J. Odenbaugh, 'What Isn't Wrong with Ecosystem Ecology' in R. Skipper et al. (eds), *Philosophy Across the Life Sciences*, Cambridge, MA: MIT Press, 2006.

22 Ibid.

23 Ibid.

Concluding remarks

Having outlined the distinction between a wide and narrow approach to the concept of 'ecosystem approach', it must be underlined by way of conclusion that this is primarily an *analytical* distinction. Additionally, there is arguably an ongoing process of convergence whereby the narrow and wide concepts, even when materially (and not only analytically) distinct, are moving closer together. This convergence is in large part a consequence of the increasing saliency in international environmental discourse of the narrative and conceptual framework of ecosystem services. The latter in fact increasingly permeates and (re-)orients the discourse of the 'ecosystem approach', although it is also true that ecosystem services are integrated *within* the context of an 'ecosystem approach' as central conceptual tools. This permeation is thus best understood as interpenetration, so that the two are increasingly interwoven, or at least 'closely linked'.[24] As we will see in detail in Chapter 9, this interpenetration is particularly evident within the context of the CBD, but it can also be observed within the context of freshwater law and fisheries law and law of the sea more broadly.

A further process which contributes to this convergence derives from the increasing interaction between Conventions; an interaction that, developed in relation to particular issues areas or themes, contributes to the development of the 'ecosystem approach'. For example, the CBD Secretariat, in a report issued under its aegis, recognizes the linkages between an 'equitable and sustainable allocation and management of' water resources and the maintenance of the 'ecological function of freshwater water ecosystems'.[25] The Ramsar Convention, in turn, has developed a series of technical handbooks for the wise use of wetlands, which can play a significant role in the interpretation of the 'ecosystem approach' within the context of the international watercourses regimes.[26]

Finally, the narrow concept is increasingly introduced explicitly or otherwise articulated in regimes originally only employing the concept of ecosystem without the specific implications of the 'ecosystem approach' as a system or method of management. This is particularly evident in the context of the law of the sea and international fisheries law.

24 McIntyre, 'The Protection', p. 92. For a critical analysis of this link, see De Lucia, 'Critical Interrogation'.

25 S. Brels, D. Coates and F. Loures, *Transboundary Water Resources Management: The Role of International Watercourse Agreements in Implementation of the CBD*, CBD Technical Series 40, Secretariat of the CBD, 2008, p. 5.

26 Ramsar Convention, *Handbooks for the Wise Use of Wetlands*, 4th edn, 2010; McIntyre, 'The Protection', p. 88, specifically speaks of the role of these Handbooks in relation to the clarification of the 'due diligence requirements of the ecosystems approach', on p. 90.

Legal-institutional clusters

Introduction

Another important element to explore when trying to locate the 'ecosystem approach', and to track its genealogy, is related to the legal-institutional clusters within which it has developed and continues to be developed. Looking at the different legal-institutional clusters within which various articulations of the 'ecosystem approach' develop arguably sheds light on the particular genealogies – in relation to the prevailing narratives, value orientations, relevant actors, institutional context, and overall legal objectives – that affect concretely the 'ecosystem approach', and dovetails with the distinction between a wide and narrow interpretation made in the previous section.

Turrell,[27] in reviewing the policy basis of the 'ecosystem approach' to fisheries, identifies what he calls 'three strands' of ocean governance development (the UNCLOS process, the UNCED process and the FAO process). This matrix has provided inspiration for a similar analytical operation within the larger context of international environmental law. However, while the idea of development strands relates primarily to the identification of linear pathways of development, my intention is rather that of underlining the institutional complexities at work in relation to the genealogical development of the 'ecosystem approach'. For this reason, I will rather speak of *clusters*. The 'ecosystem approach' has arguably evolved differently within different legal and regulatory regimes, understood in the sense of loosely subject- or theme-specific clusters of legal norms, principles and institutions. While all clusters are interrelated and communicate with one another, and increasingly so, the 'ecosystem approach' has taken and continues to take a particular orientation in each cluster and exhibits particular emphases and, importantly, genealogies. These distinctive orientations are significant enough, I submit, to warrant their organization within a historical-analytical – indeed genealogical – matrix based on different clusters. The way in which the concept of ecosystem has been articulated within what will be described as the Freshwater cluster, for example, is certainly very different from the way in which the 'ecosystem approach' to fisheries has developed within the FAO. Notwithstanding an increasing convergence across clusters, there remain legacy differences and at times incompatible orientations that are best understood, and explained, by way of a genealogical analysis. The three clusters I will focus on are the Biodiversity cluster, the Ocean and Fisheries cluster, and the Freshwater cluster.

Biodiversity cluster

The biodiversity cluster comprises a number of Conventions and other International Agreements. While the most important of these is arguably the CBD,

27 Turrell, 'Policy Basis'.

together with its two Protocols,[28] the broader regime also includes the Convention on Conservation of Migratory Species of Wild Animals (CMS),[29] the Convention on Wetlands of International Importance especially as Waterfowl Habitat (Ramsar Convention),[30] CITES, the International Treaty on Plant Genetic Resources for Food and Agriculture (Plant Treaty),[31] and the Convention Concerning the Protection of the World Cultural and Natural Heritage (Heritage Convention).[32]

The 'ecosystem approach' within this cluster has been, to a large extent, developed thanks to the work done within the context of the CBD. Traditional conservation has been found to be affected by the so-called 'pathology of natural resources management'.[33] The 'ecosystem approach' was thus introduced, not surprisingly, to overcome the limitations of classical conservation and management tools. There were identified in e.g. a narrow focus on a particular site, without consideration of the interlinkages with adjacent or otherwise connected sites; too much emphasis on species characteristics or protected areas, considering that biodiversity resides largely outside of protected areas; lack of integration with social and economic dimensions of conservation, including the problematic of market failures or distortion.[34]

During the meeting held in Jakarta in 1995, the 2nd Conference of the Parties (COP) to the CBD adopted the 'ecosystem approach' as 'the primary framework of action to be taken under the Convention',[35] following the recommendation contained in the 1st Report of the Subsidiary Body on Scientific, Technical and Technological Advice (SBSTTA).[36] SBSTTA had been

28 The Cartagena Protocol on Biosafety to the Convention on Biological Diversity, COP EM-1/1 (adopted 29 January 2000, entered into force 11 September 2003) and the Nagoya Protocol on Access to Genetic Resources and the Fair and Equitable Sharing of Benefits Arising from their Utilization, COP 10 Decision X/1 to the Convention on Biological Diversity (adopted 29 October 2010, entered into force 12 October 2014).

29 Convention on Conservation of Migratory Species of Wild Animals (adopted 23 June 1979, entered into force 1 November 1983) 1651 UNTS 333.

30 Convention on Wetlands of International Importance especially as Waterfowl Habitat (adopted 2 February 1971, entered into force 21 December 1975) 996 UNTS 245.

31 International Treaty on Plant Genetic Resources for Food and Agriculture of 2001 (3 November 2001, adopted entered into force 29 June 2004) 2400 UNTS 303.

32 Convention Concerning the Protection of the World Cultural and Natural Heritage (adopted 16 November 1972, entered into force 17 December 1975) 1037 UNTS 151.

33 C. Holling and G. Meffe, 'Command and Control and the Pathology of Natural Resource Management', 10:2 *Conservation Biology*, 1996, 328.

34 For a comprehensive list see Malawi Report, 1998, op. cit., para 14.

35 Decision II/8, 1995, op. cit., p. 55.

36 Recommendation I/3 on 'Alternative ways and means in which the Conference of the Parties could start the process of considering the components of biological diversity particularly those under threat and the identification of action which could be taken under the Convention', COP1 Report of the Subsidiary Body on Scientific, Technical and Technological Advice, 4–8 September 1995, Paris, Annex, p. 25.

tasked by COP1, with identifying alternative ways, means and actions that could be taken to implement the Convention.[37]

The COP has developed the 'ecosystem approach' in a number of subsequent decisions. The most salient[38] are decisions V/6 and VII/11. The former endorses a particular description of the 'ecosystem approach', including a set of principles (the so-called Malawi Principles, discussed in the next section) and operational guidance. The Malawi Principles will be discussed at length in Chapter 9, so I will not linger on it now, except to mention that they integrate a multiplicity of perspectives, some of which are at odds with one another, and testify to the genealogical complexity involved in the emergence of the 'ecosystem approach'. However, it can be suggested that the overall orientation is rather holistic, and aimed at overcoming some of the traditional limitations of the political epistemology of modernity, which erected and policed a strict separation between nature and culture, conservation and utilization etc.

Ocean and fisheries cluster

The cluster of ocean and fisheries is very broad and comprises two intertwined development processes, one related to the UNCLOS, and the other to the FAO. While even within this cluster there exists a plurality of articulations, the *endemic* inflection of the 'ecosystem approach' that develops within this cluster takes primarily a very specific orientation, related to its specific application to fisheries management. This 'specialized' 'ecosystem approach' to fisheries (EAF), applies the ecosystem methodology to the particular problematics of fisheries management in a sectoral manner.

The comprehensive work done within the context of the FAO represents a key contribution towards the development of the 'ecosystem approach' within the ocean and fisheries cluster, so it will be the focus of this section

37 That particular task had been marked as 'priority item', Decision I/7, 'Subsidiary Body on Scientific, Technical and Technological Advice', Report of COP1, Nassau, 28 November–9 December 1994, Annex, where the draft provisional agenda for SBSTTA was set.

38 The ecosystem approach is also mentioned in a number of 'secondary' decisions focusing primarily on other thematic and cross-cutting issues; in these, the ecosystem approach is included as the integrative framework upon which the implementation of these particular areas of biodiversity conservation is to be based; see e.g. Decision IV/4, 'Status and trends of the biological diversity of inland water ecosystems and options for conservation and sustainable use', COP4, Bratislava, 4–15 May 1998; Decision II/10, 'Conservation and Sustainable Use of Marine and Coastal Biological Diversity' in Report of the COP2, Jakarta, 6–17 November 1995 and IV/5 'Conservation and sustainable use of marine and coastal biological diversity, including a programme of work' in Report of the COP4, Bratislava, 4–15 May 1998; Decision II/16, 'Statement to the international technical conference on the conservation and utilization of plant genetic resources for food and agriculture', COP2, Jakarta, 6–17 November 1995 and Decision III/11, 'Conservation and sustainable use of agricultural biological diversity', COP3, Buenos Aires, 4–15 November 1996.

(notwithstanding other contributions).[39] The FAO's competence in relation to fisheries derives from article I of the FAO Constitution. Article I(2) requires the FAO, *inter alia*, to 'promote and where appropriate ... recommend national and international action with respect to ... nutrition, food and agriculture',[40] and the conservation of natural resources,[41] where the terms 'agriculture and its derivatives' include fisheries as well as other marine products.[42]

As far back as the early 1970s, the FAO began working towards the development of ecosystem-oriented approaches to fisheries management, in particular through a number of technical conferences aimed at widening the set of relevant variables and impacts to be considered for the further development of fisheries management.[43] These conferences laid the foundation for elaborating ecosystem-oriented fisheries management frameworks that tried to address the failure of traditional fisheries management models (TROMs) to achieve sustainable outcomes. Yet the FAO has consistently insisted upon the continuity between TROMs and its own inflection of the 'ecosystem approach', which it has aptly denominated ecosystem approach to fisheries (EAF), in order to signal its *specialized* field of application, and its sectoral orientation. Indeed, as we shall see in more details in the fourth section of Chapter 4, the FAO has explicitly distinguished EAF from *other* ecosystem approaches. The FAO in fact emphasizes how EAF is not a paradigm shift, as most of its 'principles and conceptual elements ... are already contained in a number of binding or voluntary arrangements, agreements, conventions (global or regional), codes, etc., of direct or indirect relevance to fisheries'.[44] This approach, in other words, is already part of international conservation law and practices. Its main challenges reside, according to the FAO, in ensuring coherence in this fragmented picture and in its implementation.[45] The FAO work has also other, important implications related to the particular orientation the 'ecosystem approach' takes in the context of fisheries, and how it relates to other artiulations, but the relevant points will be discussed in Chapter 4, 'Labels and terminologies'.

39 As already seen, Turrell, in reviewing the policy basis of the ecosystem approach to fisheries, identifies what he calls 'three strands' of ocean governance, the UNCLOS process, the UNCED process and a specific FAO process, Turrell, 'Policy Basis'.

40 Constitution of the United Nations Food and Agriculture Organization, 40 AJIL Supp. 76 CTS 1945/32; also in *Basic Texts of the Food and Agriculture Organization of the United Nations*, Volume I and II, 2013 edition, pp. 3–16 (hereinafter FAO Constitution), Article I(2)(a).

41 FAO Constitution, op. cit., Article I(2)(c).

42 Ibid., Article I(1).

43 Such as the FAO Technical Conference on Marine Pollution and its Effects on Living Resources and Fishing, Rome, 1970 and the FAO Technical Conference on Fishery Management and Development, Vancouver, 1972.

44 FAO, 2003, op. cit., p. 2.

45 Garcia et al., 'Ecosystem Approach', and Long, 'Legal Aspects', p. 421.,

Freshwater cluster

The freshwater cluster offers a different articulation of the 'ecosystem approach', one less focused on its methodological character and more on its more immediate legal implications. In this respect, the 'ecosystem approach' in freshwater law translates into obligations of protection and preservation of ecosystems,[46] or perhaps one could rather say that it derives from those obligations.

The first outline of an 'ecosystem approach' to transboundary water management is to be found in the context of the International Joint Commission of the United States and Canada, particularly as regards the Great Lakes Water Quality Agreement.[47] The need for an 'ecosystem approach' was outlined in a 1978 report of the Great Lakes Research Advisory Board, in which it was made clear that simply relying on 'knowledge of the chemical and physical water quality' was not enough.[48] The planning and management of 'such a priceless resource as the Great Lakes', the report continues, 'requires … understanding of the total ecosystem and the diverse interactions which occur within its chemical, physical, biological and societal components'.[49] The board then presented the case for a shift from a water quality objectives approach – the approach taken in the 1972 agreement – to an 'ecosystem approach'.[50] This report was translated into law in a 1978 revision of the Great Lakes Water Quality Agreement, which identifies, as its overall purpose, the restoration and maintenance of 'the chemical, physical, and biological integrity of the waters of the Great Lakes Basin Ecosystem'.[51]

While the Great Lakes Agreement is a bilateral agreement, the first multilateral (albeit regional) agreement to introduce the concept of 'ecosystem approach' to

46 See e.g. Article 20 of the UN Watercourses Convention, adopted by the General Assembly of the United Nations on 21 May 1997 by Resolution 51/229, annex, Official Records of the General Assembly, Fifty-first Session, Supplement No. 49. (A/51/49), entered into force 19 May 2014.

47 However, domestically ecosystem management surfaced in US legal literature from at least the late 1960s, see e.g. L. Caldwell, 'The Ecosystem as a Criterion for Public Land Policy', 10:2 *Natural Resources Journal* 1970, 203. Caldwell both recognized the need to refashion land policy on 'ecologically valid' criteria, and the fact that 'an ecosystem approach encounters resistance to the degree that it is inconsistent with the values, assumptions, institutions, and practices that shape the prevailing social arrangement which affect the custody and care of the land', p. 204.

48 Great Lakes Research Advisory Board, *The Ecosystem Approach. Scope and Implications of an Ecosystem Approach to Transboundary Problems in the Great Lakes Basin*, Special Report to the International Joint Commission, Windsor: Ontario, Canada, July 1978, p. 2.

49 Ibid.

50 Great Lakes Research Advisory Board, *The Ecosystem Approach*. For an assessment of the ecosystem approach within the context of the Great Lakes see the special issue, 'Making Sense of the Ecosystem Approach: Lessons from The Great Lakes', 20:3 *Alternatives Journal*, 1994. A seminal article on the potential implications of fully implementing an ecosystem approach in international water law is Brunnée and Toope, Environmental Security'. For a current review of the ecosystem approach in freshwater law see McIntyre, 'The Emergence'.

51 Great Lakes Water Quality Agreement, Article II.

freshwater law was the 1993 Water Convention. The Water Convention produced the Guidelines on the 'ecosystem approach' in Water Management in 1993.[52] The Guidelines were developed in order to 'assist UN/ECE Governments' towards the development and implementation of an 'ecosystem approach to day-to-day water management'.[53]

Another important step in the development of the 'ecosystem approach' within the context of international freshwater law is represented by the Watercourses Convention, the 'culmination of over 20 years of deliberation within the International Law Commission'.[54] The Watercourses Convention requires State parties to the Convention to act to protect and preserve the ecosystems of international watercourses (Article 20). The 'ecosystem approach', within the context of the Watercourses Convention, is very broadly conceived, to the extent that it generically indicates the inclusion within its protection and preservation objectives of the full range of factors that may have an impact on the relevant watercourse.[55]

There is one immediately peculiar dimension of the 'ecosystem approach' in relation to freshwater. Its development is significantly less linked to its methodological articulation aimed at the achievement of other primary goals (fisheries management; biodiversity conservation etc.) than is the case in both the biodiversity and the ocean and fisheries clusters. Within the context of freshwater law, the 'ecosystem approach' is rather more centrally linked to the substantive elaboration of *primary* obligations of protection and preservation of watercourses ecosystems, in this respect significantly expanding the obligations of States vis-à-vis the traditional transboundary harm rule.[56]

It is true that more recently the articulation of the 'ecosystem approach' in freshwater law has become more technical and methodological, thanks to the increasing linkage with the framework of ecosystem services. It is also true that even in the CBD, ecosystems, as one of the dimensions of biodiversity, are direct objects of protection, and thus of primary obligations in relation to the overall objectives of the CBD. However, the difference I have just identified, is arguably significant and resonates with the distinction discussed earlier between a wide and narrow interpretation of the concept of 'ecosystem approach'.

52 UNECE, *Guidelines on the Ecosystem Approach in Water Management*, 1993.

53 Ibid., p. 1.

54 McIntyre, 'The Emergence', p. 2.

55 McCaffrey, *The Law of International Watercourses: Non-Navigational Uses*, Oxford: Oxford University Press, 2003, p. 382.

56 Thus for example McCaffrey, *International Watercourses*, p. 394. Yet, while McCaffrey suggests that Article 20 contains an obligation to protect and preserve the ecosystems of international watercourses that is not necessarily related to transboundary effects that may arise from a failure to protect and preserve, the ultimate horizon of responsibility remains transboundary, as the effects, from a complex ecosystem perspective, are only not immediately apparent. The obligations thus remain related to other States, rather than immediately to the ecosystems of the relevant watercourses, Ibid., p. 394.

Convergence and institutional connectivity

The distinction in clusters, while analytically useful, should not be overstated. There is in fact a discernible process of convergence that is ongoing and at an increasing pace, in line with the speed at which the 'ecosystem approach' is being internalized in an increasing number of regimes. There are also many infra- and inter-cluster relations of cooperation and coordination, which make the boundaries of the clusters extremely porous. As regards infra-cluster relations, other Conventions have mostly embraced[57] the work done in the CBD on the 'ecosystem approach' *ex post*. The Ramsar Convention has re-defined the concept of wise use through the vocabulary of the 'ecosystem approach' as developed within the CBD,[58] to render it congruent with it.[59] Similarly, the Convention on Migratory Species has acknowledged the importance of the 'ecosystem approach' through cooperative arrangements with the CBD, and has recognized how the 'CBD's "ecosystem approach" should be useful as a tool for migratory species conservation and sustainable use'.[60]

57 L. Jing, *Preservation of Ecosystems of International Watercourses and the Integration of Relevant Rules*, Leiden: Brill/Nijhoff, 2014, p. 9.

58 Finlayson et al., 'Ramsar Convention', p. 176.

59 Ramsar Secretariat, 2010, *Wise Use of Wetlands: Concepts and Approaches for the Wise Use of Wetlands*, Ramsar Handbooks for the Wise Use of Wetlands, 4th edn, Ramsar Convention Secretariat, p. 28. This congruence has been reiterated in Annex A to Resolution IX.1 adopted by the 9th Conference of the Contracting Parties, Kampala, Uganda, 2005; para 22 therein in fact, updates the definition of wise use in the following terms: '[w]ise use of wetlands is the maintenance of their ecological character, achieved through the implementation of ecosystem approaches, within the context of sustainable development', footnotes omitted. Of interest is the plural reference (i.e. 'ecosystem *approaches*') as it refers in a footnote to the CBD and the joint HELCOM and OSPAR Declaration, indicating how the Ramsar Secretariat acknowledges the existence of *different* articulations of the 'ecosystem approach' in different specialized regimes. This congruence is transformed into identification by some literature to the extent that wise used is considered to be *one* form or model of an ensemble of ecosystem approaches. In the words of one study, '[w]ise use is the longest-established example among intergovernmental processes of the implementation of what have become more recently to be known as ecosystem approaches for the conservation and sustainable development of natural resources', Finlayson et al., 'Ramsar Convention', p. 177 (but the authors also speak of congruence, p. 191, again emphasizing the similarity of the substance, and underlining how the main difference is in the language used (also p. 191, where the authors recognize how wise use 'may not have been articulated as such', that is, in the language of ecosystem approach, yet it does reproduce its substantive management philosophy). Finally, as underlined by one report, the centrality of the 'ecosystem approach' within the context of the Ramsar Convention 'is reinforced by the requirement to place the wetland within a greater biogeographic region', P. Griffin, *The Ramsar Convention: A New Window for Environmental Diplomacy?* Vermont: Institute for Environmental Diplomacy & Security, University of Vermont, 2012, p. 10.

60 CMS, *Cooperation with Other Bodies CBD/CMS Joint Work Programme (2002–2005)*, Seventh Meeting of the Conference of the Parties Bonn, 18–24 September 2002, (UNEP/CMS/Inf.7), p. 11.

As for inter-cluster convergence, it may be useful to note how, given the significant role of international fisheries law for the conservation of marine biodiversity,[61] there is a significant amount of porosity, cross-fertilization and interaction between the FAO and the CBD in relation to the 'ecosystem approach' applied to marine living resources.[62] Additionally, the Ramsar Convention lies at an increasingly important intersection between the biodiversity and freshwater regimes,[63] and there is significant interaction between the CBD and the freshwater cluster, especially through the prominent role that the framework of ecosystem services has acquired for the further elaboration of the 'ecosystem approach' in both clusters.[64]

Finally, convergence is facilitated also at a broader, cross-cluster level. There are in fact several international institutions and organizations providing external connectivity between the various clusters. Some examples may help illustrate this type of convergence. First, UNEP and UNGA play an especially important role in relation to both the development of the conceptual apparatus that sustains and develops the ecosystem approach, and the construction of the overall normative framework within which the 'ecosystem approach', as a legal concept, policy tool and political objective, is situated. This connecting role is especially evident in what Turrell called the 'UNCED strand',[65] and in what the FAO calls the 'institutional path',[66] that is, a string of UN Conferences on, broadly speaking, environment and development that, since 1972, have been central to the development of international environmental law and of many of its principles and goals.

The 1972 Stockholm Conference on the Human Environment (UNCHE),[67] for example introduced the concept of ecosystem into international law and also implicitly recognized a number of the principles that form the basis of the 'ecosystem approach', such as the principles of integrated management and of stakeholder participation.[68] Moreover, the 1972 Conference led to the creation of

61 T. Henriksen, 2010, 'Conservation and Sustainable Use of Arctic Marine Biodiversity: Challenges and Opportunities', 1:2 *Arctic Review on Law and Politics* 2010, 249, p. 262.

62 See e.g. Decision II/10, 'Conservation and Sustainable Use of Marine and Coastal Biological Diversity', Report of COP2, Jakarta, 6–17 November 1995.

63 See e.g. McIntyre, 'The Protection'.

64 See e.g. McIntyre, 'Ramsar Convention' and De Lucia, 'Critical Interrogation'.

65 where UNCED stands for United Nations Conference on Environment and Development

66 FAO, 2003, op. cit., p. 75. FAO includes in this institutional path also UNCLOS and CCAMLR, and, at the onset, a number of FAO Technical Conferences where the need for new management approaches began to be discussed, see FAO Technical Conference on Marine Pollution and its Effects on Living Resources and Fishing, Rome, 1970 and FAO Technical Conference on Fishery Management and Development, Vancouver, 1972; ibid., p. 75.

67 Report of the United Nations Conference on the Human Environment, Stockholm, 5–16 June 1972, UN Doc. A/CONF.48/14/Rev.1.

68 Thus, Turrell, 'Policy Basis', p. 10. As for the principle of stakeholder participation, see Decision V/6, 2000, op. cit., Annex B 'Principles of the Ecosystem Approach'; the principle of integration takes central place in the very definition of the ecosystem approach endorsed by the COP of the CBD, 2000, op.cit, Annex A, para 1.

UNEP, a central actor in the further development of the 'UNCED strand' of the ecosystem approach.[69]

The 1992 Rio Summit (the UNCED) is also decisive, in three primary respects. First, it produces an outcome document (the Rio Declaration)[70] that, after re-affirming the principles adopted in Stockholm, sets out a number of principles that would become central to the 'ecosystem approach', such as, e.g., the principles of integration (Principle 4), of participation in environmental decision-making (Principle 10), and of precaution (Principle 15). Second, Agenda 21, a policy document ('action plan') adopted in Rio, explicitly discusses the 'ecosystem approach', most notably in Chapter 17, in relation to oceans and fisheries.[71] Finally, the CBD was open for signatures at the Rio Summit. The CBD has had a fundamental role in mainstreaming the 'ecosystem approach' in international law. Moreover, the CBD, while inserted within the biodiversity cluster, would prove a key forum for the development of the 'ecosystem approach', as we shall see in Chapter 9.

Finally, the World Summit on Sustainable Development, held in Johannesburg in 2002 is also notable in that its outcome document encouraged 'the application by 2010 of the ecosystem approach' in relation to ocean and fisheries management,[72] and that it promoted 'the wide implementation and further development of the ecosystem approach' in relation to biodiversity conservation.[73]

A last modality of convergence relates to scholarship. Scholarship in fact also significantly traverses the various clusters and contributes to this process of convergence. Indeed, in light of the elusiveness of the concept of the 'ecosystem approach', and considering its relative novelty in international environmental law, it is not surprising that scholarly discussions and doctrinal interpretations of the concept rely on a broad set of sources that traverse the clusters described here (notwithstanding the specific orientations of much scholarly literature, in particular that literature devoted to the application of the 'ecosystem approach' to fisheries).

Despite this convergence however, the distinction in clusters remains alive in both institutional settings and scholarly endeavours.[74] Moreover, it retains

69 Turrell, 'Policy Basis'

70 Report of the United Nations Conference on Environment and Development, Rio de Janeiro, 3–14 June 1992, Annex I, Rio Declaration On Environment and Development (Vol. I).

71 Agenda 21, Chapter 17, 'Protection of the oceans, all kinds of seas, including enclosed and semi-enclosed seas, and coastal areas and the protection, rational use and development of their living resources'.

72 Report of the World Summit on Sustainable Development Johannesburg, South Africa, 26 August–4 September 2002, para 30(d). Direct reference ('noting') is also made to Reykjavik Declaration on Responsible Fisheries in the Marine Ecosystem and Decision V/6, 2000, op.cit.

73 World Summit on Sustainable Development Johannesburg, 2002, op. cit., para 44(e).

74 Cluster-specific scholarship exists, but perhaps especially the CBD represents an increasingly central reference point.

arguably a clear methodological benefit, as it makes visible the genealogical pathways through which the 'ecosystem approach' has developed in each of the clusters, and the different conditions, contestations, orientations and contingencies involved. Indeed, while there is a single narrative of the 'ecosystem approach', there are multiple ecosystem approaches.

Label and terminologies

Another preliminary question that needs to be dealt with relates to the ways in which the 'ecosystem approach' is labelled. As anticipated in the introduction, there is a 'labelling knot' that must be disentangled to navigate a field affected by a disparate set of nomenclatures which offer a double layer of complexity (or confusion): in fact sometimes the same concept is given different names; some other times however the same name is used in reference to different concepts. This contributes significantly to the general, fundamental ambiguity affecting the 'ecosystem approach' and translates into a tangle of terminologies utilized in various normative contexts. Moreover, some of the terminologies are aligned with different value orientations – and in this sense, this section anticipates the main theme of the next chapter – again, however, without consistency.

Some scholarship identifies four frameworks that are encompassed by the overall concept of the 'ecosystem approach': ecosystem management; ecosystem-based management; ecosystem approach *stricto sensu*[75] and ecosystem stewardship.[76] However, while the identification of these four frameworks is based on a review of the literature, it appears to the present author to maintain a level of ambiguity even as it tries to analytically schematize the conceptual and terminological landscape. For example, as regards ecosystem-based management, Uy and Shaw suggest that while 'conceptually based on ecosystem approaches, ecosystem-based management differs from ecosystem management in that it focuses [on] an ecosystem perspective'.[77] This seems to offer little clarity.

Among the labels available in the (scientific, policy and legal) literature one finds for example: ecosystem approach; ecosystem management; ecosystem-based approach to management; ecosystem-based management; ecosystem process-oriented approach; total ecosystem management; integrated approach; integrated

75 In an important part of the literature the current expression is in fact ecosystem approach. To distinguish this legal concept with the ecosystem approach as a trajectory, which is the primary focus of this book, and as already explained in the introductory chapter, I use single quotation marks to mark the ecosystem approach as a trajectory, and that is also the expression under which I subsume the variety of labels that are used – inconsistently – in the literature. I refer then to the ecosystem approach *stricto sensu* to a more narrow delineation of the concept, one that matches the usage in the literature and in a number of legal documents and instruments.

76 Uy and Shaw, *Ecosystem-based Management*, pp. 6–10.

77 Ibid., p. 8.

agro-ecosystem approaches and holistic approach.[78] But even more elaborate labels can be found. Thus, Wang reports that

> [t]here are many other related terms, such as a bioregional approach; bioregional planning; ecoregion-based conservation; watershed management approach; holistic, intersectoral, and interactive approach; ecosystem approach that integrate the conservation of biological diversity and the sustainable use of biological resources; ecosystem approach that integrate the conservation and sustainable use of biological diversity as well as socioeconomic considerations; and precautionary ecosystem management approach.[79]

CBD COP decision VII/11 similarly describes a number of management frameworks as being in close proximity to the 'ecosystem approach', without however establishing a direct equivalence.[80] Indeed, the COP notes how a number of other 'existing approaches', such as for example sustainable forest management (whose relation with the 'ecosystem approach' is discussed at length within the context of the CBD as we shall see in Chapter 9) 'may be consistent with the application of the Convention's ecosystem approach, and support its implementation in various sectors or biomes'.[81] These are '"ecosystem based management", "integrated river-basin management", "integrated marine and coastal area management", and "responsible fisheries approaches"'.[82] The implementation of the 'ecosystem approach', suggests decision VII/11, 'can be promoted by building upon the approaches and tools developed specifically for such sectors'.[83]

One central distinction which is sometimes only terminological, but sometimes articulates a substantive difference, is the one between 'ecosystem approach' and ecosystem management. Wang, for example, conceptually separates ecosystem management and the 'ecosystem approach', the latter, he suggests, being an 'important term associated with ecosystem management'.[84] Wang however concedes that '[t]he meanings of these two terms partly overlap'.[85] And so does the terminology, which is in many respects difficult to disentangle. Hatcher and Bradbury for example, while

78 Some of which are articulated in particular within the context of thematic or cross-cutting programmes of work under the CBD, J. Stadler and H. Korn, 'The "Ecosystem Approach" in the Light of COP Decisions and Background Papers' in H. Korn et al. (eds), *Report on the Scientific Workshop on "The Ecosystem Approach – What Does it Mean for European Ecosystems?* Bonn: BfN–Bundesamt für Naturschutz, 2004, and Wang, 'Ecosystem Management', p. 43.

79 Ibid., p. 43, fn 24 (the corresponding footnote text from which the quotation is taken is located at p. 63).

80 Decision VII/11, 2004, op. cit., para 8.

81 Ibid.

82 Ibid.

83 Ibid.

84 Wang, 'Ecosystem Management', p. 43.

85 Ibid., p. 43.

endorsing criticisms of the 'variable use of the term',[86] refer to, in the same article,[87] 'marine ecosystem management',[88] 'ecosystem-based approaches to management',[89] 'integrated management',[90] 'ecosystem approaches',[91] 'ecosystem approach to marine management'[92] and 'ecosystem-based management' (with and without the qualifier 'marine'),[93] before concluding that '[i]n short, ecosystem management is part and parcel of what we also call integrated coastal zone management and good oceans governance'.[94] A recent WWF report also presents entangled terminologies, considering that the title of the report is '*Ecosystem-Based Management* in Multilateral Environmental Agreements: Progress towards Adopting the *Ecosystem Approach* in the International Management of Living Marine Resources'.[95]

Another relevant dimension relates to the fact that the different terminologies, which may arise from significant conceptual differences, may also, more trivially, represent semantic habits which eventually crystallize. The label 'ecosystem approach' which emerged within the context of the CBD seems to be an example of the latter. The report of the Malawi Workshop, where nomenclature is explicitly discussed, recorded that '[i]t was debated whether the term "ecosystem approach" was not preferable to "ecosystem-based approach"'. The report suggested at first that 'the term "ecosystem-based approach" would reflect better the particular type of reasoning and analysis to tackle the objectives to implement the Convention',[96] particularly as it would not imply 'a focus on any particular ecosystem', but rather the deployment of principles and methods to be used to manage 'particular areas of land and water'.[97] 'However', the workshop report concludes, 'as the term "ecosystem approach" has been used throughout the discussions within the Convention, it was felt that it was advisable to continue to use this term'.[98]

One significant distinction relates to the *object* of 'management'. Labels such as 'ecosystem approach' or ecosystem-based management – as opposed to, for

86 Hatcher and Bradbury, 'Marine Ecosystem Management', p. 206.
87 And it is not always obvious to the reader (or at least *this* reader) whether the different nomenclatures are a stylistic variation or imply specific reference to different conceptual and management frameworks.
88 Hatcher and Bradbury, 'Marine Ecosystem Management' (this particular expression is contained in the title of the article)
89 Ibid., p. 206.
90 Ibid., p. 205.
91 Ibid., p. 206.
92 Ibid., p. 207.
93 Ibid., p. 221.
94 Ibid., p. 225.
95 Currie, *Ecosystem-Based Management*, emphases mine.
96 Report of the Workshop on the Ecosystem Approach, 1998, op. cit., para 12.
97 Ibid.
98 Ibid.

example, ecosystem management – are supposed to incorporate the notion that it is 'human activities which are being managed, rather than the ecosystem within which they take place'.[99] The label 'ecosystem approach' is in fact generally considered to have superseded the label ecosystem management, more common in North American usage and practice: 'ecosystem management has been largely abandoned because of the twin realizations that we do not govern nature, we govern people, and that ecosystems are much too large and complex realities to try to manage in any direct sense'.[100] However, the wording used in scientific reports, policy documents and legal texts is not always cognizant of such differences. In the text of the Malawi Principles, the object of management is ecosystems: 'ecosystems should be managed for …';[101] 'need to … manage the ecosystem in an economic context';[102] 'ecosystems must be managed …'.[103]

UNEP provides another example, as its own programme for the promotion of the utilization of the 'ecosystem approach' is called 'Ecosystem Management Programme'.[104] Different labels are also often used interchangeably in the same official documents,[105] to the point that confusion is inevitable. UNEP provides another illustrative example, through its usage of both ecosystem management and 'ecosystem approach' in its publications, sometimes even in the mixed form of 'ecosystem management approach'. EU regulations also often refer, in the same legal text, to different labels, such as, for example, the regulation establishing a Programme to support the further development of an Integrated Maritime Policy,[106] which uses the expressions 'ecosystem approach',[107] 'ecosystem-based management'[108] and 'ecosystem-based approach to the management of …' interchangeably.[109]

99 As reported in the Report of the Reykjavik Conference on Responsible Fisheries in the Marine Ecosystem, Reykjavik, Iceland, 1–4 October 2001, at Section H, para 7 of the Observations from the Scientific Symposium by the Co-chairs, Mr Michael Sinclair and Mr Jóhann Sigurjónsson. This conclusion was then incorporated in ICP-7, 2006, op. cit., para 29: 'Many delegations underlined that ecosystem approaches should address the management of human activities affecting oceans and seas, and not the management of ecosystems per se'.

100 D. Wilson, 'The Paradoxes of Transparency Science and the Ecosystem Approach to Fisheries Management in Europe', MARE Publication Series 5, Amsterdam: Amsterdam University Press, 2009, p. 167. However, some doctrine still considers that 'ecosystem approaches to marine management universally identify the ecosystem as the object of management., Hatcher and Bradbury, 'Marine Ecosystem Management', p. 207. Emphasis mine.

101 Malawi Principles, 2010, op. cit., Principle 1, in Decision V/6, 2000, op. cit., Section B.

102 Ibid., Principle 4.

103 Ibid., Principle 6.

104 See UNEP, Six Priority Areas Factsheets - Ecosystem management, 2010.

105 Such is the case with ICP-7, 2006, op. cit. where throughout one can find 'ecosystem approach', 'ecosystem approaches' and 'ecosystem-based management'.

106 European Parliament and Council Regulation (EU) 1255/2011 of 30 November 2011 establishing a Programme to support the further development of an Integrated Maritime Policy Text with EEA relevance [2011] OJ L321/1.

107 Ibid., Preamble, para 11.

108 Ibid., Article 3(2)(b).

109 Ibid., Article 3(3)(c).

CBD COP Decision V/6 also discusses both ecosystem management and the 'ecosystem approach'. However, the relationship between the two is conceptualized in terms of how the 'ecosystem approach' can facilitate ecosystem management, hence distinguishing the two concepts and establishing the 'ecosystem approach' as a broader framework that serves to orient the other.[110] The language is further jumbled, however, when, under the operational guidance for the 'ecosystem approach', the decision states that '*ecosystem management* has to be carried out even in the absence of [certain] knowledge', and to this purpose, the '*ecosystem approach* can facilitate practical management by *ecosystem managers*'.[111] This linkage, or connection, between the 'ecosystem approach' and ecosystem management resonates with at least some literature discussing explicitly 'an 'ecosystem approach' to 'ecosystem management'.[112] Indeed, this literature suggests that 'ecosystem management be based on the ecosystem approach', itself defined as an approach 'based on complex systems theory'.[113] It is only *this* 'ecosystem approach' to ecosystem management that 'truly represent a paradigm shift in the management of ecological systems'.[114] Yet in other cases the two are considered equivalent.[115] Others still, speak of an ecosystem-based management approach.[116]

This terminological confusion, which is at least in part a legacy of an evolutionary conceptual process (a genealogical 'descent'), may also be read as a symptom of conflicting views trying to find space within a novel discourse (conflicting views which will be discussed in more detail in the next chapter). In this latter respect, naming is an important process in any attempt to impose particular closures. Yet again, some of the terminological and conceptual differences may be traced to different understandings of the concept of ecosystem.

The focus on the management of ecosystems seems to imply their recognition as the object of protection. This implication seems clearly established in CBD Article 8, dedicated to 'in-situ conservation'. Article 8(d) identifies 'the protection of ecosystems'[117] as one of the forms through which *in-situ* conservation is to be achieved. Less clear is the extent to which ecosystem management implies a specific spatial ontology of ecosystems, considering the explicitly functional understanding of ecosystems within the CBD. Indeed,

110 Decision V/6, 2000, op. cit., Annex B 'Operational Guidance', Guideline 1.

111 Ibid., emphasis mine.

112 A. Crober, The Ecosystem Approach to Ecosystem Management, Senior Honours Thesis, Department of Geography, Faculty of Environmental Studies, University of Waterloo. Unpublished, 1999.

113 Crober, 'Ecosystem Approach', p. 3.

114 Ibid.

115 'Much discussion about ecosystem management, or taking an ecosystem approach, emphasizes the need to …', J. Kay, H. Regier, M. Boyle and G. Francis, 'An Ecosystem Approach for Sustainability: Addressing the Challenge of Complexity', 31:7 *Futures* 1999, 721, p. 722.

116 R. Siron et al., 'Ecosystem-Based Management in the Arctic Ocean: A Multi-Level Spatial Approach', 61:1 *Arctic* 2008, 86, p. 87.

117 CBD, Article 8(d). See also Decision V/6, 2000, op. cit., Section A, para 3.

the definition of ecosystems contained in the CBD does not indicate any particular spatial, material ontology, pointing rather to 'dynamic complex[es]' and 'functional unit[s]'.[118] That is, and by contrast to the terms 'biome' or 'ecological zone', an ecosystem is defined in relation to the 'problem being addressed', and it may well be 'a grain of soil, a pond, a forest, a biome or the entire biosphere'.[119] The only naturalistic spatial concept defined in the CBD is the concept of 'habitat', defined as 'the place or type of site where an organism or population naturally occurs'[120] (and yet, provisions relating to *in-situ* conservation, which include ecosystems as one of the three dimensions of biological diversity, would argue for a spatial framing of ecosystems). The conceptual distinction between ecosystem and habitat is clearly marked in the definition of 'in-situ conservation' contained in Article 2, as well as in Article 8(d), both of which indicate that conservation is to include both 'ecosystems and natural habitats', hence underlining the distinction.

By contrast, the term 'ecosystem approach' refers more explicitly to a methodological approach.[121] The implication is that an 'ecosystem approach' in the methodological sense may very well be implemented sectorally, or focus on a specific or narrow target or objective (such as a species, or an environmental target) which is to say, without the application of a fully holistic framework.

Such is the case of the FAO's 'ecosystem approach' to *fisheries* management or the EU's ecosystem-based *fisheries*[122] and *marine*[123] management. In this respect, the interesting question is whether the different terminologies refer to substantive differences in the understanding of the concept. As already mentioned, one significant distinction relates to the object of 'management'. Labels such as the 'ecosystem approach' or ecosystem-based management – as opposed to, for example, ecosystem management – supposedly incorporate the notion that it is 'human activities that are being managed, rather than the ecosystem within which they take place'.[124]

118 CBD, Article 2.
119 Decision V/6, 2000, op. cit., Section A, para 3.
120 CBD, Article 2.
121 See, for example, Recommendation IX/6, 'Ecosystem approach: further elaboration, guidelines for implementation and relationship with sustainable forest management' SBSTTA 9, Montreal, 10–14 November 2003, Annex I, Section A, para 4, where the 'ecosystem approach' is defined as a 'methodological framework for supporting decisions in policy-making and planning', and as an implementation tool.
122 European Parliament and Council Regulation (EU) 1380/2013 of 11 December 2013 on the Common Fisheries Policy [2013] OJ L354/22.
123 Council Directive 2008/56/EC of 17 June 2008 establishing a framework for community action in the field of marine environmental policy (Marine Strategy Framework Directive) [2008] OJ L164/19.
124 As reported in the Report of the Reykjavik Conference on Responsible Fisheries in the Marine Ecosystem, Reykjavik, Iceland, 1–4 October 2001, at Section H, para 7 of the Observations from the Scientific Symposium by the Co-chairs, Mr Michael Sinclair and Mr

Within the context of fisheries management, the FAO has attempted to systematize the terminological landscape, and has mapped it onto 'two different but related [and converging, in the FAO's view] paradigms':[125] ecosystem management and fisheries management. According to the FAO's analysis – and here a novel, important differentiation emerges – the former takes an ecocentric perspective,[126] to the extent that it aims at the protection and conservation of 'the structure, diversity and functioning of ecosystems', especially through protected areas. The latter, aimed as it is at meeting 'the goals of satisfying societal and human needs for food and economic benefits through management actions that focus on the fishing activity and the target resource',[127] is more anthropocentric. The FAO, importantly, identifies sustainable development as the unifying element, as the framework that may facilitate a convergence between the two.[128] The FAO then distinguishes *both* between terminologies and between the underlying represented concepts, or paradigms.[129]

Considering that the next chapter discusses the conflicting views and the competing narratives that traverse the concept of the 'ecosystem approach', this last distinction delineated by the FAO seems an appropriate way to conclude the last substantive section of this chapter.

Conclusions

This chapter has offered some preliminary, yet important, analytic clarifications. Both the distinction between a wide and a narrow concept of the 'ecosystem approach', and the distinction in clusters serve the general purpose of providing a backdrop for the more detailed analyses that are carried out in the next two chapters aimed at exploring the genealogical complexity underlying the 'ecosystem approach' and the competing narratives within which it is situated. The distinction in clusters, in particular, serves the purpose of emphasizing how the different legal-institutional clusters within which the 'ecosystem approach' has developed, contributes to track its complex genealogies, to the extent that such complexity is in a number of respects linked to the differential genealogical paths – and the different types of contestations – at work in the different clusters (notwithstanding the ongoing process of convergence and the important role that particular institutions and organizations, as well as scholarship and doctrine, fill in relation to connecting the different articulations).

Jóhann Sigurjónsson. This conclusion was then incorporated in ICP-7, op. cit., para 29: 'Many delegations underlined that ecosystem approaches should address the management of human activities affecting oceans and seas, and not the management of ecosystems per se'; in this sense also Dir. 2008/56/EC, op. cit.

125 FAO, 2003, op. cit., p. 11.

126 Ibid.

127 Ibid.

128 Ibid.

129 Garcia et al., *Ecosystem Approach*; 'Terminology and Paradigms' is the title of Chapter 1.

A final section of this chapter also dealt in detail with questions of terminology. The 'labelling knot', as I call it, is an important question that on the one hand testifies to the confusion affecting scholarly inquiries into the 'ecosystem approach'; on the other hand, terminological plurality and instability is also linked to significantly different orientations that the 'ecosystem approach' may take, orientations that are, however, only sometimes reflected in the different terminologies deployed in the literature. In other words, the different terminologies and the related semantic and conceptual overlaps, contradictions and ambiguities are also a clear sign of the genealogical complexities underlining the descent and emergence of the concept.

The chapter has also shown, especially in the section entitled 'Legal-institutional clusters', how the 'ecosystem approach', to use the vocabulary of this book, represented in all three clusters a nominal problematization of traditional resource management practices. Yet the articulations within the different clusters have taken distinct orientations: holistic in the biodiversity cluster; sectoral and avowedly anthropocentric in the oceans and fisheries cluster; normative rather than methodological in the freshwater cluster.

In conclusion, this chapter, while providing this preliminary set of necessary clarifications, has also made visible what will be more centrally explored in the next chapters, namely the competing narratives and the complex genealogies within which the 'ecosystem approach' is situated.

5 Underneath the 'ecosystem approach'

Ecology as a framework of ambiguity

Introduction

As we have seen in the preceding chapter, attempting to provide an answer to the question 'what is an ecosystem approach' is a difficult task from both a scientific and legal perspective.[1] This chapter begins to explore the reasons why such difficulty is not only present and evident, but also, to a large extent inevitable and irreducible. This chapter focuses on the problematics that arise from recognizing that ecology is a framework of ambiguity, both at the most general level of ecology as a field of knowledge, and at the more particular level of the concept of ecosystem, the central naturalistic underpinning of the 'ecosystem approach'. Such recognition moreover, rather than prompting us to seek the most reliable or true articulation of ecology, alerts us to the irreducible genealogical nature of this field of knowledge, in a way that further corroborates the usefulness of the genealogical approach taken in this book, particularly in relation to the 'ecosystem approach'.

This chapter first discusses ecology as a framework of ambiguity in relation to its complex genealogy (section two) and in relation to its double epistemic role – as a science and as a worldview (section three). Section four considers in more detail a number of the complexities and competing perspectives that traverse ecology – at the intersection of its double epistemic role – complexities that articulate in diverse ways the oscillation between the competing paradigms of order and chaos. Section five, finally, addresses these complex genealogies as they specifically operate in relation to the concept of ecosystem.

The complex genealogies of ecology

Ecology as a science is traversed by disciplinary conflicts between different theoretical perspectives, as a number of philosophers of science have illustrated.[2]

1 Long, 'Legal Aspects'.

2 D. Keller and F. Golley (eds), *The Philosophy of Ecology: From Science to Synthesis*, Athens, GA: University of Georgia Press, 2000; M. Sagoff, 'The Plaza and the Pendulum: Two Concepts of Ecological Science', 18:4 *Biology and Philosophy* 2003, 529; deLaplante and Odenbaugh, 'What Isn't Wrong'; M. Loreau, *From Populations to Ecosystems: Theoretical Foundations for*

Many of its central concepts, rather than being the result of rational scientific practices, are rather the result of a series of social, political, institutional and economic contingencies.[3] Ecology is, in this respect, the result of a complex genealogy, never beyond the 'play of forces', never beyond the reach of contingency and conflict; never, furthermore, a complete, neat, singular, linear, coherent field of knowledge unravelling closer and closer to the truth; always susceptible, rather, to the impurities and lacerations of genealogical histories. Indeed, ecology can be understood constitutively as 'the borderland between the biological and social sciences',[4] regardless of the more recent disappearance of this dimension, in light of the attempts of ecologists to transform ecology into a 'hard science'.[5]

This complex genealogy is first visible in the multiplicity of competing theoretical frameworks operating under the label of ecology. The first great divide in this respect is that between community and ecosystem ecology. Each having developed its own concepts, theories and methodologies,[6] the divide remains significantly in place, despite recent work attempting to arrive at a synthesis.[7] While some measure of disagreement and theoretical competition is to be expected, some of the debates traversing the science of ecology and some of the controversies affecting the relationship between competing theoretical perspectives are best characterized precisely as lacerations. The controversy about the contribution of (bio)diversity to the stability of ecosystems is one example: the conflict has been so heated that some media commentators have spoken of 'full war among ecologists'.[8] Another example, as we shall see in more detail later, is the range of conflicting views – and the degree of incompatibility among them – regarding the concept of ecosystem.

a New Ecological Synthesis, Princeton: Princeton University Press, 2010; K. deLaplante and V. Picasso, 'The Biodiversity-Ecosystem Function Debate in Ecology' in K. deLaplante, B. Brown and A. Peacock (eds), *Handbook of the Philosophy of Science. Volume 11: Philosophy of Ecology*, North Holland: Elsevier, 2010.

3 Thus e.g. F. Golley, *A History of the Ecosystem Concept in Ecology: More than the Sum of its Parts*, New Haven, CT: Yale University Press, 1993, as regards the concept of ecosystem. See also J. B. Hagen, *An Entangled Bank: The Origins of Ecosystem Ecology*, New Brunswick, NJ: Rutgers University Press, 1992, and G. Mitman, *The State of Nature: Ecology, Community, and American Social Thought, 1900–1950*, Chicago and London: University of Chicago Press, 1992.

4 Mitman, *State of Nature*, p. 1.

5 Ibid., p. 1. See also T. Allen, A. Zellmer and C. Wuennenberg, 'The Loss of Narrative', in B. Beisner and K. Cuddington (eds), *Ecological Paradigms Lost: Routes of Theory Change*, North Holland: Elsevier Academic Press, 2005, which for example observes how '[t]hese are times of an unrelenting mechanistic binge in ecology', p. 33.

6 Loreau underlines how community and ecosystem ecology 'provide two different perspectives on the same material reality', M. Loreau, *From Populations to Ecosystems Theoretical Foundations for a New Ecological Synthesis*, Princeton: Princeton University Press, 2010, p. ix; deLaplante and Picasso similarly state that 'Population/community ecology and ecosystem ecology present very different perspectives on ecological phenomena', deLaplante and Picasso, 'Biodiversity-Ecosystem Function', p. 169.

7 Keller and Golley, *Philosophy of Ecology*; Loreau, *From Populations.*

8 deLaplante and Picasso, 'Biodiversity-Ecosystem Function', p. 169.

The even greater, and apparently insurmountable, divide between the narratives of order and chaos (with the latter paradigm having spawned an alleged paradigm shift in environmental law),[9] continues to fuel debate in the philosophy and history of science literature,[10] and is the source of conflicting views in the legal literature.[11] From yet another perspective, the sharp distinction between individualist and collectivist approaches is equally controversial.[12] This distinction consciously or unconsciously informs the entire history of the science of ecology, which then is understood as either a normative model from which to draw social and political insights, or as a field of knowledge to be normatively aligned with the prevailing social and political ideas.[13]

Ecology between science and worldview

Ecology however lies, in a deeper sense, at the centre of several epistemological and ontological intersections and informs several particular knowledge domains, such as philosophy, law, economics, ethics, as well as the practices of social movements and of subsistence communities. Besides being a scientific discipline, ecology inspires in fact what can be described as an ecological *worldview*. As Gregg Mitman emphasizes:

> Ecology as a subject and as a discipline has come to symbolize many things … the environmental movements of the 1960s, … population growth control, … Aldo Leopold's 'land ethic'[,] a scientific discipline that studies the interaction between organisms and their environments.[14]

9 See e.g. Tarlock, 'Nonequilibrium Paradigm', and D. Botkin, 'Adjusting Law to Nature's Discordant Harmonies', 7:1 *Duke Environmental Law and Policy Forum* 1996.

10 D. Worster, 'The Ecology of Order and Chaos', 14:1/2 *Environmental History Review, (Conference Papers)* 1990, 1; C. Eliot, 'The Legend of Order and Chaos: Communities and Early Community Ecology' in K. deLaplante, B. Brown and A. Peacock (eds), *Handbook of the Philosophy of Science. Volume 11: Philosophy of Ecology*, North Holland: Elsevier, 2010.

11 See e.g. the so called Pardy-Ruhl Dialogue on Ecosystem Management, in which one of the central questions was the merit of the conceptual framework of the New Ecology, whose main axiom is that nature is in a state of constant flux, hence relegating the notion of 'balance of nature' to the realm of myth (thus in particular D. Botkin, *Discordant Harmonies: A New Ecology for the Twenty-first Century*, 1st edn, Oxford: Oxford University Press,.1990. The dialogue comprises five papers published in *Pace Environmental Law Review*: B. Pardy, 'Changing Nature: The Myth of the Inevitability of Ecosystem Management', 20:2 *Pace Environmental Law Review*, 2003, 675; J. Ruhl, 'The Myth of What Is Inevitable under Ecosystem Management: A Response to Pardy', 21:2 *Pace Environmental Law Review* 2004, 315; B. Pardy, 'Ecosystem Management in Question: A Reply to Ruhl', 23:1 *Pace Environmental Law Review* 2006, 209; J. Ruhl, 'The Pardy-Ruhl Dialogue on Ecosystem Management, Part IV: Narrowing and Sharpening the Questions', 24:1 *Pace Environmental Law Review* 2007, 25; B. Pardy, 'The Pardy-Ruhl Dialogue on Ecosystem Management Part V: Discretion, Complex-Adaptive Problem Solving and the Rule of Law', 25:2 *Pace Environmental Law Review* 2008, 341.

12 See e.g. Eliot, 'The Legend'.

13 Hagen, *Entangled Bank*; Mitman, *State of Nature*.

14 Mitman, *State of Nature*, p. 1.

One of the central characteristics of ecology and of the ecological worldview in particular, is its malleability, its susceptibility to being deployed within any number of articulatory practices. In this sense ecology provides an illuminating example of a field of discursivity which cannot be 'sutured',[15] but remains continuously and irreducibly contested and the object of struggles.

Ecology, then, is in many ways a placeholder whose meaning and content remain suspended between a number of competing senses. Its first and more immediate sense refers to ecology as a science. *This* ecology is firmly located within the paradigm of modern science. The term ecology however evokes a second field of meaning. Ecology in this second sense is a source of normativity. It operates not as a science, but as a philosophical and ethical framework, which is most commonly deployed with the normative aim of (re)orienting the relationship between humanity and the non-human world.[16] In this sense then, ecology is not simply a field of knowledge, but rather more broadly underpins and sustains an entirely new worldview. Yet, deLaplante shows how there is not *one* ecological normativity, from which it is possible to derive clear normative implications. Indeed, he argues that the opposite is true: it is impossible to derive '*unique* philosophical consequences'[17] from ecological science. To a significant extent, deLaplante observes, this is due to the fact that there is a multiplicity of ecological sciences (or theoretical articulations thereof), each of which is in turn able to support a multiplicity of ecological metaphysics and epistemologies.[18] The best that can be done, deLaplante suggests, is to narrow the range of possible inferences.[19]

Both horizons – the science and the worldview – are in this sense often inextricably intertwined. Moreover, they are affected by mutually reverberating internal ambiguities and contestations. As regards ecology as a science, at least some of its theoretical articulations are contested in relation to their scientific pedigree due to their soft or weak predicting abilities, as Sagoff, among others, underscores.[20] Moreover, ecology is increasingly described as postmodern[21] or

15 This is the term employed by Laclau and Mouffe when arguing that no field of discursivity can be ultimately fixed; for example, they suggest that '[t]here is no single underlying principle fixing – and hence constituting – the whole field of differences'. Thus every totality remains always incomplete, it can never be 'sutured and self-defined', E. Laclau and C. Mouffe, *Hegemony and Socialist Strategy: Towards a Radical Democratic Politics*, 2nd edn, London: Verso, 2001, p. 111.

16 As deLaplante explains, 'ecological science has been viewed … as a foundational source for the development of metaphysical, epistemological and normative views' regarding the relationship between humanity and nature, deLaplante, 'Environmental Alchemy', p. 361.

17 Ibid.

18 Ibid., pp. 361–80.

19 Ibid.

20 Sagoff, M., 'The Plaza and the Pendulum: Two Concepts of Ecological Science', *Biology and Philosophy*, 18:4, 2003, 529'.

21 Allen, T., Zellmer, C. and Wuennenberg, C., 'The Loss of Narrative', in K. Cuddington and B. Beisner (eds), *Ecological Paradigms Lost: Routes of Theory Change*, Burlington, MA: Elsevier Academic Press, 2005.

post-normal,[22] in light of its epistemological pluralism, its inevitable entanglement with values, and the central role assumed by uncertainty and complexity.

From this (postmodern) point of view, science can no longer provide certainty regarding statements of facts or assessments of risks, as ignorance acquires a crucial epistemic role.[23] Moreover, in the context of 'truth pluralism'[24] – that is of competing claims to truth which cannot be adjudicated objectively – science is fully revealed to be 'normative science'.[25] This, in a certain sense, is not new: knowledge and values are entangled on both objective and subjective grounds.[26] Every decision, then, carries within it specific normative, ethical and political commitments arising from both scientific and legal processes.[27] These considerations partly displace the problematic of deriving normative implications from science,[28] as normative choices are *always already* embedded in science itself. Ecology, in this sense, is *always already* a worldview.

Ecology then, as we will see throughout this chapter, is traversed irreducibly by an entanglement of epistemologies and of value orientations which effectively makes it a carrier of specific philosophical commitments that align with different social and political projects. The literature distinguishes – in relation to the contributions of ecology to different philosophical and ethical orientations – two different positions, called environmentalism and ecologism[29] or shallow and

K., deLaplante, 'Is Ecosystem Management a Postmodern Science?' in K.E Cuddington. and B.E Beisne (eds), *Ecological Paradigms Lost: Routes of Theory Change*, Burlington, MA: Elsevier Academic Press, 2005.

22 deLaplante, 'Is Ecosystem Management'; deLaplante specifically discusses Ravetz and Funcowitz' framework of post-normal science in relation to the ecosystem approach, see S. Funtowicz and J. Ravetz, 'Three Types of Risk Assessment and the Emergence of Postnormal Science', in D. Golding and S. Krimsky (eds), *Social Theories of Risk*, Westport, CT, Praeger, 1992; and S. Funtowicz and J. Ravetz, 'Science for the Postnormal Age', 25:7 *Futures*, 1993, 739.

23 Thus for example Tallacchini, 'Legal Framework'.

24 S. Gutwirth and E. Naim-Gesbert, 'Science et droit de l'environnement: réflexions pour le cadre conceptual du pluralism de vérités', 34 *Revue interdisciplinaire d'études juridiques* 1995, 33.

25 Ibid., p. 61: 'Le pluralisme de vérités montre bien que ce rôle absolu, extra-politique, extra-juridique et indiscutable accordé à la science et ses verités est inacceptable'. See also, on normative science, R.T., Lackey, 'Appropriate Use of Ecosystem Health and Normative Science in Ecological Policy' in D. J. Rapport et al. (eds), *Managing for Healthy Ecosystems*, Boca Raton, FL: CRC Press, 2002.

26 See Tallacchini, 'Legal Framework', p. 1096; K. Shrader-Frechette, *Risk and Rationality. Philosophical Foundations for Populist Reforms*, Berkeley: University of California Press, 1991; K. Shrader-Frechette, 'Methodological Rules for Four Classes of Scientific Uncertainty' in J. Lemons (ed.), *Scientific Uncertainty and Environmental Problem Solving*, Oxford: Blackwell Science, 1996.

27 Tallacchini, 'Legl Framework', p. 1095.

28 See for example deLaplante, 'Environmental Alchemy', pp. 361–80, where he observes how it is highly problematic to derive *unique* philosophical consequences from the science of ecology.

29 Thus, Tallacchini, *Diritto*.

deep ecology.[30] Particularly with regards to the latter pair, the distinction intends to provide a demarcation line which separates *shallow* liberal environmentalism – the prevalent ideological orientation in environmental law, and environmental discourse more generally, today[31] – from *deep* ecological critiques, the latter emphasizing the need for a rupture with the entire prevailing cultural paradigm.[32]

Daniel Worster speaks in this respect of the 'moral ambivalence' of ecology;[33] an ambivalence evident in its two main traditions: an 'imperial' one, which expresses an anthropocentric, engineering vision of the world; the other 'arcadian', drawing from ecology a philosophical and ethical vision, aligned with the values of romanticism, holism and ecocentrism.[34] This moral ambivalence or ambiguity of ecology is an important point, which will be discussed further later in the book, in relation to anthropocentrism and ecocentrism, and their limitations as critical concepts (Chapter 7), and in relation to biopower and biopolitics (Chapter 8). Now, however, I will address in some detail some of the ways in which ecology as a science slips into and intersects with ecology as a worldview.

Multiple ecologies

This section will offer two illustrations of the ways in which ecology has been used towards different, indeed opposed, political projects, or has been stretched between competing visions. The first example will discuss the competing paradigms of order and chaos, the second the competing visions of the social implications of ecology, and their polarization in either social competition or cooperation.

Between order and chaos

In its formative period, the science of ecology was dominated by a paradigm of order and equilibrium, usually referred to as the 'balance of nature' paradigm. The American ecologist Frederic Clements, a key figure in the history of ecological sciences, was the leading figure of this paradigm. While Clements focused on a *dynamic* ecology which tracked all the changes implicated by the processes of vegetation succession, he ultimately viewed ecosystems as inevitably reaching

30 Thus, A. Næss, 'The Shallow and the Deep, Long-Range Ecology Movement. A Summary', 16:1 *Inquiry* 1973, 95–100. See also Curry, *Ecological Ethics*.

31 See, *ex pluribus*, Wilkinson, 'Using Environmental Ethics' and, especially, Bernstein, *The Compromise*.

32 Næss' original distinction hinged on the fact that shallow environmentalism fights instead 'against pollution and resource depletion' and that its 'central objective [is] the health and affluence of people in the developed countries', Næss, 'The Shallow and the Deep', p. 95.

33 Worster, *Nature's Economy*.

34 Ibid.

what he called a 'climax state'.[35] Nature, in Clements' view, was oriented towards stability in ways that could be understood and predicted. As Eliot summarizes, Clements' view hinges at its most basic on two claims: 'that vegetation develops … in a way comparable to, or literally identical with, the development of an individual organism',[36] what Clements called a 'superorganism', due to the particularly close integration and level of self-organization achieved by a given community of plants;[37] and that 'the development of vegetation in an area necessarily culminates in a particular type of vegetation, called that area's "climax" which is determined by its climate'.[38]

From this perspective, it is clearly possible to identify, ontologically, an entity – a thing, something – that can be protected and preserved, insofar as it is 'real and orderly'.[39]

This paradigm, based on an allegedly deterministic progression of natural communities towards a climax state of equilibrium, has been replaced by a paradigm of contingency and chaos, in which each individual component entertains only the most tenuous and contingent relationship with its neighbours within a natural community,[40] and in which the overall organizing principle is constant change, rather than equilibrium.

This shift began to appear already in the 1940s, when Clements' theory of the climax still dominated the field of ecology (at least in the United States). Then,

> a few scientists began trying to speak a new vocabulary. Words like 'energy flow', 'trophic levels', and 'ecosystem' appeared in the leading journals, and they indicated a view of nature shaped more by physics than botany. Within another decade or two nature came to be widely seen as a flow of energy and nutrients through a physical or thermodynamic system.[41]

In the words of one of the most prominent voices of the new paradigm, Daniel Botkin, the old paradigm of balance and order was nothing but a myth, a myth superseded by a revolution: '[t]here has been a revolution in environmental sciences. At the heart of this revolution is a shift from the old idea of the constancy of Nature which is part of the ancient myth of the

35 F. Clements, 1916, *Plant Succession: An Analysis of the Development of Vegetation*, Washington DC: Carnegie Institution of Washington, 1916.
36 Eliot, 'The Legend', p. 65.
37 Worster, 'Order and Chaos', p. 4.
38 Eliot, 2010, op. cit., p. 65.
39 Ibid., p. 66.
40 See, for example, Worster, 'Order and Chaos'; C. Lévêque, *Ecology from Ecosystem to Biosphere*, Enfield, NH: Science Publishers, 2003; Eliot, 'The Legend'.
41 Worster, 'Order and Chaos', p. 4.

Balance of Nature.'[42] Botkin identifies three key features of the Balance of Nature myth:

> [f]irst, Nature, undisturbed by human influences, achieves a permanency of form and structure that persists indefinitely. Second, this permanent condition is the best condition for Nature: best for other creatures, best for the environment, and best for humans. Third, when disturbed from this natural state, Nature is capable of returning to it.[43]

It is useful to quote Daniel Botkin at length, since he not only advocates the new paradigm on scientific grounds, but also articulates its normative implications:

> [h]aving altered nature with our technology, we must depend on technology to see us through to solutions. The task before us is to understand the biological world to the point that we can learn how to live within the discordant harmonies of our biological surroundings, so that they function not only to promote the continuation of life but also to benefit ourselves: our aesthetics, morality, philosophies, and material needs.[44]

Botkin further argues – something which is extremely significantly in relation to its prescriptive implications – that '*[n]ature in the twenty-first century will be a nature that we make*; the question is the degree to which this molding will be intentional or unintentional, desirable or undesirable'.[45]

Three themes are evident in these quotations from Botkin: first, the emphasis on the role of technology; second, and relatedly, the emphasis on the manufacture of nature in accordance with human beings' will and desires; and third, that both depend on science and knowledge. This resonates strongly with Bacon's project as outlined in the New Atlantis.[46] It also, however, importantly resonates with a biopolitical framework of care, control and optimization, as will be amply discussed in Chapter 8.

This same dichotomy is represented in the field of environmental law. As Botkin himself complained, despite the ecological revolution 'our laws and policies are still based on outmoded concepts'.[47] His warnings are echoed by legal

42 Botkin, D., 'Adjusting Law to Nature's Discordant Harmonies', 7:1 *Duke Environmental Law and Policy Forum*, 1996, 25.

43 Ibid., p. 26.

44 Botkin, D., *Discordant Harmonies: A New Ecology for the Twenty-first Century*, Oxford: Oxford University Press, 1st edition, 1990, p. 191.

45 Ibid., p. 193. Emphasis mine.

46 And despite the fact that the method employed by contemporary scientists is not the empirical-inductive method dear to Bacon, but rather the hypothetico-deductive one dear to Descartes. See e.g. Ost, F., *La Nature hors la loi. L'Écologie à l'épreuve du droit*, Paris: La Découverte, 2003.

47 Botkin, 'Adjusting Law', p. 25.

scholars. Dan Tarlock for example, emphasizes how the balance of nature paradigm informed early US federal environmental legislation,[48] which now needs to be replaced. In 1994, when Tarlock described the 'partial unravelling' of environmental law due to the emergence of the new ecological paradigm, environmental law had 'just [begun] to address this paradigm shift'.[49] Today, some 20 years later, the new paradigm has penetrated much more profoundly to become the ontological basis for the juridification of nature.[50]

Between competition (individualism) and cooperation (collectivism)

The contestations described above between order and chaos spill over into another major area which has historically been the terrain of debates and contestations in the field of ecological science: the relation between competition and cooperation – or, to use terms usually referring to a social domain, between individualism and collectivism.[51] Indeed, it has been the case that ecology has been deployed as a supporting argument in both directions. In the early days of ecology, the debate aligned with the theories of Clements (which we have seen) and Gleason. Gleason, in particular, developed what he called the 'Individualistic Concept of Ecology'.[52] This theory fundamentally rests on the view that

> individual plants disperse and establish themselves independently of others, so that plant communities are merely unstructured aggregates of independent plants. Such unstructured ... communities would have dynamics rather unlike those of the integrated collectives attributed to Clements which develop as units towards definable climaxes.[53]

But the debate traverses the entire history of biological and ecological sciences. Mitman for example, in his history of the Chicago School of biology shows how

48 Tarlock identifies the National Environmental Policy Act of 1969, the Endangered Species Act of 1973, the Wilderness Act of 1964, and parts of the Clean Water Act, such as section 404, as having been the outcome of the paradigm of the balance of nature in ecology. He then goes on to say that '[t]wenty-five years after this paradigm was incorporated into law, it – and thus the basis for the core of biodiversity protection law – is now unraveling. In the twenty-five years since it has been enshrined in environmental law, the equilibrium paradigm has been rejected in ecology and replaced with a complex, stochastic non equilibrium one', Tarlock, 'Nonequilibrium Paradigm', 1122–3, footnotes omitted.

49 Ibid., p. 1123.

50 Much of this new paradigm also underlies the new model of ecosystem management, as evident in the work of, for example, see e.g. Ruhl, J., 'The Myth of What Is Inevitable under Ecosystem Management: A Response to Pardy', 21:2 *Pace Environmental Law Review* 2004, 315; and Ruhl, J., 'The Pardy-Ruhl Dialogue on Ecosystem Management, Part IV: Narrowing and Sharpening the Questions', 24:1 *Pace Environmental Law Review* 2007, 25.

51 See for a review of the founding debate on these issues Eliot, 'The Legend'.

52 Ibid., p. 65.

53 Ibid., p. 66.

social and political events crucially oriented and re-oriented the theoretical preferences of ecologists. The experience of World War I led ecologists, for example, to adopt a perspective in which competition was downplayed in favour of cooperation, the former seen as leading to unavoidable conflict whose effects, as had just been proven, were catastrophic.[54] In this respect, '[i]n contrast to a view of nature as the scene of a struggle for life, they [the ecologists at Chicago] proposed one where nature was based on interaction and cooperation whereby a stable order was achieved'.[55] Nature was seen as normative, and biology as a model for human sociality.[56]

However, in a post-World War II context, ecologists were now 'fearful of totalitarianism and captivated by cybernetics discourse of the organism/machine'.[57] The 'aggregation ethics [and] totalitarianist biology'[58] conducted at Chicago came under increasing attack, as it threatened to 'undermine the importance of the individual at the root of democracy'.[59] A nominalist orientation regained full ground: the group was seen to possess no 'distinctive properties of its own; it was a mere amalgamation of individuals'.[60] Competition between individuals thus regained 'central importance as a creative and stabilizing force',[61] until it became the 'primary explanation of community structure'.[62]

Regardless of the new disciplinary focus on individualism however, ecologists maintained the same tendency, i.e. to look to nature and biology as a way of 'healing human society' and as a source of normativity.[63] Importantly,

> competition became the guiding principle in postwar population ecology … the principles of energy flow and thermodynamics became the basis for ecosystem ecology, which usurped community studies at the forefront of ecological research. Eugene Odum's 'Fundamentals of Ecology', published in 1953, was oriented around the ecosystem concept, and rapidly displaced 'Principles of Animal Ecology' as the leading ecology textbook. Ecology [with Odum's studies being funded by the Atomic Energy Commission] not only appropriated military funds, it also appropriated the cybernetics narrative that emerged from military research on aircraft-missile guidance systems. The ecosystem blurred the distinction between inorganic and organic by reducing everything to energy as the common denominator.[64]

54 Mitman, *State of Nature.*
55 Ibid., p. 7.
56 Ibid.
57 Ibid., p. 5.
58 As one critic described the biology done at Chicago, Mitman, *State of Nature*, p. 6.
59 Ibid.
60 Ibid.
61 Ibid., p. 208.
62 Ibid.
63 For example, Mitman recalls how both Lorenz's ethology and Wilson's sociobiology see 'biology and nature as offering the normative guidelines along which to construct human society', Ibid., p. 209.
64 Ibid., p. 210.

In many ways this shift also translated into a shift in the role of ecologists as scientists. As nature became 'a system of components that could be managed, manipulated, controlled', the role of ecologists 'increasingly became that of environmental engineers', that is, 'ecologists were to be professional managers who could monitor and fix the environmental problems created by human society'.[65] In this sense, ecologists' new role turned decisively biopolitical.

Biology, ecology and the ethics of conservation

The preceding debates, still traversing ecology and conservation biology, have taken on a particular ethical tinge in an internal struggle affecting conservation biology, a crucial branch of biology behind the development of the concept of biodiversity and a significant, even decisive, normative influence on environmental policy. This debate, characterized by some commentators as a 'battle for the soul of conservation science',[66] is primarily linked to the ethics of conservation, as especially captured by the question: *why* should we conserve nature? The binary anthropocentrism/ecocentricsm is a key boundary in this debate. Yet a call for unity in conservation biology appeared in *Nature* in November 2014.[67] The comment takes up the 'age-old conflict around [the] seemingly simple question … of why … we conserve nature'.[68] The authors of the comment – and 238 additional signatories – propose 'a unified and diverse conservation ethic; one that recognizes and accepts all values of nature, from intrinsic to instrumental, and welcomes all philosophies justifying nature protection and restoration, from ethical to economic, and from aesthetic to utilitarian'.[69]

This is, in many ways, a recognition of the epistemological plurality that characterizes ecology as a postmodern science. Yet there are two main competing approaches to conservation. One is attuned to the tradition of protected areas and ecological integrity, strongly advocating that nature should be conserved for its 'intrinsic value' (rather than merely for instrumental purposes) and finds its most prominent representative in Michael Soulé, one of the scientists behind the concept of biodiversity. The other approach, whose leading advocates are Peter Kareiva, chief scientist for The Nature Conservancy (a leading American environmental organization operating worldwide) and Michelle Marvier, acknowledges the end of nature in the Anthropocene. Therefore, it favours a pragmatic embrace of artificial nature and human management in order to

65 Ibid.

66 K. Kloor, 'The Battle for the Soul of Conservation Science', 31:2 *Issues in Science and Technology* 2015, http://issues.org/31-2/kloor/ (Accessed on 26 September 2018).

67 H. Tallis and J. Lubchenco, 'A Call for Inclusive Conservation', 515 *Nature* 2014, 27.

68 Ibid., p.27.

69 Ibid.

conserve a nature beneficial to humans – a discourse in tune with Botkin[70] – through what is called 'new conservation'.[71]

Against the traditional separation between humans and nature upheld by traditional conservation biology measures such as protected areas, the 'new conservation' shifts perspective, towards a '[f]orward-looking conservation' that 'protects natural habitats where people live and extract resources and works with corporations to find mixes of economic and conservation activities that blend development with a concern for nature'.[72] Kareiva and Marvier's argument hinges on two primary reasons. First, protected areas, the primary tool of traditional conservation, are insufficient, and under siege.[73] Second, linking nature conservation and human benefits (through the concept of ecosystem services) allows us to prioritize conserving areas that deliver important benefits to human communities (especially poor ones) which 'is imperative for conservation effectiveness'.[74] In this respect, Marvier and Kareiva urge conservation to become pluralist, i.e. to adopt a larger toolbox, rather than rely only on protected areas. Yet in response, Soulé argues that such an approach would only hasten 'ecological collapse globally, eradicating thousands of kinds of plants and animals'.[75]

So on the one hand, we have a biocentric, or ecocentric approach, which, however, aims at upholding the modern fence between humanity and nature – epistemologically, and, crucially, practically. Indeed, this vision of nature reflects the modern threshold separating the sacred from the abject. On the other hand, we have a new ecology or new conservation attuned to a postmodern, 'Anthropocenic' view of nature, which, while moving beyond the anthropocentric/ecocentric binary, is also still in line with the modern trajectories of mastery and domination, especially when understood, as we shall see later in the book, in a biopolitical key. Both of these positions, moreover, find expression in the 'ecosystem approach', in particular as it is developed within the context of the CBD, where the relationship between protected areas and the 'ecosystem approach' is formally discussed and developed by the Conference of the Parties. Yet the 'ecosystem approach' also tends to envelop both within a larger biopolitical frame, as we shall see in Part III of the book. For the moment, it is important

70 For a full account of this debate see Kloor, 'Battle for the Soul'.

71 See e. g. M. Marvier, 'New Conservation is True Conservation', 28:1 *Conservation Biology* 2013, 1; M. Marvier and P. Kareiva, 'The Evidence and Values Underlying "New Conservation"', 29:3 *Trends in Ecology and Evolution* 2014, 131.

72 Kareiva and Marvier, quoted in Kloor, 'Battle for the Soul'.

73 Marvier and Kareiva, 'Evidence and Values', p. 131.

74 Marvier, 'New Conservation', p. 1.

75 Soulé as quoted in Kloor, 'Battle for the Soul'. Soulè, Kareiva and Marvier have been prominently debating conservation in the past few years; see e.g. M. Soulé, 'The New Conservation', 27:5 *Conservation Biology* 2013, 895; Marvier, 'New Conservation'; Marvier and Kareiva, 'Evidence and Values'; and P. Kareiva, 'New Conservation: Setting the Record Straight and Finding Common Ground', 28:3 *Conservation Biology* 2014, 634.

to note how the irreducible contestations characterizing conservation biology carry significant problems into the 'ecosystem approach'.

At this point it is time to shift attention to the concept of ecosystem, central to both ecology – in all its articulations – and to the 'ecosystem approach', so as to further explore the contingencies and contestations characterizing the genealogy of the 'ecosystem approach'.

The concept of ecosystem

So far, we have seen the general background against which the 'ecosystem approach' needs to be read. The framework of ambiguity we have traced, clearly and inevitably has important effects on the concept of the 'ecosystem approach'. However, what makes the 'ecosystem approach' a contested concept, but also facilitates its success, can also be more immediately linked to its specific naturalistic underpinning – the concept of ecosystem – itself an ambiguous and contested concept.

The concept of ecosystem is one of the central organizing ideas within the science of ecology. In fact, it has had a significant role in shaping the entire discipline in its initial development phases. Yet, the concept of ecosystem has emerged under the influence of contingent dynamics of interaction and competition – individually and institutionally – within the scientific communities involved.[76] And while it has proved to be an enduring concept, it was only *one* among several other similar concepts which emerged in the early days of ecology in order to describe similar phenomena, processes or objects, which have now been abandoned.[77]

This complex genealogy, which is perhaps still evident in the many definitions of ecosystem available,[78] has a direct bearing on the observation that the 'term 'ecosystem' has multiple meanings and uses'.[79] Within this multiplicity, according to philosopher of ecology deLaplante, some theorists consider ecosystems to be well defined objective entities,[80] while others maintain that '[e]cosystems are only mental constructs, not real, discrete, or living things on the landscape'.[81] There is, in other words, no objective entity comprised of communities and capable of reaching a stable equilibrium, hence any 'attempt to preserve any such thing [i.e. ecosystems]' is misguided.[82]

76 Golley, F., *A History of the Ecosystem Concept in Ecology: More than the Sum of its Parts*, New Haven, CT and London: Yale University Press, 1993.

77 Ibid., p. 168. Some of these other concepts are: biogeocenosis, microcosm, epimorph, elementary landscape, microlandscape, biosystem etc. In particular as regards biogeocenosis, its meaning and its history see ibid., pp. 169–74.

78 According to Hatcher and Bradbury, for example, there exist at least forty different definitions of ecosystem, Hatcher and Bradbury, 'Marine Ecosystem Management', p. 208.

79 deLaplante, 'Ecosystem Management', p. 398.

80 M. Soulé, 'The Social Siege of Nature', in M. Soulé and G. Lease (eds), 1995, *Reinventing Nature? Responses to Postmodern Deconstruction*, Washington, DC: Island Press.

81 A. Fitzsimmons, 1999a, 'Ecosystem Management: An Illusion?', *PERC Reports* 17:5, 1, 4.

82 Fitzsimmons, quoted in Eliot, 'The Legend', p. 68; see also Fitzsimmons, 'Ecosystem Management'.

A brief genealogy of the concept of ecosystem

The actual word 'ecosystem' was first used as a formal scientific concept in 1935 by Alfred Tansley.[83] However, the central ecological question of the interactions between species and their physicochemical environment had been part of the conceptual toolset of botanists and zoologists since the eighteenth and nineteenth centuries.[84] It was already clear then that 'the individual organism and its environment [must] be seen as a unity'.[85] Tansley however expressed this existing knowledge in a new way, offering the first formal definition of an ecosystem as 'a biotic assemblage and its associated physical environment in a specific space'.[86] It was only in the 1950s though, after the publication of Eugene Odum's *Fundamentals of Ecology*,[87] that the concept, until then one of several scientific concepts which aimed to describe the interactions between biota and their physicochemical environment, acquired a dominant position.[88] The success of the expression is relatively independent of the continued use of its main conceptual components, which in fact were common to many of the other terms and concepts. The success of the expression itself may, at least in part, be explained by its ability to capture metaphors important to the western scientific mind. As Golley underlines, the word ecosystem 'emphasizes the system, which is both a modern and a technical concept; and it conveys the idea of an ecological machine'.[89] The machine metaphor has been a crucial organizing metaphor of western modernity at least since Descartes, who considered 'the human body as a machine',[90] implying that there is no conceivable 'difference between the machines made by craftsmen and the various bodies that nature alone composes'.[91]

This metaphoric success also coincided with a shift in the perception of the role of ecologists, from conservation and ecological stewards, to environmental engineers and managers of a nature conceptually transformed into 'a

83 See, e.g. Golley, *History*; S. Bocking, 'Visions of Nature and Society. A History of the Ecosystem Concept', 20:3 *Alternatives* 1994, 12.

84 For a brief overview of the development of ecology in the eighteenth and nineteenth centuries, an overview which won't be necessary to report here, see Lévêque, *Ecosystem to Biosphere*, Chapter 2.

85 Ibid., p. 22, quoting Haldane. See also Golley, *History*.

86 A Tansley, 'The Use and Abuse of Vegetational Concepts and Terms', 16:3 *Ecology* 1935, 284.

87 E. Odum, *Fundamentals of Ecology*, Philadelphia: W. B. Saunders Company, 1953.

88 Golley, in his history of the concept of ecosystem, mentions several other concepts, such as biogeocenosis, microcosm, epimorph, elementary landscape, microlandscape, biosystem and several others; see Golley, *History*, p. 168, and throughout for the specific histories.

89 Ibid.

90 As quoted in F. Capra, *The Turning Point: Science, Society, and the Rising Culture*, Toronto: Bantam Books, 1984, p. 42.

91 As quoted in T. Hayward, *Ecological Thought. An Introduction*, Cambridge, UK: Polity Press, 1995, p. 17.

system of components that could be managed, manipulated and controlled'.[92] As we have already seen the concept of ecosystem 'blurred the distinction between inorganic and organic' and reduced 'everything to energy as the common denominator',[93] hence fully responding to the trajectories of modernity illustrated earlier, whereby nature is thoroughly disembodied and abstracted.

However, another important element that may explain the success of the ecosystem concept is that its apparent linguistic neutrality may have endowed it with an integrating capacity, enabling it to include narrower approaches and to transcend dichotomies.[94] This explanation is supported by the fact that the idea of ecosystems proved attractive to both the environmental movement of the 1960s and 1970s, keen to utilize the morally inspired metaphorics of holism and Mother Earth, and to managers and industrialists, keen to deploy its technical and mechanistic dimensions with a view to manage and control natural systems, in a Baconian – and then biopolitical – fashion.[95]

But what is an ecosystem? Tansley's definition emphasizes three elements: living organisms; physical environment, and a bounded space. In fact ecosystems are generally understood as discrete fragments of the biosphere displaying a relative autonomy from other fragments – neighboring ecosystems – where biotic and abiotic components interact.[96] Stressing the spatial dimensions, others describe ecosystems as 'spatially explicit units of the earth that [include] all of the organisms, along with all components of the abiotic environment within its boundaries'.[97] From this perspective, a key characteristic of an ecosystem is that it is a geophysical entity, bounded and discrete, within which certain interactions take place. Thus 'a dung pile or a whale carcass are ecosystems as much as a watershed or a lake'.[98]

In the early period of ecology, the main regulative ideas associated with ecosystems were teleology, stability and equilibrium. As noted above, Clements, while never using the term ecosystem, understood the successional processes of natural communities as eventually reaching a final stage of stable equilibrium, which he called 'climax'. Clements further conceived of ecosystems as quasi super-organisms, insofar as he saw them as self-organizing communities exceeding the mere aggregation of their component parts. Tansley too considered equilibrium to be a central element of his ecosystem concept,[99] even though he

92 Mitman, *State of Nature*, p. 210.
93 Ibid.
94 Golley, *History*, p. 190.
95 Ibid.,
96 Lévêque, *Ecosystem to Biosphere*, p. 33.
97 Likens, quoted in ibid., 2003, p. 33.
98 Ad Hoc Committee on Ecosystem Management, Ecological Society of America, as quoted in A. Fitzsimmons, *Defending Illusions. Federal Protection of Ecosystems*, Lanham, MD: Rowman & Littlefield, 1999b, p. 23.
99 It must be noted that other precursor concepts, such as the notion of 'trophic ecology' developed by Stranciskii in Russia, also were organized around the central idea of equilibrium. For further details on Stranchinskii's theory, see Golley, *History*.

tried to distance his definition from the Clementsian concept of super-organism.[100] Ecosystems, in this sense, are specific spatial or geographic units, which tend to display stability.

Ecologists today, however, tend to think of ecosystems not so much as amenable to representation as a map, but rather as a flowchart, since they focus primarily on processes, energy flows and nutrient cycling. In this respect, Eugene Odum, another central figure in the history of ecological sciences, had already begun to move the focus from communities and places to energy flows. According to Odum, an ecosystem[101] is 'any unit that includes all of the organisms (i.e. the "community") in a given area interacting with the physical environment so that a flow of energy leads to clearly defined trophic structures, biotic diversity, and material cycles (i.e. exchange of materials between living and nonliving parts) within the system'.[102] Odum however still conceived ecological processes as eventually leading to order, reaching a state of equilibrium which he called 'homeostasis'. As Worster points out, while Odum had dropped Clements' terminology, he still maintained a very similar understanding of nature as ordered and stable. Accordingly, Odum cautioned against anthropogenic changes to ecosystems and urged ecologists to reach out to the general public in order to influence public opinion and public policy.[103]

100 Ibid.

101 Concept which, according to Worster, occupied the central role of the sovereign in Odum's ecology; see Worster, 'Order and Chaos', p. 4.

102 Odum's definition is taken from ibid., pp. 4–5.

103 And this is an attitude replicated multiple ways in multiple contexts. As regards to the concept of biodiversity for example, Meffe and Viederman recommend scientists to 'take a clear stand that biodiversity is good, that functioning and intact ecosystems are good, that continued evolutionary change and adaptation are good, and that diversity and variation in general is good. Scientists cannot and should not remove themselves from these usually unstated value judgments' (as quoted in Lackey, 'Appropriate Use', p. 12). Normative science is a rather widespread phenomenon, particularly in the context of complex social questions, such as those relating to the environment. Lackey summarizes thus: 'Some argue that normative science is desirable, even essential, for implementing ecological policy. Scientists, they assert, have an obligation to incorporate policy value judgments into ecology, even to the point that such "science" concepts as ecosystem health should be adopted as the cornerstone of ecological policy (Callicott, 1995). Some scientific disciplines and professions (e.g., conservation biology, restoration ecology) unapologetically embrace normative science postulates as the core of their trade (e.g., biological diversity is inherently good, extinction of populations and species is inherently bad, ecological complexity is inherently good, evolution is good, biological diversity has intrinsic value) (Soule 1985)', ibid., p. 19. James Hansen offers another example within the context of climate change. Hansen has been a very vocal public speaker using his scientific credential to legitimize his public advocacy. Furthermore, he embedded his normative scientific efforts within his role as grandfather, charting his responsibilities – in a personal tone – for his grandchildren, in J. Hansen, *Storms of My Grandchildren: The Truth about the Coming Climate Catastrophe and Our Last Chance to Save Humanity*, New York: Bloomsberg Publishing, 2009.

Today, however, the prevailing idea of ecosystems is as fluid entities that may be apprehended conceptually and materially at very different scales, depending on perspective, focus and goals; their boundaries, indeed, are more often than not established *ad hoc*.[104] However, there are different views on ecosystems, to a large extent aligned with disciplinary perspectives.[105] Some suggest in fact that the boundaries or spatial delimitation of ecosystems is a 'mental construct', as ecological systems are not, in and of themselves, 'discrete, identifiable units', but rather lack 'boundaries in space or time'.[106] It is the human gaze that selects, largely in relation to a relevant problem, a particular segment of the Earth and declares it an ecosystem.

Yet some theorists think of ecosystems (or rather, ecoregions) in terms of place and scales, emphasizing 'the role of place, scale, and both natural and social elements'.[107] Others 'tend to focus on environmental processes with special attention given to energy flows and nutrient cycling'.[108] The notion of ecosystem is then able to accommodate spatial views, where ecosystems are best represented by a map, as well as processual views, best represented by a flowchart, or an organigram.[109] What remains significant is that there are, in fact, *different* perspectives, so that the very concept of ecosystem remains ambiguous and contested.

This difference in conceptualization of ecosystems – and thus, ultimately, of nature – is perhaps dependent on the fact that the term 'ecosystem' is applied in many different ways, in many different contexts, and in order to describe many different things. In this respect deLaplante urges us to distinguish between at least three major usages of the concept of ecosystem, aligned with three major research traditions in ecology: (1) ecosystem as object (2) ecosystem theory and (3) ecosystem methodology.[110]

Ecosystem as object is the concept captured by Tansley's definition: the complex set of biotic elements and the physico-chemical processes within a particular spatial region. This concept is mostly used in theoretical contexts and is the subject of many debates regarding its metaphysics and ontology. *Ecosystem theory* is linked to trophic-dynamic theory and focuses on the biogeochemical cycling of energy in ecosystems, i.e. its aim is to describe and explain

104 '[Ecosystems boundaries are] arbitrarily defined to suit particular purposes', W. J. Christie et al., 'Managing the Great Lakes Basin as a Home', 12:1 *Journal of Great Lakes Research* 1986, p. 4. See also, Lévêque, *Ecosystem to Biosphere*, p. 33: 'the limits of an ecosystem depend largely on the objectives pursued', and are most usefully 'defined operationally as a function of the processes studied or the questions posed'.

105 T. Loveland and J. Merchant, 'Ecoregions and ecoregionalization: geographical and ecological perspectives', 34:1 *Environmental Management* (Supplement) 2004, 1.

106 R. Ricklef, quoted in Fitzsimmons, 'Ecosystem Management', p. 24.

107 Loveland and Merchant, 'Ecoregions', in the abstract.

108 Ibid.

109 Lévêque, *Ecosystem to Biosphere*, p. 33.

110 See deLaplante and Odenbaugh, 'What Isn't Wrong'; see also deLaplante, 'Ecosystem Management', pp. 397–416, where the same distinctions are made.

the flow of energy and matter in an ecosystem. It may focus on matter, chemicals, energy, or on the organizational and developmental properties of ecosystems as wholes. In this latter case it is normally described through systems ecology. Finally, *ecosystem methodology* is the application of an ecosystem methodology to a variety of other fields. It is, as deLaplante describes it, a 'style of research' characterized by the qualifier 'an "ecosystem approach" to....', and it may be equally applied to environmental management, to public education, to human well-being, to fisheries, or to urban development. Its main feature is that it entails situating the investigation within its broader environmental context, on the assumption that at least some of its properties depend on interactions and relations with the environment within which the investigation is situated.[111] Significantly, a number of different ecological sub-disciplines operate within each conceptualization of ecosystem.

Jax similarly observes how the 'ecosystem concept is used in various meanings and is thus strongly ambiguous'.[112] In fact, he continues, '[t]he ecosystem is used both as a concrete object (unit) and as a perspective (denoting a 'holistic' approaches transcending disciplinary and administrative boundaries). Different value dimensions are attributed to the ecosystem'.[113] He suggests, however, that '[a]lthough these characteristics cause problems (e.g. unclear goal systems and an unreflected mix between facts and values) they can be made productive ... if considered explicitly',[114] with specific reference to the deployment of the concept within the context of the 'ecosystem approach'.

Golley more broadly emphasizes how the concept of ecosystem has assumed a variety of meanings: a philosophical idea; a physical object; a scientific paradigm, structuring scientific organization and research.[115] And it is this broad and diverse usage which produces an almost irreducible ambiguity. For this reason, the very usefulness of the concept of ecosystem has been challenged, whether for its association with the Clementsian concept of super-organism; or for its association with determinism; or for the utilization, in ecosystem studies, of inductive, rather than hypothetical-deductive methodologies.[116]

Ultimately, it is this very ambiguity and the multiplicity of senses in which the concept can be used that has determined the success of the idea of ecosystem. The concept has proved attractive, as already mentioned, to both the environmental movement, for the morally inspired holistic metaphors the concept resonates with, and to industrial capitalism for its technical and mechanistic

111 All three descriptions draw from deLaplante and Odenbaugh, 'What Isn't Wrong'.

112 K. Jax, 'On the Different Perceptions of the Ecosystem Concept within the Ecosystem Approach: Problems and Potentials', in H, Korn, R. Schliep and J. Stadler (eds), *Report of the International Workshop on the 'Further Development of the Ecosystem Approach'*, Bonn: BfN - Skripten, 2003, 78, p. 36.

113 Ibid.

114 Ibid.

115 Golley, *History*, p. 200.

116 Ibid., p. 200ff.

dimensions, aligned with the goals of managing and controlling natural systems.[117] The concept of ecosystem then emerges from the same complex genealogies, and is affected by the same 'moral ambivalence' affecting ecology in general.[118]

Conclusions

This chapter has shown that the entire scientific scaffolding underpinning the 'ecosystem approach' – that is ecology and the concept of ecosystem – is best understood as a framework of ambiguity. Its complex genealogical emergence, and the competing approaches – theoretical and normative – makes the construction of a stable, uniform, concept of the 'ecosystem approach' problematic. By contrast, these complex genealogies help situate the 'ecosystem approach' in what is inevitably a field of discursivity immersed in a genealogical 'play of forces'. Both these elements will be revisited in the last chapter, where I will try to draw out a number of productive ambiguities from this genealogical complexity.

117 Ibid., p. 3. Similarly, A. Bell, 'Non Human Nature and the Ecosystem Approach. The Limits of Anthropocentrism in Great Lakes Management', 20:3 *Alternatives Journal* 2004, 20, p. 23.

118 Worster, 'Order and Chaos'.

6 The 'ecosystem approach' between competing narratives

Introduction

The competing narratives and the complex genealogies underlying both ecology in general and the concept of ecosystem more specifically, are also central to the 'ecosystem approach'. Situated within a space of conflicting values, the 'ecosystem approach' is *one* of the terrains where irreducible contestations over the values underlying and informing ecological science and environmental law take place. These contestations are usually framed within a narrative space where anthropocentrism and ecocentrism represent opposite poles along a gradient of intermediate positions (such as biocentric or weak anthropocentric positions). This chapter explores these aspects in order to further demonstrate its complex genealogical character. The first section will outline, in broad terms, the conflicting values and the competing narratives involved, in order to describe the field of tensions within which the 'ecosystem approach' finds articulation(s). The remainder of the chapter illustrates how each narrative may find expression in a particular articulation of the 'ecosystem approach'.

Conflicting values, competing narratives

As Lackey very aptly put it already in 1998, 'defining ecosystem management is itself a highly value-laden process'.[1] The 'ecosystem approach' in fact 'strikes at the core of our values, ethics, and moral philosophy'.[2] Lackey raises a number of important questions, which he considers fundamental to the 'ecosystem approach':

> What rights, if any, should be granted to the non-human world? How important is this generation's material well-being compared with that of our children? Are our children more important than the natural world? How are the benefits of ecological resources to be distributed within society? … Is the separation between society and nature real or imagined?'[3]

1 Lackey, 'Ecosystem Management', p. 109.
2 Ibid.
3 Ibid.

These, argue Lackey, 'are not science issues but reflect deeply held moral and religious views'.[4]

It is thus arguable that the 'ecosystem approach', and any process which leads to a definition of its parameters, core elements and objectives, is caught between conflicting values and visions of society and nature. Indeed, as has been noted, the 'widespread support' for the 'ecosystem approach' and its vocabulary, 'masks a fundamental philosophical conflict between ecocentric and anthropocentric attitudes toward nature'.[5] Some further examples will help better illustrate these axiological polarizations. Stanley distinguishes between an anthropocentric and what he calls a 'biocentric' view of ecosystem management.[6] The latter considers resources utilization secondary to the primary – and constraining – goal of the maintenance of ecological integrity, while the former – the anthropocentric version – has as its primary goal that of optimizing human resources use, although ecological considerations are included in its management horizon.[7] Yaffee's distinction is more nuanced, as he identifies three models of ecosystem management, each infused with specific meanings reflecting different interests, values and knowledges. He calls these three models, 'environmentally sensitive multiple use'; 'ecosystem approach to resource management'; and 'ecoregional management'.[8] 'Environmentally sensitive multiple use' has an anthropocentric outlook and its goal is to allow multiple human uses in an environmentally rational manner. The 'ecosystem approach to resource management' has a biocentric outlook and incorporates holistic thinking and the recognition of the complexity of ecological systems. 'Ecoregional management', finally, has an ecocentric outlook, and 'shifts management focus toward ecosystem processes and away from biota'.[9]

These conflicting and competing views are relevant in the context of international law. This can be seen in the Report on the work of the United Nations Open-ended Informal Consultative Process on Oceans and the Law of the Sea at its Seventh Meeting (ICP-7). ICP-7 reports how some commentators understand the 'ecosystem approach' as enabling 'the integrated management of human activities based on the best available science about ecosystems, ecological interactions and the precautionary approach, in order to achieve the sustainable use of goods and services and the maintenance of ecosystem integrity (i.e., structure and function)'.[10] Others understand the 'ecosystem approach' in a different way. Thus, in the report the 'ecosystem approach' is described as a framework for

4 Ibid.
5 Bell, 'Non Human Nature', p. 23.
6 T. Stanley, 'Ecosystem Management and the Arrogance of Humanism', 9:2 *Conservation Biology* 1995, 255.
7 Ibid., p. 256.
8 S. Yaffee, 'Three Faces of Ecosystem Management', 13:4 *Conservation Biology* 1999, 713.
9 Ibid., in abstract.
10 ICP-7, 2006, op. cit., para 43.

> protecting of the ecosystems in their entirety: all species, communities and ecosystem functions in a given area, and [for] acting with caution where knowledge [is] lacking. [I]mpact assessment [is] needed in advance to avoid undue harm to the environment. Activities [should] only be allowed to proceed if no harm were caused.[11]

Within the context of fisheries the polarization of conflicting values is even more evident. The discussion that took place during the twenty-fifth meeting of the FAO Committee on Fisheries (COFI),[12] report for example Tudela and Short, 'clearly demonstrated that two opposed visions of [the ecosystem approach] are struggling against each other'.[13] The two views, or 'visions',[14] are indeed radically different. One vision considers the 'ecosystem approach' as 'a radical change in the paradigm underlying ocean management (including fisheries)'.[15] Ecosystem complexities, in this vision, make precaution a central element of an 'ecosystem approach' to management. Moreover, 'the preservation of ecologically viable populations of ecosystem components is key', both in relation to 'species' functional redundancy' and in relation to ensuring 'ecosystem resilience and sustained provision of ecosystem services'.[16]

The other vision considers ecosystems 'from a purely mechanistic perspective and seeks to maximize human profits by engineering the food web and other ecosystem manipulations'.[17] The underlying logic here is, according to Tudela and Short, entirely different, as it aims to mould ecosystems 'that are tailored for supplying the optimum harvest of the desired species, e.g. in the case of fisheries'.[18]

The FAO itself situates the 'ecosystem approach' (to fisheries) within the context of these competing narratives, which it explicitly links to anthropocentrism and ecocentrism.[19]

It appears thus that the 'ecosystem approach' is both compressed and stretched between these two competing narratives, each trying to impose a particular closure on the concept. While the effect of these competing narratives has already been briefly mentioned when discussing terminology (Chapter 4), this chapter discusses two paradigmatic articulations of each of these narratives in more details.

11 Ibid., para 43.

12 The discussion followed the presentation of the following publication: FAO, 2003, op.cit., Tudela and Short, 'Paradigm Shifts'.

13 Ibid., p. 282. Tudela and Short refer more specifically to 'ecosystem-based management'. Issues regarding use of different labels in different contexts have been discussed (and clarified) in Chapter 4.

14 Ibid.

15 Ibid.

16 Ibid.

17 Ibid.

18 Ibid.

19 FAO, 2003, op. cit. p. 11.

Ecocentric articulations of the ecosystem approach

Ecocentric articulations of the 'ecosystem approach' purport to offer a fundamental critique of the anthropocentrism of legal modernity. As mentioned, the strong narrative of the paradigm shift identifies the 'ecosystem approach' as a 'quite radical'[20] or 'fundamental'[21] shift that entails a challenge to the philosophical and legal framework that dominates legal modernity.

It is conservation biologist Robert Grumbine who has provided the clearest ecocentric articulation in a now classic essay taking stock of the emerging paradigm of ecosystem management,[22] so this will be our starting point. Grumbine suggests that one of the primary goals of the 'ecosystem approach' is that of accommodating human use and occupancy *within* ecosystem constraints. This implies that '[a]long with defining the ecosystem management approach as a new policy framework there appears to be a parallel [need] of redefining the fundamental role of humans in nature'.[23] Grumbine emphasizes that the key question is one of 'sufficiency', yet, as he further notes, 'many green economy, stewardship, and "sustainability" models suffer from concentrating on efficient management instead of sufficient management'.[24] This requires the re-definition of how human activities and uses should be contained within appropriate ecological limits. In this sense, Grumbine observes, the 'ecosystem approach' is a response to the resourcism of traditional resource management.[25]

A number of other scholars share this perspective, and consider the 'ecosystem approach' to be a shift in paradigm, either in progress or necessary: for Keiter and Boyce the 'transition to ecosystem management manifests a willingness to accept nature largely on its own terms and to control incompatible human uses';[26] Goldstein suggests that 'using ecology to redefine land management implies an ethical reorientation – the ecosystem is valued as an object of respect and admiration', that is, ecocentrically;[27] for Noss and Cooperrider '[b]iodiversity conservation ultimately requires a rejection of humanism or anthropocentrism … It requires a biocentric [a term which is arguably used here as an equivalent to ecocentric] embrace of all life'.[28]

20 J. Vallentyne and A. Beeton, 'The 'Ecosystem' Approach to Managing Human Uses and Abuses of Natural Resources in the Great Lakes Basin', 15:1 *Environmental Conservation* 1988, 58.

21 Wallace et al., 'Moving Toward', p. 1.

22 Grumbine, 'What is Ecosystem Management?' Grumbine uses the expression 'biocentric', but for our purposes the distinction between ecocentric and biocentric is not especially relevant.

23 Ibid., p. 28.

24 Ibid., p. 35.

25 Ibid., p. 34.

26 R. Keiter and M. Boyce, *The Greater Yellowstone Ecosystem: Redefining America's Wilderness Heritage*, New Haven, CT: Yale University Press, 1994, p. 404.

27 B. Goldstein, 'Can Ecosystem Management Turn an Administrative Patchwork into a Greater Yellowstone Ecosystem?', 8:2 *Northwest Environmental Journal* 1992, 285.

28 R. Noss and A. Cooperrider, *Saving Nature's Legacy: Protecting and Restoring Biodiversity*, Washington, DC: Island Press, 1994, p. 328.

The key element in all these accounts is that for all these authors a crucial move implicated and required (by *this* articulation of the 'ecosystem approach') is the reconstruction and re-imagination of nature, so that the 'fence' which has segregated humans from nature can be dismantled.[29] Moreover, all these accounts reflect a modal shift in the understanding of reality, and of nature: away from the atomistic and towards the systemic and relational. Ecosystems are thus apprehended as wholes, and all participants – humans and non-humans alike – are connected in a relational field comprised of places, processes, individuals and ecological communities. This shift has potentially significant effects on law: representations of reality mediate crucially between 'the world of natural objects and the world of legal objects';[30] in return, legal notions convey 'a particular partitioning of the world'.[31]

Relatedly, sometimes explicitly and sometimes only implicitly, all these accounts suggest that legal subjectivity be attached to non-human entities, in light of the 'goal of protecting ecological integrity', considering that 'owls and gentians [also] have evolutionary needs as do humans'.[32] While there is a clear echo of the seminal article of Christopher Stone discussing this theme,[33] from which Grumbine in particular draws direct inspiration, what is important to emphasize is how one such understanding of the 'ecosystem approach' places it squarely within current debates that not only imagine nature as a legal subject, but that also, more radically, problematize the very notion of legal subjectivity, and its hegemonic liberal genealogy. In this sense, the 'ecosystem approach' may operate as a conceptual picklock for a radical reconstruction of a legal subjectivity no longer centred on abstract rational agency, but on a plurality of embodied, vulnerable agencies.[34]

Anthropocentric articulations of the ecosystem approach

Introduction

Other articulations of the 'ecosystem approach' are framed through an anthropocentric narrative. As I have argued elsewhere,[35] this is the prevailing

29 Grumbine, 'What is Ecosystem Management?', p. 35.

30 Smith as quoted in Tallacchini, 'Legal Framework', p. 1092.

31 Ibid., p. 1093.

32 Grumbine, 'What is Ecosystem Management?', p. 35.

33 C. Stone, *Should Trees Have Standing? Law, Morality, and the Environment*, 3rd edn, Oxford: Oxford University Press, 2010, and A. Grear (ed.), *Should Trees Have Standing? 40 Years on*, Cheltenham: Edward Elgar, 2012b.

34 N. Naffine, 'Who are Law's Persons? From Cheshire Cats to Responsible Subjects', 66:3 *Modern Law Review* 2003, 346; A. Grear, 'Law's Entities: Complexity, Plasticity and Justice', 4:1 *Jurisprudence* 2013b, 76; Cullinan, *Wild Law*; P. Burdon, *Exploring Wild Law: The Philosophy of Earth Jurisprudence*, Kent Town, S.A.: Wakefield Press, 2011a.

35 De Lucia, 'Competing Narratives'.

narrative within which the 'ecosystem approach' is framed in international law. This is in large part due to the fact that, as we have seen in Chapter 1, international environmental law is fundamentally and predominantly aligned to an anthropocentric frame of reference.[36] This is especially evident in the central role that the concept of sustainable development has acquired as both a legal concept and as a broad legal narrative. It is therefore important to situate the 'ecosystem approach' within the larger legal narrative within which it finds expression.

There are two primary reasons why this is important. First, and in general, the conceptual and normative meaning of the 'ecosystem approach' can be fully appreciated only when placed within the broader normative context of sustainable development, to the extent that the latter operates as a quasi-constitutional principle providing specific normative guidance and pull.[37] Sustainable development is, in other words, a central legal narrative that must be considered in order to understand the 'ecosystem approach' in international environmental law. Indeed, regardless of the formal legal status of sustainable development, its normative force emerges from its ability to provide coherence and a clear normative horizon that forms and informs the manner in which interpreters operate.[38]

Second, the usefulness of reading the 'ecosystem approach' in relation to sustainable development derives from the particular relationship that *specifically* links the two legal concepts together. This relationship is probably best understood in terms of *instrumentality*. The 'ecosystem approach' is in fact often characterized in policy and normative documents as a 'tool' to achieve the objective of sustainable development.[39] The next two sections will deal with each of these points in turn.

The 'ecosystem approach' and the sustainable development narrative

The 'ecosystem approach' has from very early on developed in parallel to what would become a central narrative in international law and politics: sustainable development. This is particularly relevant in the context of biodiversity and natural resources management, as the holistic orientation of the 'ecosystem approach', as we shall see, has been re-oriented, as it were, so as to fit within the broader normative paradigm dominated – 'quasi-constitutionally' – by the concept of sustainable development. This concept – itself contested – has developed within the context of international law with a clear and evident anthropocentric orientation.[40] Indeed, the foundational texts of international environmental law

36 Gillespie, *International Environmental Law*.
37 Lowe, 'Sustainable Development'.
38 In this sense, e.g., ibid.
39 For a longer discussion of this relationship see De Lucia, 'Competing Narratives'.
40 Thus for example Gillespie, *Illusion of Progress*.

illustrate how the anthropocentric vision is the prevailing, hegemonic approach to environmental protection. The 1972 Stockholm Conference contains this orientation in its very title – Conference on the *Human* Environment[41] – and explicitly grounds itself in an anthropocentric worldview to the extent that Man is the barycentre of the entire Declaration,[42] and all references to ecosystems and the Earth are functional to the well-being of 'present and future generations' (Principles 1 and 2) or 'all mankind' (Principle 5), or to the production of resources (Principle 3). The need for an 'efficient synthesis'[43] of environmental and developmental concerns was an important element of the debate at Stockholm,[44] and the attention to ecosystems was, already there, counter-balanced and re-calibrated in terms of the instrumentalization of their protection for humanity's well-being. The Rio Declaration consolidates this synthesis, enshrining the concept of sustainable development as the organizing principle and as the ideological narrative for the entire environmental legal landscape.[45] In fact, Principle 1 states unequivocally that '[h]uman beings are at the centre of concerns for sustainable development'.[46]

An important element of differentiation hinges on the relation of the 'ecosystem approach' with the notion of ecological limits. While Grumbine, in his articulation of the 'ecosystem approach', orients attention towards sufficiency, sustainable development, and the articulation of the 'ecosystem approach' aligned to its narrative, emphasizes efficiency.[47] To be sure, the concept of ecological limits is recognized even within the context of sustainable development as an important and meaningful constraint. Yet, and this is the crucial difference, while '[t]he concept of sustainable development does imply limits', such limits are nevertheless 'not absolute … but limitations imposed by the present state of

41 Emphasis mine.

42 The Declaration proceeds from the consideration of a need to identify a 'common outlook and for common principles to inspire and guide the peoples of the world in the preservation and enhancement of the human environment', Declaration of the United Nations Conference on the Human Environment (UN Doc. A/Conf.48/14/Rev. 1, 1973), *incipit.*

43 See D. Leary and B. Pisupati, *The Future and International Environmental Law*, Tokyo: United Nations University, 2010, p. 5 which quotes from the Founex Report, in which this efficient synthesis was theorized and promoted, see M. O. de Almeida, *Environment and Development: The Founex Report on Development and Environment*, New York: Carnegie Endowment for International Peace, 1972.

44 Particularly fuelled by the concerns of developing countries, see Bernstein, *Compromise.*, p. 32ff.

45 In its broadest sense. See Lowe 'Sustainable Development', which constructs the international legal space as a field open to the influence of cultural narratives and political contestations through the notion of interstitial norms, of which the concept of sustainable development is a primary example.

46 The second sentence states that human beings also 'are entitled to a healthy and productive life in harmony with nature', Rio Declaration On Environment and Development, (U.N. Doc. A/Conf.151/26 (Vol. I)), Principle 1.

47 On the fundamental differences between the two concepts see, e.g. T. Princen, *The Logic of Sufficiency*, Cambridge, MA: MIT Press, 2005.

technology and social organization on environmental resources and by the ability of the biosphere to absorb the effects of human activities'.[48] Limits, are, then, mobile and entirely, or at least largely, contingent on human technical ingenuity. In this respect, sustainable development is a strategy aimed at securing a 'new era of economic growth'.[49] This is really the conceptual basis for the ideology of a 'green economy', understood as a new development path that is 'based on sustainability principles and ecological economics',[50] which is currently being mainstreamed.[51]

The green economy, which took centre stage at the recent Rio + 20 UN Conference on Sustainable Development held in Rio de Janeiro in 2012 (one of its two themes was in fact the green economy in the context of sustainable development and poverty eradication)[52] is not so much a re-formulation of the framework of sustainable development, but a sharpening of its (economic) focus. Indeed, as UNEP underlines, 'the concept of a "green economy" does not replace sustainable development, but', the report suggests, 'achieving sustainability rests almost entirely on getting the economy right'.[53] The green economy moreover, which is further understood as 'one of the several mutually complementary constructions that have emerged in recent years to enhance convergence between the different dimensions of sustainable development',[54] promises to be 'a new economic growth paradigm that is friendly to the earth's ecosystems and can also contribute to poverty alleviation'.[55]

48 World Commission on Environment and Development (WCED), *Our Common Future: Report of the World Commission on Environment and Development*, 4 August,1987, p. 24.

49 Ibid., p. 28 (but the reference is used repeatedly throughout the report).

50 UNEP, *Restoring the natural foundation to sustain a Green Economy. A century-long journey for Ecosystem Management*, International ecosystem management programme (IEMP) UNEP-IEMP, Policy Series paper, 2011b, p. 1, HTTP: http://mahenvis.nic.in/Pdf/Report/report_eco_Restoring%20the%20natural%20foundation.pdf (Accessed 26 September 2018).

51 UNEP, 2011b, op. cit.; UNEP, *Towards a Green Economy. Pathways to Sustainable Development and Poverty Eradication*, A Synthesis for Policy Makers, 2011c; UNEP, 2011a.

52 UNGA resolution 'Implementation of Agenda 21, the Programme for the Further Implementation of Agenda 21 and the outcomes of the World Summit on Sustainable Development', 31 March 2010, UN Doc. A/RES/64/236, para 20(a).

53 UNEP, 2011a, op. cit., p. 2.

54 UNGA, Progress to Date and Remaining Gaps in the Implementation of the Outcomes of the Major Summits in the Area of Sustainable Development, as well as an Analysis of the Themes of the Conference Report of the Secretary-General, (UN Doc. A/CONF.216/PC/2), para 43.

55 UNEP, *The Transition to a Green Economy: Benefits, Challenges and Risks from a Sustainable Development Perspective*, Report by a Panel of Experts to Second Preparatory Committee Meeting for United Nations Conference on Sustainable Development (Prepared under the direction of: Division for Sustainable Development, UN-DESA United Nations Environment Programme UN Conference on Trade and Development), 2012b, p. 3, HTTP: https://plagiarism.repec.org/trica-papuc/trica-papuc2.pdf (Accessed 26 September 2018).

Sustainable development (with its corollary of sustainable economic growth, and its most current re-formulation in terms of green economy) thus provides a normative horizon for environmental protection in general and for the 'ecosystem approach' in particular, as will be more evident from the next section.[56] This is further evident in the perspective offered by UNEP. A recent UNEP report suggests (as one of its key findings) that the 'Ecosystem Management Approach plays a critical role in addressing substantial issues of Green Economy development, including promoting the sustainable use of natural capital and providing cost-effective environmentally friendly approaches'.[57]

The 'ecosystem approach' as a tool

At this point is useful to show how the relationship between the 'ecosystem approach' and the broader principle – and narrative – of sustainable development takes a *specific* form. Indeed, the 'ecosystem approach' is often described in terms of a *tool* to achieve sustainable development. This is evident in the CBD, and reinforced by the Plan of Implementation of the World Summit on Sustainable Development, where the 'ecosystem approach', in the context of biodiversity protection,[58] is presented as one of the key tools to achieve sustainable development.[59] Even more explicit is the ICES, which states that

> [t]he 'ecosystem approach' is *embedded* in the concept of sustainable development, which requires that the needs of future generations are not compromised by the actions of people today. The 'ecosystem approach' puts emphasis on a management regime that maintains the health of the ecosystem alongside appropriate human use of the marine environment, for the benefit of current and future generations.[60]

Moreover, the converse seems also to be true: 'sustainable forest management, as developed within the framework established by the Rio Forest Principles, can be considered as a means of applying the 'ecosystem approach' to forests'.[61] The FAO similarly suggests that the 'ecosystem approach' (to fisheries) 'is, in effect, a way to implement sustainable development in a fisheries context'.[62]

56 Similarly, albeit within the particular context of the Great Lakes, Bell, 'Non Human Nature'.
57 UNEP, 2011b, op. cit., p. 2.
58 Plan of Implementation of the World Summit on Sustainable Development, 4 September, 2002, para 44 begins thus: 'Biodiversity, which plays a critical role in overall sustainable development and poverty eradication…').
59 Plan of Implementation of the World Summit on Sustainable Development, 4 Septembter, 2002, op.cit., para 44(e).
60 ICES, 2005, op. cit., p. 4, para 4.1, emphases mine.
61 Decision VII/11, 2004, op. cit.
62 FAO, 2003, op. cit., p. 11.

Particularly explicit is the European Commission:[63] the 'ecosystem approach' is, in the view of the Commission, 'an instrument to pursue sustainable development in its three dimensions'.[64] A report of the Secretary General to the General Assembly of the United Nations at its fifty-ninth session, however, provides the clearest evidence of such a deep instrumental linkage: 'The "ecosystem approach" is the backbone of sustainable development'.[65]

Articulating the 'ecosystem approach' anthropocentrically

With particular reference to fisheries management, and within the context of the EU, the European Commission considers that an ecosystem approach 'to fisheries management is about ensuring goods and services from living aquatic resources for present and future generations ... such fishery management will strive to ensure that benefits ... are high while direct and indirect impacts ... on marine ecosystems are low.[66]

The Commission explicitly recalls a prior FAO definition, which reads as follows:

> the purpose of an 'ecosystem approach' to fisheries is to plan, develop and manage fisheries in a manner that addresses the multiple needs and desires of societies, without jeopardizing the options for future generations to benefit from the full range of goods and services provided by marine ecosystems.[67]

The FAO is indeed, entirely aligned with an anthropocentric articulation of the 'ecosystem approach', and is very explicit about its adoption of an anthropocentric perspective. Moreover, while it recognizes two strands of the 'ecosystem approach' – one ecocentric and one anthropocentric, it also considers that the two are merging (indeed are 'required' to merge) under the broader frame of sustainable development.[68]

But the general thrust of both the EU and the FAO's purpose-oriented definitions are also evident within the context of OSPAR. The OSPAR Commission states that 'the essence of the 'ecosystem approach' is to allow sustainable exploitation of natural resources while maintaining the quality, structure and functioning of marine ecosystems'.[69] Furthermore, the joint OSPAR-HELCOM

63 For the purposes of this article included under the rubric of international law.

64 European Commission Communication, *The Role of the CFP in Implementing an Ecosystem Approach to Marine Management*, COM 2008, 187, (Brussels 11 April 2008), p. 3.

65 Oceans and the law of the sea, Report of the Secretary-General, Addendum, 18 August 2004, para 244.

66 European Commission Communication, 2008, op. cit., p. 3.

67 FAO, 2003, op. cit., p. 14.

68 Ibid., p. 11.

69 OSPAR Commission, *Quality Status Report*, 2010, p. 9.

definition,[70] also adopted by ICES, offers a perfect example of how these conflicting values are simultaneously at work; it hence bears repeating. The 'ecosystem approach', in this respect, is defined as

> a comprehensive integrated management of human activities based on the best available scientific knowledge about the ecosystem and its dynamics, in order to identify and take action on influences which are critical to the health of the marine ecosystems, thereby achieving sustainable use of ecosystem goods and services and maintenance of ecosystem integrity.[71]

Two themes, or narratives, traverse all these definitions of the 'ecosystem approach'; narratives that are arguably in tension with one another. While this tension will be thoroughly explored in Chapter 9, it is necessary to anticipate how they align with the broad narrative of sustainable development. These two narrative horizons are, on the one hand the anthropocentric framework of ecosystem services; and on the other the ecocentric horizon implied by concepts such as ecosystem integrity and ecosystem health.

As already mentioned, the different emphases, as Brooks et al. observe, reflect the fact that particular epistemic communities or interest groups infuse the 'ecosystem approach' with a local, institutional and sometimes parochial.[72]

Yet, this double register of articulation that suggests a polarity opposing anthropocentrism and ecocentrism, ecosystem services and ecosystem integrity, is a central *focus* for the critical biopolitical analysis that I carry out in Part III of the book, in relation to both the critique of the modern anthropocentrism/ecocentrism binary and to the increasing interpenetration of the concepts of the 'ecosystem approach' and of ecosystem services.

Conclusions

In this chapter I have explored in some detail the two main competing narratives that shape and orient the field of discursivity within which the 'ecosystem approach' is situated. Each tries to impose its closure, to fix a specific meaning of the 'ecosystem approach' in what, however, remains an open space of competition and contestation. Indeed, the 'ecosystem approach', this chapter has shown, is caught between conflicting values and visions of society and nature. Hence, the widespread support for its language should not lead us to believe that there is broad agreement as to what it is. On the contrary, as has already

70 Statement on the Ecosystem Approach to the Management of Human Activities, 'Towards an Ecosystem Approach to The Management of Human Activities', 2003, op.cit., Agenda item 6, ANNEX 5 (Ref. §6.1), para 5

71 ICES, 2005, op. cit., p. 4, para 4(1), emphases mine.

72 R. O. Brooks, R. Jones and R. A. Virginia, R. A., *Law and Ecology: The Rise of the Ecosystem Regime*, Burlington: Ashgate, 2002, pp. 268–9.

been noted, this widespread support crucially hides fundamental philosophical conflicts, conflicts readily aligned with the anthropocentric/ecocentric threshold.

This chapter also concludes the second part of the book, where I have articulated a genealogical reconstruction of the 'ecosystem approach', of its complexities and of the competing narratives within which it is situated. In Part III I will shift the theoretical register and start by questioning the very conclusion of Part II, namely the distinction between anthropocentrism and ecocentrism, as a prelude to the articulation of a biopolitical analysis of the 'ecosystem approach'.

Part III

A biopolitical reading of the ‘ecosystem approach’

7 Beyond anthropocentrism and ecocentrism

Introduction

Part II has shown how the 'ecosystem approach' is the result of complex genealogies, how it is inserted in the midst of a 'play of forces'. Moreover, we have seen how competing narratives try to impose closure on what is ultimately a discursively open concept. I have explored these competing narratives and the way they affect the articulations of the 'ecosystem approach' in some detail, and it has emerged that they map with the binary anthropocentrism/ecocentrism. I have shown that the anthropocentric articulation is aligned with the broad narrative of sustainable development and constructs natural ecosystems as service providers, while the ecocentric one is concerned with the dismantling of the Cartesian fence separating humans and nature, and with the ascription of subjectivity to natural entities.

From the preceding analysis one may then draw a polarized image of the field of discursivity within which the 'ecosystem approach' is situated. This is indeed the impression that one can infer from the literature, where the polarization between anthropocentric and ecocentric articulations is common. While attempting to reconstruct the genealogies of the 'ecosystem approach', the terminology and the conceptual semantics of these key referents have remained largely unproblematized. However, in line with the most recent scholarship,[1] it is now time to problematize these conceptual categories, for a number of reasons and in relation to a number of interlinked purposes.

First, and at a very general level, one such polarized framework of analysis, one moreover that inevitably demarcates the good from the bad in an unrealistically neat manner, is incapable of fully capturing the profound genealogical ambiguities underlying the 'ecosystem approach'. This, indeed, is an utterly modern framing, to the extent that modernity rests decisively on the

1 A. Philippopoulos-Mihalopoulos, 'Actors or Spectators? Vulnerability and Critical Environmental Law', 3:5 *Oñati Socio-Legal Series* 2013, 854–76; Grear, 'New Horizon'; A. Grear, 'Deconstructing Anthropos: A Critical Legal Reflection on "Anthropocentric" Law and Anthropocene "Humanity"', 26:3 *Law Critique* 2015, 225; De Lucia, 'Beyond Anthropocentrism'.

establishment of thresholds.[2] Establishing exact thresholds – whether terminological or conceptual – erases the ambiguities that, however, irreducibly inhabit both camps. A polarized critique tends to aim at what can be described as a 'righteous and destructive closure'. My goal, as discussed in Chapter 1 (especially in the final section), is rather that of opening up, of fully exploring the ambiguities in every direction. This also means being able to capture the productive ambiguities that may be available to a careful, nuanced reading – this will be a central theme to which I will return in the last substantive chapter of this book. This, indeed, may turn out to be the *central* and *decisive* characteristic of the 'ecosystem approach', that it is a thoroughly post-normal concept that cannot be grasped through modern binary categories.

The first problem that arises when analysing the 'ecosystem approach' through this binary is that, while it simplifies an analytically complex situation and hence makes it in many ways intelligible, it also remains too coarse to be able to capture all the crucial nuances that a thoroughly critical deconstruction of the 'ecosystem approach' must expose. The binary situates the 'ecosystem approach' along a narrative that is unrealistically *linear*. Such linearization (along a scale which, in true modern fashion, progresses from chaos to enlightenment) unduly simplifies a complex, non-linear, utterly conflictual – i.e. irreducibly genealogical – discourse. Moreover, an analysis based on this binary fails fully to recognize the complex genealogies that emerge from the recognition of ecology as a framework of ambiguity.

Moreover, both terms of the binary conceal as much as, or indeed more than they reveal. While certainly useful in relation to (a) the identification of the different ideological projects trying to articulate the 'ecosystem approach', and (b) proving the point that the 'ecosystem approach' is fundamentally contested and located between competing narratives, this binary is, I argue, ultimately unsatisfying, or rather, incapable of capturing the complexities involved.

Finally, an important point emerges from the problematization of the anthropocentrism/ecocentrism binary, one that requires us to rotate the angle from which the research question is approached. Both terms of the binary – that is, both anthropocentrism and ecocentrism – remain within the trajectories of modernity, so that the distinction between an anthropocentric and an ecocentric articulation of the 'ecosystem approach' offers too little critical purchase in relation to the starting point of my enquiry, namely, whether and to what extent the 'ecosystem approach' entails a paradigm shift.

2 See, for a detailed discussion of the relation of modernity with the establishment of thresholds, V. De Lucia, *Re-embodying Law: Transversal Ecology and the Commons* in R. Thomas-Pellicer, De Lucia V. and Sullivan S. (eds), *Law, Philosophy, Ecology: Exploring Re-embodiments*, London: Routledge, 2016, and V. De Lucia, 'Semantics of Chaos. Law, Modernity and the Commons', 12:2 *Pólemos Journal of Law, Literature and Culture*, 2018b.

A crucial step in this book, then, is to thoroughly problematize the binary frame anthropocentrism/ecocentrism, in order to move on to a biopolitical framework of critique which, I argue, is better equipped to explore and deconstruct the nuanced, ambiguous and paradoxical concept – and articulations thereof – of the 'ecosystem approach'. This chapter contains two main sections, the first problematizing anthropocentrism and the second problematizing ecocentrism. The central aim of problematizing ecocentrism is to prepare the terrain for a biopolitical critique of the 'ecosystem approach', a critique that will be equally aimed at anthropocentric and ecocentric articulations of the concept. In this sense, the answer to the main research question no longer hinges on the secondary instrumental question of whether the 'ecosystem approach' is anthropocentric or ecocentric, but rather, on the more problematic one of whether, to what extent and how the 'ecosystem approach' – in both anthropocentric and ecocentric articulations – enables and enacts a biopolitical form of power.

However, the concepts of anthropocentrism and ecocentrism will not be entirely discarded. They will retain a *residual* utility, especially in relation to their ability to capture synthetically the polarities of the mainstream discourse, and more specifically, the competing narratives that are explicitly discussed in

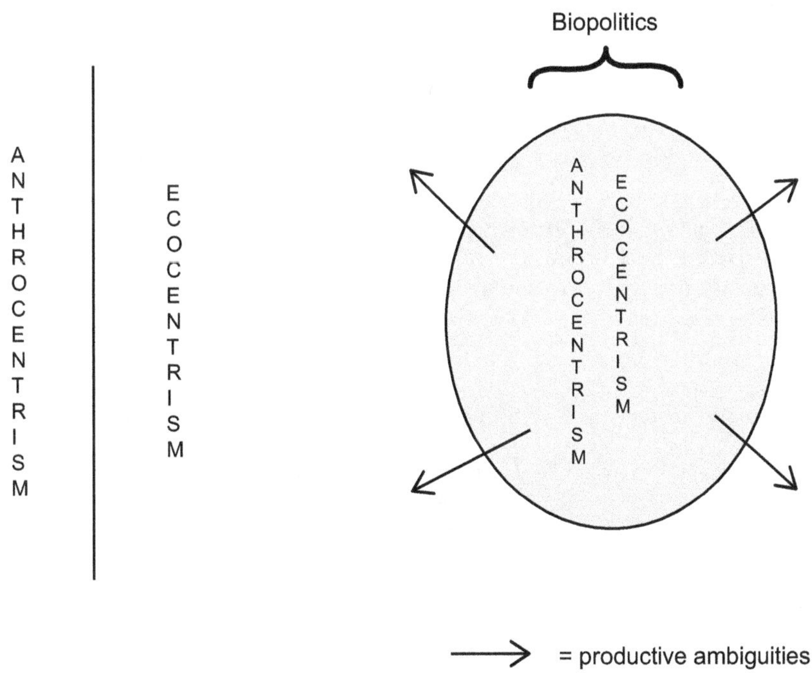

Figure 7.1

the literature concerned with the 'ecosystem approach'. Rather than entirely discarding the two concepts, the aim of this section is to show that they cannot be usefully deployed as the *central* critical framework of analysis, but must be inserted within a biopolitical horizon. The residual utility of these two concepts will be further specified in the conclusion.

Before proceeding with the problematization of anthropocentrism and ecocentrism, it might be useful to offer a visualization of the perspectival shift (Figure 7.1), a shift that re-articulates the problematic from a binary threshold to a circular enfolding where both critique and resistance are 'inside and against', an expression that indicates how critique is immanent and does not hinge or depends on an external, transcendental frame of reference that ultimately upholds the binary (I will return on this point at length in Chapter 10).

Problematizing anthropocentrism

Introduction

As we have seen in Chapter 1, anthropocentrism is premised on the Cartesian separation between mind and matter.[3] This separation, organized in a particular hierarchical form, gave rise to the modern epistemology of mastery according to which humankind, through the development of science and technology, is able to dominate and exploit an objectified nature, devoid of reason and only capable of responding to mechanistic natural laws.[4] Humanity, in this view, is the centre around which the rest of the world 'whirls like a frilly skirt';[5] the latter's value is only instrumental and contingent on the benefits it may provide to human beings.[6] Notwithstanding the diversity of legal cultures and traditions, anthropocentrism has fundamentally informed not only the way modern law constructs, categorizes and orders nature,[7] but also the manner in which law *protects* nature, as Gillespie has shown in relation to international environmental law.[8] However, the conceptual semantics

3 See, *ex pluribus*, Grear, 'Vulnerable Living Order'; Tallacchini, *Diritto*; and Godden, 'Nature as Other'.

4 Tallacchini, *Diritto*; W. Leiss, *The Domination of Nature*, Montreal and Quebec: McGill-Queens University Press, 1994; C. Merchant, *The Death of Nature: Women, Ecology, and the Scientific Revolution*, New York: HarperCollins, 1990; L. Code, 'Doubt and Denial: Epistemic Responsibility Meets Climate Change Scepticism', 3:5 *Oñati Socio-Legal Series* 2013, 838.

5 Philippopoulos-Mihalopoulos, 'Towards a Critical', p. 22.

6 Curry, *Ecological Ethics*; Gillespie, *International Environmental Law*.

7 See e.g. Tallacchini, *Diritto*; Godden, 'Nature as Other'; Burdon, *Exploring Wild Law*; Grear, *Vulnerable Living Order*.

8 Gillespie, *International Environmental Law*. See also K. Bosselmann, *The Principle of Sustainability: Transforming Law and Governance*, Aldershot: Ashgate, 2010; and Bosselmann, 'Losing the Forest'.

of the term anthropocentrism may conceal more than they reveal. Indeed, they do conceal, as I will try to illustrate in the following sections, important things. This concealment is due, in part, to the unwarranted subsumption, under one term, of a great variety of articulations that are only distant relatives. In part (and this is the most important operative concealment perhaps) it is due to the fact that the central term of reference – the Anthropos – makes undifferentiated reference to humanity, and in so doing it masks through abstraction the very concrete, material differences that 'anthropocentric' law facilitates and helps re-produce. I will now address each in turn.

Multiple anthropocentrism(s)

A first problem arising in the use of the term anthropocentrism is that anthropocentrism is not a monolithic concept. There are many inflections of anthropocentrism, whose conceptual core and practical effects vary significantly. This question is discussed, for example, by Patrick Curry, who concludes by using anthropocentrism to refer specifically to the 'unjustified privileging of human beings ... at the expense of other forms of life'.[9] Yet anthropocentrism can refer to a number of *justified* ethical positions, a case in point being weak anthropocentrism (to which I will return later in this section). Moreover, all those stances that can be gathered under the heading of stewardship (albeit itself not a univocal concept) arguably retain an anthropocentric outlook,[10] to the extent that stewardship 'originated, both as a practice and a concept, in patriarchal, elitist, and anthropocentric social systems and ideologies'.[11] Some stances however also internalize a significant element of care and empathy with the non-human world, so that some authors consider stewardship as also conceivable in non-anthropocentric terms.[12] Yet, in a further complicating circularity, the 'managerial', resource-oriented orientation may ultimately remain embedded in every articulation of stewardship.[13] One definition of stewardship that is thus oriented reads as follows:

9 Curry, *Ecological Ethics*, p. 55.

10 Thus in particular a 'classic' critique of stewardship: Palmer, 'Stewardship'.

11 J. Welchman, 2012, 'A Defence of Environmental Stewardship', 21 *Environmental Values* 2012, 297, p. 298. For the original argument see Palmer, 'Stewardship'.

12 J. McIntyre-Mills, *Systemic Ethics and Non-Anthropocentric Stewardship. Implications for Transdisciplinarity and Cosmopolitan Politics*, Switzerland: Springer International Publishing Switzerland, 2014.

13 Thus Palmer, 'Stewardship', which at p. 66 argues that the strongest resonance of stewardship is with financial management, through a linkage she establishes with church campaigns of 1950s and 1960s, campaigns aimed at gathering more resources, particularly (although not only) in the form of money. Stewardship, in this context, was considered good management of resources which, ultimately, belonged to God.

> Stewardship is the responsible use (including conservation) of natural resources in a way that takes full and balanced account of the interests of society, future generations, and other species, as well as of private needs, and accepts significant answerability to society.[14]

While this definition of stewardship is not ecocentric as such, it is evidently far removed from Curry's definition of anthropocentrism, and in this sense sufficiently illustrates the point. Its emphatic reference to other species, however, as already indicated, still arguably proceeds from the privileged position humans are implicitly accorded at the centre of the world. Thus, while some scholarship understands stewardship as a *tertium genus* vis-à-vis both anthropocentrism and ecocentrism, it seems to the present author that stewardship, as a form of 'responsible management',[15] remains a form of anthropocentrism yet one where responsibility and empathy for the non-human world has an important place.[16]

Another articulation that upsets a monolithic understanding of anthropocentrism is weak anthropocentrism. This position, articulated by Bryan Norton, is particularly noteworthy in that it has provided significant inspiration for at least some environmental legal theory[17] and therefore requires some discussion. Norton's starting point is that there is no need to locate value intrinsically in nature (responding to those criticizing anthropocentrism and promoting its opposite, ecocentrism). His aim is to rescue anthropocentrism as a valuable ethical framework for environmentalism, one that can be efficacious and more easily defended than ecocentrism. To this purpose, Norton distinguishes between strong and weak anthropocentrism. Strong anthropocentrism is, in Norton's account, characterized by 'felt preferences', that is 'any desire or need of a human individual that can at least temporarily be sated by some specifiable experience of that individual'.[18] Weak anthropocentrism, on the other hand, is characterized as being based on what he calls 'considered preferences',[19] that is

14 R. Worrell and M. Appleby, 'Stewardship of Natural Resources: Definition, Ethical and Practical Aspects', 12:3 *Journal of Agricultural and Environmental Ethics* 2000, 263, p. 269.

15 See e.g. W. Cunningham, M. Cunningham and B. Woodworth Saigo, *Environmental Science: A Global Concern*, 12th edn, New York: McGraw-Hill, 2012, especially Chapter 2.

16 T. Derr, quoted in S. Sharpe, *Redeeming the Time: A Political Theology of the Environment*, London: Bloomsbury Publishing, 1998, p. 32.

17 Particularly in Italy, see e.g. Tallacchini, *Diritto* and Tallacchini, 'Legal Framework', which explicitly refers to and advocates weak anthropocentrism; see also M. Cecchetti, 'La Disciplina Giuridica della Tutela Ambientale come Diritto dell'ambiente', *federalismi.it, Rivista di Diritto Pubblico Italiano, Comunitario e Comparato*, https://federalismi.it/nv14/articolo-documento.cfm?artid=6610 (Accessed 26 September 2018), who, like Tallacchini, also embraces a weak anthropocentrism understood as an 'eco-compatible revision of anthropocentrism' (p. 56; see also Tallacchini, *Diritto*, p. 160)

18 B. Norton, 'Environmental Ethics and Weak Anthropocentrism', 6:2 *Environmental Ethics* 1984, 131, reprinted in D. Clowney and P. Mosto (eds), *Earthcare: An Anthology in Environmental Ethics*, Lanham, MD: Rowman & Littlefield, 2009 (to which I will refer), p. 161.

19 Ibid.

preferences that are formed through deliberative processes and are 'rationally defensible in the context of a defensible worldview'.[20]

Deliberation and rationality, Norton further argues, would prevent preferences from remaining impermeable to criticism and objections, insofar as they are adopted within the bounds of a 'rational worldview ... which includes fully supported scientific theories and a metaphysical framework interpreting those theories'.[21] In this sense, argues Norton, weak anthropocentrism provides the basis for a 'criticism of value systems which are purely exploitative of nature',[22] if they are 'not consonant'[23] with such a rational worldview. Weak anthropocentrism, in this sense, would be *justifiable*, and hence outside of Curry's definition of anthropocentrism. Moreover, it seems that this articulation of anthropocentrism can be easily superimposed on the general framework of sustainable development, which also rests on a 'rational worldview' and on 'fully supported scientific theories', and in this respect is not a 'purely exploitative' system of environmental governance.

What Norton underlines as the crucial distinction is consequently not that between anthropocentrism and non-anthropocentrism, but rather the distinction between individualism and non-individualism.[24] This last point, that is, demarcating value systems along an individualist/non-individualist line, is certainly attractive and useful.

The aim of this section however, is not to discuss each individual articulation of anthropocentrism, and its merits (for example, Norton's argument seems to offer a rather unproblematized account of rationality), but rather to show how anthropocentrism, if deployed in an unqualified manner, conceals more than it reveals, given the great variety of articulations that exist. This concealing effect is even more evident, and more pernicious, if one deconstructs the conceptual core of anthropocentrism, the *Anthropos.*

Deconstructing the Anthropos[25]

The primary semantic reference of anthropocentrism is man, the Anthropos, understood as a symbolic placeholder for humanity. Indeed, Curry's definition refers directly to the privilege accorded to humanity vis-à-vis nature. Yet anthropocentrism, as Anna Grear insightfully suggests, is 'far removed from being a concern with human beings in any rich and inclusive sense'.[26] Anthropocentric law, in its

20 Ibid., p. 159.

21 Ibid.

22 Ibid.

23 Ibid.

24 'While the development of a non-anthropocentirc axiology committed to intrinsic value for nonhuman natural entities remains an interesting philosophical enterprise, the dichotomy on which it is based has less importance than I usually thought', ibid., p. 164.

25 This particular formulation is a quotation from Grear, 'Deconstructing Anthropos'.

26 Grear, *Should Trees*, p. 82.

imbrication with capitalism,[27] traverses humanity and selects only *certain* human beings as the beneficiaries of current regimes of ecological accumulation.[28] As some critical (legal[29] and non-legal[30]) scholarship has begun to emphasize, anthropocentrism, with its undifferentiated reference to humanity or mankind, masks very differentiated realities within which different human individuals and communities live. As Grear explains cogently and at length for example,[31] the *Anthropos* that is put up as the universal representation of humanity, is in fact a placeholder for a very narrow instantiation of it, namely the white, male, rational agent. Anthropocentrism then, is not a reference to humanity, but to a particular subset thereof that historically has colonized, exploited and plundered both other human communities and the non-human world, through what Grear calls 'vectors of oppression linking intra- and inter-species hierarchies'.[32]

Malm and Hornborg similarly deconstruct the Anthropocene for its inability, or unwillingness, to account for the role of power and power relations in the pervasive transformations that determined the geological shift under discussion.[33] They argue that the standard Anthropocene narrative is premised 'for reason of logical necessity' on a species narrative and on the notion of human nature (and its species-properties), which, in a certain sense, teleologically, underlies the Anthropocene. 'Anything less', Malm and Hornborg suggest, would turn the Anthropocene from a geology of *mankind* to a geology of *only a few*.[34] Yet, and this is precisely Malm and Hornborg's point, the industrial revolution (the most common reference point for the onset of the Anthropocene) with its reliance on the steam engine and on the large-scale introduction of fossil fuels into the human economy, is 'historically predicated on highly inequitable global processes from the start'.[35] Indeed, uneven distribution is a condition for the '*very existence* of modern, fossil-fuel technology', and hence for the Anthropocene.[36] Malm and Hornborg remind us thus that 'a significant chunk of humanity is not party to the fossil economy at all'.[37] Grear similarly

27 This imbrication is a central feature of modern law; see e.g. De Sousa Santos, *New Common Sense*.

28 For an argument about environmental regimes as regimes of ecological accumulation, particularly in relation to climate change, see M. Paterson, 'Legitimation and Accumulation in Climate Change Governance', 15:3 *New Political Economy* 2010, 345.

29 Grear, 'Deconstructing Anthropos'; De Lucia, 'Beyond Anthropocentrism'.

30 See especially A. Malm and A. Hornborg, 'The Geology of Mankind? A Critique of the Anthropocene Narrative', 1:1 *The Anthropocene Review* 2014, 62.

31 Grear, 'Deconstructing Anthropos'.

32 Ibid., p. 9 (page numbers missing on the document, so I refer to the page number as evinced from the PDF reader).

33 A. Malm and A. Hornborg, 'The Geology of Mankind? A Critique of the Anthropocene Narrative', 1:1 *The Anthropocene Review*, 2014, 62.

34 Ibid., p. 63.

35 Ibid.

36 Ibid., p. 64. See also, at length, Hornborg, *Power of the Machine*.

37 Ibid., p. 65.

emphasizes how 'intra-species hierarchies' are central to 'histories of eco-violation', histories which 'are directly interwoven with well-practiced, patterned and predictable distributions of egregious intra-species injustice'.[38]

The concept of anthropocentrism, and the narrative associated with it and with the emerging frame of the Anthropocene, conceals, then, very specific mechanics of exclusion and of power relations (operating, crucially in and through law) through which *both* other humans *and* ecosystems are 'systematically disadvantaged'[39] by legal modernity.[40] The semantics of the Anthropos – whether in the form of anthropocentrism or through the new horizon of the Anthropocene – operates according to the same logic, embedded in the concepts of the legal subject or legal person, and manifesting through the mechanics of private property and public sovereignty (in which international law has had a central role).[41]

Problematizing ecocentrism

Introduction

Ecocentrism, which in its most general sense implies the recognition of the intrinsic value of nature, equally requires 'particular critical attention',[42] especially in relation to the risk it runs of remaining merely a specular image of its opposite, functioning within a mechanics of reversal which fails to *exit* from the dialectic of modernity, obsessed with '-centrisms'.[43]

Indeed, there are a number of ways in which ecocentrism can be problematized. We can group these under two main headings. One directly problematizes ecocentrism on its own terms, providing some form of critique *internal* to it. A second perspective, which will be discussed first, takes an external approach and problematizes ecocentrism to the extent that it either denies its very possibility (given that all valuation is, of necessity, human), or its usefulness, on pragmatic and strategic grounds.

External critiques

As anticipated, the external critique criticizes ecocentrism on two separate questions. The first relates to the very possibility of ecocentric approaches, insofar as any act of valuation, even allegedly ecocentric ones, proceed from human

38 Grear, 'Deconstructing Anthropos', p. 231.
39 Grear, *Should Trees*, p. 78.
40 English literature speaks more often of hegemonic liberal articulations of modern law; see in this respect e.g. Grear, 'Vulnerable Living Order', and Wilkinson, 'Using Environmental Ethics'.
41 See e.g. A. Anghie, 2004, *Imperialism, Sovereignty and the Making of International Law*, Cambridge: Cambridge University Press, from which also Grear 'Deconstructing Anthropos' draws.
42 Grear, *Should Trees*, p. 82.
43 Philippopoulos-Mihalopoulos, 'Actors or Spectators?'.

judgement, and hence are irreducibly anthropocentric. The second question relates to the pragmatic and strategic shortcomings that ecocentric approaches may have.

Is Non-anthropocentrism possible?

Whether a non-anthropocentric stance is possible is a recurrent question in the literature, a question with discordant answers. The question has two dimensions, hinging respectively on the source of value and on its location. As regards the *source* of value, a number of authors submit that the human perspective is always necessarily embedded in value judgements, hence human values are inevitably anthropocentric. Some suggest that 'the attempt to provide a genuinely non-anthropocentric set of values or preferences [is] a hopeless quest'.[44] Frederik Ferré underlines, for example that a 'perspectival anthropocentrism' is 'inescapable', insofar as 'humans have no choice but to think as humans'.[45] Tim Hayward, discussing 'what is not wrong with anthropocentrism', similarly emphasizes how humans' view of the world is inevitably the result of their particular perspective, which places them 'in many real respects … at the centre of it'.[46] In fact, Hayward observes that '[t]here is an ineliminable element of anthropocentrism in ethics as such'.[47] From a legal-pragmatic perspective, Christopher Stone, in his famous essay exploring whether trees should have legal standing, observes that perhaps 'at the level of praxis, an earth justice system will be unavoidably anthropocentric'.[48] This seems a reasonable approach, at least prima facie, in that it underscores how either directly or indirectly humanity remains the ultimate source of value.

This critique however, is easily rebutted. As Curry observes, authors who subscribe to the inevitability of anthropocentrism make the 'common mistake of inferring that values must be anthropo*centric* … from the fact that when humans are the valuers … those values are indeed necessarily anthropo*genic*'.[49]

Pragmatic problems with ecocentrism

Another problematic aspect of ecocentrism, according to the literature, relates to pragmatic considerations. Tim Hayward, for example, maintains that 'blanket

44 William Grey, quoted in A. Schillmoller, and A. Ricketts, 'Recognising Rights for Nature: A Nego-tiation of Principle and Pragmatism', 21 *Transformations* 2012, 1–14.

45 F. Ferré, 'Personalistic Organicism: Paradox or Paradigm?', in R. Attfield and A. Belsey (eds), *Philosophy and the Natural Environment*, London: Royal Institute of Philosophy Supplement, 36, Cambridge: Cambridge University Press, 1994, p. 72.

46 Hayward, 'Anthropocentrism', p. 51.

47 Ibid., p. 55.

48 C. Stone, *Should Trees*, p. 4.

49 P. Curry, 'Nature Post-Nature', 26 *New Formations* 2008, 51, p. 54. Emphasis in the original.

condemnations of "anthropocentrism" not only condemn some legitimate human concerns; they also allow ideological retorts to the effect that criticisms of anthropocentrism amount to misanthropy'.[50] This concern is echoed by Andrew Light, who emphasizes that arguments from non-anthropocentric philosophical positions are 'naïve and imprudent',[51] to the extent that such positions lead one to 'negotiate environmental priorities from the point of view of an irreconcilable and intractable moral view opposing human interests'.[52] From the pragmatic standpoint Light speaks from, '[m]aking questions of human benefit and satisfaction irrelevant'[53] would simply and invariably ensure the irrelevance of non-anthropocentric environmentalism at the negotiating table. Bryan Norton, too, justifies his weak anthropocentric position on the basis that it yields better results in the area of public policy.[54]

This set of critiques, however, rests largely on a straw man, as there is no necessary or logical connection between a non-anthropocentric value orientation and an opposition to human values (also bearing in mind the distinction carved out in the previous section between anthropocentric and anthropogenic). Moreover, the deconstruction of the monolithic notion of 'humanity' has shown how anthropocentrism is not equally concerned with all humans. Dismissing then non-anthropocentric stances because of their opposition to human values amounts to a neutralization of legitimate political concerns in the name of the pragmatics of policy-making.

Internal critiques

Introduction

While external critiques are relatively easy to reject or overcome, internal critiques are far more problematic. From this internal perspective, ecocentrism is unable to achieve its stated goal of a radical shift and it rather remains an internal articulation of the dialectic of modernity. There are several reasons for it. Ecocentrism shares the modern obsession with a centre;[55] ecocentric articulations rarely problematize the otherwise problematic, unstable and contested concept of nature;[56] finally, the most purportedly radical legal trajectory associated with ecocentrism remains thickly *inside* modernity, to the extent that, despite its

50 Hayward, 'Anthropocentrism', p. 49. Curry also discusses this problematic aspect of ecocentrism (and in particular whether humans are, or ought to be included or not), though he arrives at a different conclusion, see Curry, 'Nature Post-Nature', p. 57ff.

51 A. Light, 'Contemporary Environmental Ethics from Metaethics to Public Philosophy', 33:4 *Metaphilosophy* 2002, 426, p. 439.

52 Ibid.

53 Ibid., quoting E. Katz and L. Oechsli, 'Moving beyond Anthropocentrism: Environmental Ethics, Development, and the Amazon', 15:1 *Environmental Ethics*, 1993, 49, p. 58.

54 Clowney and Mosto, *Earthcare*.

55 Philippopoulos-Mihalopoulos, 'Actors or Spectators?'.

56 Grear, A. 'Law's Entities: Complexity, Plasticity and Justice', 4:1 *Jurisprudence* 2013b, 76.

pragmatic advantages, it furthers an ethical and legal framework premised on *individual rights.*[57]

The Problem of '-centrism'

In a first sense, the main critique that can be raised against ecocentrism relates to the fact that ecocentrism reproduces the modern obsession with the centre. Ecocentrism, which in its most general sense implies the recognition of the intrinsic value of nature, requires 'particular critical attention',[58] especially in relation to the risk it runs of filling the role of an opposite yet analytically dependent complement to anthropocentrism, of which it remains merely a specular refraction. Ecocentrism would thus merely function within a mechanics of reversal which fails to *exit* from the paradigm of modernity, obsessed with '-centrisms'[59] and, conversely, with peripheries.[60] It is in this sense that Philippopoulos-Mihalopoulos speaks of a 'tyranny of the centre'[61] that affects both ecocentrism and anthropocentrism, and that leads to epistemological 'delusions', in the sense of the construction of a '*centre*' from which to know.[62] Indeed, the epistemology of the centre, suggests further Philippopoulos-Mihalopoulos, ensures that there is no available alternative: 'However much we battle between extremes, we have no choice but to remain faithful to the soothing idea that there must be a centre'.[63]

This mere reversal, as noted in the literature, also, and importantly, maintains the 'radical split' between humanity and nature.[64] And if this is especially true for those ecocentric approaches that would rather exclude humans from the scope of concern, and as such might be rightfully described as either misanthropic, strategically counterproductive, or both,[65] the split is also upheld to the extent that a clear distinction is demarcated by the alternative conceptual and

57 Thus for example the prevailing legal strategy employed by scholars affiliated with the emerging legal philosophy called Earth Jurisprudence, see De Lucia, 'Towards an Ecological Philosophy',, and A. Schillmoller and A., Pelizzon, 'Mapping the Terrain of Earth Jurisprudence: Landscape, Thresholds and Horizons', 3:1 *Environmental and Earth Law Journal* 2013, 1 – 32.

58 Grear, 'Law's Entities', p. 82.

59 Philippopoulos-Mihalopoulos, 'Actors or Spectators?'

60 A similar reasoning based on the notion of dichotomization is articulated by Hasley, and in relation to the same logic of reversal, albeit he discusses it in Deleuzian language, see M. Hasley, *Deleuze and Environmental Damage. Violence of the Text*, Aldershot: Ashgate, 2006, pp. 35–6.

61 A. Philippopoulos-Mihalopoulos, 'Epistemologies of Doubt' in A. Grear and L. Kotzé (eds), *Research Handbook in Human Rights and the Environment*, Cheltenham, UK: Edward Elgar, 2015, p. 29.

62 Ibid., p. 28.

63 Philippopoulos-Mihalopoulos, 'Actors or Spectators?', 857.

64 Curry, *Ecological Ethics*, p. 57.

65 Ibid.

terminological framing. In this sense, the binary anthropocentrism/ecocentrism operates as the ethical articulation of the ontological binary man/nature.[66]

Unproblematized nature

Another problematic risk that affects any version of ecocentrism is that the idea of nature often remains unproblematized in ecocentric articulations of law. Taking Earth Jurisprudence as one example of a legal philosophy decisively driven by what can be called an 'ecocentric turn' will help concretize this point. Earth Jurisprudence is founded on what Berry calls the Great Law (and Cullinan the Great Jurisprudence): a set of timeless and unified principles, or laws, 'manifest in the Universe itself',[67] which serve as a standard against which human laws are to be assessed. As I have argued elsewhere,[68] this approach resembles modern natural law theory.[69] Drawing normativity from nature, by contrast, arguably demands to openly problematize the ambiguities afflicting the idea of nature and other important ecological concepts, such as 'ecosystem', 'ecosystem health' and 'ecosystem integrity'. This is particularly true if one accepts the idea that the concept of nature (and I underline, the *concept*, not its material reality) is socially constructed, historically situated and politically negotiated.[70] In other words, the concept of nature is a methodological device that captures and continuously re-enacts the political epistemology of modernity. This ambiguity operates as an inevitable inner tension that remains under-problematized (at best) or even unproblematized (at worst), by much Earth Jurisprudence scholarship,[71] whose focus has primarily been the re-articulation of the humanity – nature relationship, rather than the exploration of 'the ontological and epistemological issues associated with the *idea* of nature'.[72] Indeed, a problematic, ambiguous concept of nature, as Schillmoller and Pelizzon

66 Philippopoulos-Mihalopoulos, 'Epistemologies of Doubt', 31.

67 Cullinan, *Wild Law*, pp. 76–7. See also T. Berry, *The Great Work: Our Way into the Future*, New York: Bell Tower, 1999.

68 De Lucia, 'Towards an Ecological Philosophy'.

69 A resemblance and affinity openly advocated by at least some scholars working within the context of Earth Jurisprudence. See in this respect Burdon, 2014, op. cit.

70 See N. Evernden, *The Social Creation of Nature*, Baltimore, MD: The Johns Hopkins University Press, 1992, and B. Latour, *The Politics of Nature*, London: Harvard University Press, 2004.

71 See in this respect Schillmoller and Pelizzon, 'Mapping the Terrain'.

72 Ibid., p. 16, emphasis theirs. Schillmoller and Pelizzon represents an exception, as they openly acknowledge (from *within* the legal and epistemological perspective of Earth Jurisprudence) how nature is a 'contested concept with negotiable meaning' (p. 3), that will require 'continuing negotiation' (p. 7). It must be further noted that at least some Earth Jurisprudence scholars are aware of these issues. Cullinan for example cautions that '[t]he diversity of nature and our limited knowledge of it allow ample scope for everyone from fascists to nihilists and beyond to claim nature supports their theories', C. Cullinan, *Voices of Earth Jurisprudence: An Introduction to Earth Jurisprudence: Guiding Principles and Wild Law Possibilities*, Center for Earth Jurisprudence, 2011, p. 5.

observe, bring home the 'problematic tension between the requirement of a concept of nature upon which to ground action, and an awareness of the impossibility of settling upon a definitive version of what nature is'.[73]

It would be thus wise to approach nature as a genealogical concept, openly acknowledging it as a terrain of contestation and negotiation. Problematizing nature in this sense however, also inevitably entails a problematization of ecocentrism. Ecocentrism itself then is suddenly inserted in a genealogical 'play of forces', further destabilizing the binary framework of analysis that neatly and univocally identify the two keys of the binary, and linearly demarcates the bad (anthropocentrism) from the good (ecocentrism).

Ecocentrism and subjective rights

A final problem with ecocentrism is that it may extend, rather than radically criticize, legal modernity, in light of its tight linkage with the idea of extending legal subjectivity to nature (the reader may recall that this was one of the central ideas in ecocentric articulations of the 'ecosystem approach' such as Grumbine's). This is particularly true in relation to Earth Jurisprudence, whose prevailing strategy for effecting an ecocentric turn in law is a rights-based approach.[74] In a manner similar to the problematic -centrism of ecocentrism, a rights-based strategy risks maintaining the framework of legal analysis within a horizon premised on a subject – object grammar. This is a particularly problematic critique if one approaches the question of radical change from the perspective illustrated, for example, by Pieraccini, in which the task of legal scholars is to create a *new* language disentangled from the constraints and commitments (philosophical, conceptual and legal) of modernity, and responsive to the requirements of the Anthropocene.[75]

73 Schillmoller and Pelizzon, 'Mapping the Terrain', p. 8.

74 Ibid., p. 3 describe for example EJ as generally advocating 'the recognition of rights of nature'. However, there is no necessary relationship between EJ as a legal philosophical framework and a rights-based approach. In fact, a number of EJ scholars and theorists are sceptical or critical of rights discourse. E. Fitz-Henry, 'Between Rights and Services', Earth Jurisprudence: Building Theory and Practice, from Third Wild Law Conference, Griffith University, 16–18 September 2011; cautions for example against uncritical embrace of rights discourse (particularly as regards the role nature rights may play towards further marginalization of already marginalized communities and the risk of corporate appropriation of rights discourse, leading to further commodification of nature). See also, among others, Bosselmann, *The Principle*, particularly Chapter 4 (Bosselmann cautions against overextending the usefulness of a rights approach); Schillmoller and Ricketts, 'Recognising Rights'; Schillmoller and Pelizzon, 'Mapping the Terrain'. Even Burdon recognized recently the limitations of a rights discourse, P. Burdon, 'Environmental Protection And The Limits Of Rights Talk', *Right Now*, 2012, http://rightnow.org.au/topics/environment/environmental-protection-and-the-limits-of-rights-talk/, (Accessed 26 September 2018).

75 M. Pieraccini, 'Reflections on the Relationship between Environmental Regulation, Human Rights and beyond – with Heidegger' in A. Grear and E. Grant (eds), *Thought, Law, Action and Rights in the Age of Environmental Crisis*, Cheltenham: Edward Elgar, 2015.

This is not the place for a thorough critique of the rights-based approach, nor for an evaluation of its merits – pragmatic and/or theoretical.[76] The intent of this section was merely to *highlight* the problems involved in an uncritical deployment of the narrative frame of ecocentrism, and how they in turn affect and destabilize the binary framework of analysis that is the focus of my critique.

Conclusions

This chapter has problematized both anthropocentrism and ecocentrism on a number of grounds. While still useful as a descriptive matrix, particularly in relation to their widespread use in the relevant literature, I have argued that the problems that affect them render them unable to offer a strong and sufficiently nuanced critical framework of analysis. This conclusion hinges on the fact that the 'ecosystem approach' is a genealogically complex concept that defies neat polarizations or linear accounts of its development. The choice between the two terms of the binary, both reproducing a 'centric' dialectic, remains arguably undecidable for two reasons. First, and most immediately, the utter core of the 'ecosystem approach' is empty, there is no true version, but its meaning is genealogical, i.e. inserted inevitably within a 'play of forces'. Second, both '-centrisms' remain within the horizon of sense of modernity. This is exactly the juncture where the biopolitical key, I argue, proves to be the best interpretive register available to a critical legal engagement with the 'ecosystem approach'. A biopolitical analysis, in fact, is genealogical. It does not concern itself with any 'inexorable' unravelling of values 'towards their perfection',[77] but rather wishes to arrive at the 'lacerations' implicated in the descent and emergence of every concept.[78] The same competing narratives, conflicting values and relations of forces not only traverse the 'ecosystem approach' in its different articulations, but are also at work *even* in the same definition. Thus, the polarization of the 'ecosystem approach' in two distinct camps (anthropocentric and ecocentric) does not exhaust the contestations that are reproduced even within specific articulations of the concept (and this is exactly where a biopolitical frame of analysis will be able to offer, I argue, important insights). On the contrary, one such binary framework conceals the biopolitical problematics that are internal to both narratives of the 'ecosystem approach'. It is then now time to delve into biopower and biopolitics, and to articulate in full a biopolitical framework of analysis.

76 Critique that I have partially attempted in De Lucia, 'Towards an Ecological Philosophy'.

77 Douzinas and Gearey, *Critical Jurisprudence*, p. 54.

78 Descent and emergence are the two elements of genealogy, the reader may recall, which Foucault draws out from Nietzsche, the former linked to the an evolutionary, yet not linear, concept of history, and the latter to the notion of history as the result of constant 'play of forces', M. Foucault, 'Nietzsche, Genealogy, History', in D. F. Bouchard (ed.), *Language, Counter-Memory, Practice: Selected Essays and Interviews*, Ithaca, NY: Cornell University Press, 1977. For a more detailed account I will refer the reader to Chapter 1, where I have discussed genealogy as a methodology.

8 A biopolitical framework

Introduction

After having problematized the concepts of anthropocentrism and ecocentrism, and after having outlined their limited critical purchase, this chapter shifts the critical register towards a biopolitical framework of analysis. Here the analytics of biopolitics come into full view and to full use. An analytics of biopolitics in fact, allows us to simultaneously problematize *both* narratives from the same critical perspective, to the extent that each, albeit differently, constructs, operationalizes and enforces, through similar mechanics, the dual, bivalent biopolitical *dispositif*. Ecocentric articulations of the 'ecosystem approach', from a biopolitical perspective, are not immune from critique. They are not insulated from being re-presentations of nature that functionalize it towards its regularization along decisively biopolitical trajectories. They do not, importantly, abandon the perspective of the centre, a perspective which, as we have seen, is one of the crucial signals of the inadequacy of environmental law. On the other hand, anthropocentric orientations are not necessarily uniformly aligned to negative articulations of the 'ecosystem approach'. In fact, anthropocentrism, once critically unpacked, does not offer a reliable frame of analysis, as there are different humanities, and in relation to some of these humanities even 'centrisms' may find an affirmative declension, linked to the inevitability of a human epistemology, but *not* to an epistemology of mastery and to a praxis of domination. Furthermore, and importantly, an analytics of biopolitics disallows and deactivates analytical frames that neatly separate the two narratives. On the contrary, a biopolitical frame suggests, encourages and even demands that privileged focus be given to the genealogical complexity, the inevitable slippery margins, the constitutively porous interfaces, and the irreducibly intertwined discourses and practices that traverse, join, effect and continuously re-orient the narratives involved.

Another key advantage of using this analytics of biopolitics rather than the binary anthropocentric-ecocentric framework, is that a biopolitical reading of the 'ecosystem approach' allows us to *simultaneously* read the 'ecosystem approach' negatively and positively, to the extent that life, situated at the 'moving margins' of intersection and tension between biology and history,[1] is both enhanced and subjugated by power in the same gesture.[2] Furthermore, a biopolitical key allows law to be understood in its multiple and complementary registers of operation.[3] The question of sovereignty, always oscillating between a conceptual obsolescence (advocated by progressive lawyers) and its material re-assertion (underlined by realists), finds also a useful re-framing through the analytics of biopolitics. Finally, the biopolitical perspective through which the original research question is discussed, inevitably re-calibrates the question itself.

While the history of biopolitics, with its multiple lines of elaboration, is complex to trace, a usual starting point is the work of French political philosopher Michel Foucault. In Foucault's thought, biopolitics is one of two modes of operation of a new form of power that seizes life under its purview: biopower.[4] Biopower, in brief, is the mode of power that seizes life under its political purview, and is bipolar in character. Biopower in fact envelops under its matrices and mechanics of control life as individual bodies and as populations. It is arguably the most advanced articulation of power today, and the broad horizon of sense within which the entire tradition of modernity can be organized. As such, biopower is a key vantage or perspective through which modernity can be understood,[5] and finds its most decisive intensification precisely in the Anthropocene, and in relation to not only human populations, but also life in the broadest sense. A key point I will illustrate at length in this chapter is in fact how biopower and biopolitics seize life in its entirety under their purview: all ecosystems, ecosystem processes and living organisms, that is, biodiversity.

Biopolitics, which represents the polarity of biopower most relevant in this book, has a varied and long history; a history that unfolds along multiples lines of elaboration, and shows, arguably, no necessary theoretical coherence.[6] In its most general sense, biopolitics looks at man as a species with the objective of achieving 'overall states of equilibration or regularity' in relation to its biological processes.[7] I will however apply biopolitics to nature, and in that respect a crucial point is that, from the perspective of biopolitics, nature is no longer simply an object of exploitation,

1 Esposito, *Bios*, p. 31.
2 Ibid., p. 37.
3 For other applications of the biopolitical framework of analysis to environmental law more broadly, and to invasive alien species, see, respectively, V. De Lucia, 'Critical Environmental Law and the Double Register of the Anthropocene: A Biopolitical Reading' in L. Kotzé (ed.), *Environmental Law and Governance in the Anthropocene*, Hart Publishing, 2017b, and V. De Lucia, 'Bare Nature. The Biopolitical Logic of the International Regulation of Invasive Alien Species', *Journal of Environmental Law*, 2018c, Advance Online
4 M. Foucault, *The History of Sexuality. Volume I: An Introduction*, Pantheon Books, 1978.
5 As Esposito himself, a leading advocate of the intrinsic biopolitical foundation of modernity, recognizes (as we shall see later in this chapter), Esposito, *Bios*.
6 For the genealogy of the concept see, e.g. Esposito, *Bios*.
7 M. Foucault, *Society Must Be Defended: Lectures at the Collège de France 1975-1976*, London: Penguin Books, 2004, p. 246.

but becomes the recipient of a series of interventions that aim at its optimization, and at the enhancement of its productive forces.[8]

French philosopher Michel Foucault usually represents the starting point of any discussion of biopolitics and, consequently, of this chapter. Foucault, however, does not exhaust biopolitical thinking. As this chapter shows, the concept of biopolitics has been subjected to a number of sophisticated elaborations that significantly stretch the concept. This shows how biopolitics – and in this respect biopolitics and biopower intersect in a multiplicity of contradictory, even confusing ways – has become a key register of political theory today, yet one that is integrated within the context of different political projects and intellectual trajectories.

It is perhaps useful to note at the outset that I wish to emphasize *a particular set* of interlocking aspects that pertain to biopower and biopolitics and that traverse the theorizations of biopolitics from Foucault to Esposito, through Agamben and Negri, rather than claim any loyalty to Foucault, or to any original, or true, articulation of biopolitics. Moreover, I will articulate what, following Lemke, can be described as an 'analytics of biopolitics'.

This chapter begins by outlining the intertwined concepts of biopower and biopolitics in the thought of Michel Foucault. The next section briefly traces the ways in which the concept of biopolitics has been developed by subsequent philosophers, in order to highlight the intimate continuity between sovereignty and biopolitics, a continuity downplayed, if not rejected outright by Foucault, and yet central to the argument developed in this book. Finally, and crucially, the last part of the chapter transposes biopolitics, originally aimed at exploring how power encompasses life in relation to human populations, to the field of ecology, and offers first a specifically *ecological* declension of it and subsequently an analysis of the relation between sovereignty and biopower.

Biopower and biopolitics

Bipolar biopower: discipline and regulatory controls

Biopower is a complex and composite notion. Exploring it usually means starting from Michel Foucault.[9] It is complex to the extent that Foucault's

8 Lemke, *Biopolitics*, p. 68, also more generally M., Hardt, and A., Negri, Empire, *The New World Order*, Cambridge, MA: Harvard University Press, 2000.

9 However, recent scholarship is increasingly tracing the concept to other philosophers that precede Foucault, such as Heidegger and Arendt, even if they did not explicitly deploy a biopolitical language, thus e.g. C. Wolfe, 2013, *Before the Law: Humans and Other Animals in a Biopolitical Frame*, Chicago, IL: University of Chicago Press, 2013; T. Campbell and A. Sitze (eds), *Biopolitics. A Reader*, London: Duke University Press, 2013; or to other social theorists who deployed a prodromic language of biopolitics, even if their conceptual apparatus was not fully developed in a biopolitical sense, thus e.g. Esposito, *Bios*.

theorizing is not coherent, linear or complete, but, rather, 'turbulent';[10] and because a number of philosophers have expanded, stretched and recalibrated the concept. It is composite, because biopower is a composition of two forms, or strategies, of power, both responding to the overall transformation of the mechanics and goals of power, but which take power in significantly different directions. In this book, I am concerned with only one of the two forms, namely biopolitics, and, it should be noted at the outset, the two terms – biopolitics and biopower – are used somewhat interchangeably (as is indeed often the case even in Foucault). It is, however, important to begin with an outline of biopower in general, its relationship with sovereign power, and the ways in which this relationship is articulated, ambiguously, in Foucault. In a later section, I account for how the concept of biopower (and biopolitics) has been analysed and further elaborated by other political philosophers and scholars who significantly developed the concepts of biopower and biopolitics.

The core of biopower lies in what Foucault characterizes as a transition from 'the power of the sovereign to the power over life'.[11] Sovereign power, argues Foucault, primarily takes the form of a 'deduction' or 'subtraction',[12] to the extent that it operates negatively by way of removing, taking away, extracting, or appropriating resources ('a portion of the wealth, a tax of products, goods and services, labour and blood') away from a subject.[13] In the last chapter of the first volume of his *History of Sexuality*,[14] and also in the lectures gathered under the title 'Society must be defended',[15] Foucault further notes that sovereign power finds its paradigmatic expression in a right to kill, so that sovereignty is 'power of life and death'.[16] The sovereign's mode of operation is, then, its right to seize life and dispense death. The relationship between power and life, from the sovereign perspective, is paradoxical, insofar as, Foucault further notes, 'it is at the moment when the sovereign can kill that he exercises his right over life'.[17]

Sovereign power, however, is overtaken by a new form or mode of power,[18] which is the opposite of sovereign power, insofar as, rather than entailing the

10 Campbell and Sitze speak of the 'turbulence' of Foucault's text, *Biopolitics*, p. 7.

11 Foucault, *Society Must be Defended*, p. 239.

12 Foucault, *History of Sexuality*, p. 136.

13 Ibid.

14 Ibid. However the very first outline of biopower was originally contained in *La volonté de Savoir*, whose third part was entitled 'Right of Death and Power over Life', a part which was subsequently reproduced in English translation in the *History of Sexuality*, vol. I., see e.g. Campbell and Sitze, *Biopolitics*, pp. 3–4.

15 Foucault, *Society Must be Defended.*

16 Ibid., p. 136.

17 Ibid., p. 240.

18 Foucault has a precise timeline within which he inserts the appearance of the biopower. However, as in this book I operate according to trajectories and theoretical frameworks, rather than precise periodizations and historical coordinates, I do not see it as necessary to reproduce Foucault's timeline, as it adds nothing to the argument presented in this chapter and book.

right to let live which is ultimately exercised through the power to kill (or, rather, through its restraint), it is a power to 'make live and let die'.[19] At its most general level, biopower, subsuming life under its care, 'distribut[es] the living in the domain of value and utility',[20] and, unlike sovereign power which becomes manifest 'in its murderous splendour', biopower qualifies, measures, appraises and hierarchizes.[21]

This power has two declensions, one focused on individual bodies – disciplinary power, or anatomopolitics – and the other focused on humans as a living species – biopolitics. Both introduce a shift from a politico-juridical to a technical-administrative exercise of power (a shift that moreover resonates strongly with a number of accounts of the development of international environmental law).[22] In the next section, I outline in some detail the differences between the two, before moving on to speak primarily of biopolitics. Until then however, there is some terminological overlap, resulting from the corresponding blurriness in both Foucault's works and in those of later commentators (with particular regard to what sometimes appears to be the near equivalent and indifferent usage of the terms biopower or biopolitics).

This new form of power over life is, as Foucault puts it, 'positive': it aims at fostering life. Biopower is however 'bipolar' and evolves in 'two basic forms'[23] that represent 'two poles of development linked together by a whole intermediary cluster of relations'.[24] The two forms are disciplines and regulatory controls.[25] The key differences are their focus and mode of operation. Their aim however is the same, that is, they are forms of power that 'do not limit, but, on the contrary, increase and multiply (certain) human abilities and forces in order to produce docile bodies, obedient citizens, normal individuals, healthy populations … and rich societies'.[26] The fact that biopower aims at the improvement

19 Foucault, *Society Must be Defended*, p. 241.

20 Foucault, *History of Sexuality*, p. 144.

21 Ibid.

22 The trend towards the scientific management of the environment started, according to Kuokkanen, in 1972, on the occasion of the Stockholm Conference on the Human Environment. Further, 'the incorporation of scientific and technical knowledge into policy-making [engendered] a process of professionalization. The integration of epistemic communities entailed a shift from traditional diplomacy and politics toward expertise. The new type of policy coordination having recourse to scientific knowledge began to frame a new approach to environmental issues', T. Kuokkanen, *International Law and the Environment: Variations on a Theme*, Leiden/Boston: Martinus Nijhoff, 2002, p. 254.

23 See in particular Foucault, *History of Sexuality* and *Society Must be Defended*. For some secondary literature see e.g. P. Rabinow and N. Rose, 'Biopower Today', 1:2 *BioSocieties* 2006, 195 and Lemke, *Biopolitics*.

24 Foucault, *History of Sexuality*, p. 139.

25 See in particular Foucault, *History of Sexuality* and *Society Must be Defended*. For some secondary literature see e.g. Rabinow and Rose, 'Biopower'.

26 L. Bernini, 'Sovereignty and Biopolitics in Hobbes' *Leviathan*', working paper, 2011, p. 2, https://www.academia.edu/3721237/Sovereignty_and_Biopolitics_in_Hobbes_Leviathan_May_2011_(Accessed 26 September 2018), originally published in Italian as L. Bernini, 'Il

of *certain* capacities of a population (in its specific biopolitical articulation) that are, moreover, aligned to what will be argued are capitalist aims of wealth production, is an important point that the reader should bear in mind, as I will return to it in several occasions.

The first pole of biopower relates to the emergence of technologies of power 'essentially centered on the body'[27] and strategically aimed at 'its disciplining, the optimization of its capabilities, the extortion of its forces, the parallel increase of its usefulness and its docility, its integration into systems of efficient and economic controls'.[28] These technologies, aimed at the 'separation, alignment, serialization, and surveillance' of individual bodies,[29] are explored by Foucault in a number of studies[30] and understood as *disciplines*, or as an '*anatomopolitics of the human body*'.[31] Foucault, however, identifies a further pole of this biopower in which another set of technologies offer themselves as an addition to disciplinary techniques operating on the body, which he calls biopolitics.

Biopolitics

This mode of power shifts its gaze and operative mode(l), and begins focusing not on man as a body, but rather on man as a population, as a species. In other words, while disciplinary power 'tries to rule a multiplicity of men to the extent that their multiplicity can and must be dissolved into individual bodies that can be kept under surveillance, trained, used and, if need be, punished',[32] biopolitics has an entirely different orientation.[33] Biopolitics is no longer interested in the multiplicity of men *qua* individual men or individual bodies. This multiplicity becomes interesting only 'to the extent that they form … a global mass' which can be understood in terms of broad biological processes requiring 'regulatory controls'.[34] Foucault calls this technology of power a 'biopolitics of the human race',[35] or alternatively, a 'biopolitics of the population'.[36] Biopolitics focuses

Cristallo Biopolitico di Hobbes' in L. Bernini (ed.), *Apocalissi Queer. Elementi di Teoria Antisociale*, Pisa: Edizioni ETS, 2013, pp.146–58.

27 Foucault, *Society Must be Defended*, p. 242.
28 Foucault, *History of Sexuality*, p. 139.
29 Foucault, *Society Must be Defended*, p. 242.
30 See e.g. M. Foucault, *Discipline and Punish: The Birth of the Prison*, London: Vintage Books, 2nd edn, 1995, and Foucault, *History of Sexuality*.
31 Ibid., p. 139. Emphasis in the original.
32 Foucault, *Society Must be Defended*, p. 242.
33 It must be noted however how, as recognized in the literature, Foucault's terminology is not always clear and precise; in this sense e.g. Rabinow and Rose, *Biopower*, p. 197.
34 Foucault, *History of Sexuality*, p. 139.
35 Foucault, *Society Must be Defended*, p. 243.
36 Foucault, *History of Sexuality*, p. 139. On this and further points related to the genealogy of biopolitics see also M. Foucault, *The Birth of Biopolitics*. Lectures at the Collège de France 1978–1979, New York: Palgrave-Macmillan, 2010.

on processes such as 'the ratio of births to deaths, the rate of reproduction, the fertility of the population, and so on'.[37] The central 'objective of knowledge' of biopolitics is precisely these processes and the economic and political problems that they pose.[38] These are 'the targets [biopolitics] seeks to control'.[39]

The aim of biopolitics (which eventually becomes the prevailing form of biopower so that Foucault almost identifies biopolitics with biopower and refers to disciplinary techniques as disciplinary power)[40] is, then, the regularization of life. Biopolitics[41] comprises a set of 'regulatory mechanisms'[42] or 'regulatory controls'[43] that aim at establishing 'an equilibrium, maintain an average, establish a sort of homeostasis, and compensate for variations within this general population and its aleatory field'.[44] Furthermore, 'security mechanisms have to be installed around the random element inherent in a population of living beings so as to optimize a state of life'.[45]

Biopolitics, in other words, looks at humankind as a species with the objective of achieving 'overall states of equilibration or regularity' in relation to its biological processes.[46] The field of operation of biopolitics includes permanent factors affecting a population's productivity and capacity to work (such as illnesses, age etc.) and further includes the relation between humanity as a species and its environment (climate, geography, hydrography), insofar as it may have effects on a population's productivity.[47] As Soirila notes, '[t]hrough the application of statistics, medicine and biology, ... biopower concentrates on shaping the population as a whole in order to maximize its capacities, effectivity and productivity'.[48] The complementarity with disciplinary power is evident, insofar as both contribute to the development of capitalism, which 'would not have been possible without the controlled insertion of bodies into the machinery of production and the adjustment of the phenomena of population to economic processes'.[49] Economic productivity is, in fact, a central concern of biopower and biopolitics.[50]

37 Foucault, *Society Must be Defended*, p. 243.
38 Ibid.
39 Ibid.
40 Ibid. Rabinow and Rose, among others, underline Foucault's terminological imprecision, Rabinow and Rose, 'Biopower', p. 197.
41 From now on I will use biopolitics and biopower interchangeably, even though biopolitics does not exhaust biopower, but is merely its most central form or mode, especially for the purposes of this book.
42 Foucault, *Society Must be Defended*, p. 246.
43 Foucault, *History of Sexuality*, p. 139.
44 Foucault, *Society Must be Defended*, p. 246.
45 Ibid.
46 Ibid.
47 Ibid., pp. 244–5.
48 U. Soirila, 'The European Court of Human Rights, Islam and Foucauldian Biopower', 2 *Helsinki Law Review* 2011, 365, p. 372.
49 Foucault, *History of Sexuality*, p. 141.
50 Soirila, 'European Court', p. 375, footnotes omitted.

Another important aspect is that biopolitics deals with problems which are aleatory, unpredictable at the individual level, occur over an extended period of time and can therefore only be managed at the population level. Thus, the 'mechanisms introduced by biopolitics include forecasts, statistical estimates, and overall measures'.[51]

Another important point, especially with respect to ecosystem approaches to environmental management, is that the purpose of biopolitics is not that of modifying 'any given phenomenon … but essentially, to intervene at the level at which these general phenomena are determined'.[52] Biopolitics then aims at controlling or reducing mortality and, by converse, increasing productivity and life expectancy. It is thus a matter of 'using overall mechanisms and acting in such a way so as to achieve overall states of equilibration or regularity; it is, in a word, a matter of taking control of life and the biological processes'.[53] This 'taking control of life' is a central concern of science, ethics and law in the Anthropocene, as we have seen in Chapter 6 when describing the implications of the 'new ecology' for environmental conservation policy and law. Foucault emphasizes that biopower is 'continuous' and 'scientific';[54] unlike sovereign power, biopower 'has no control over death, but it can control mortality'.[55] What happens with biopower and biopolitics is that 'power … takes life under its care'.[56] Ultimately, these processes cannot be disciplined, but ought to be regularized, in order to become calculable and predictable.

Finally, and crucially important, as we shall see, biopolitics elicits a new subject: the population. The population is at once 'a scientific and political problem … a biological problem and … a power's problem'.[57]

A key to the elaboration and enactment of biopolitical technologies of power is knowledge, and the relation between power and knowledge is a crucial dimension of Foucault's work. Biopower, in both in its modalities, is the result of this interpenetration between knowledge and power. Foucault famously joined the two terms in the expression power-knowledge, which captures precisely the combinatorial effect of power and knowledge. Certain knowledges, observes Foucault, enable certain technologies of power and conversely, power is in the position to normalize, discipline or disqualify incompatible or otherwise competing knowledges.[58] Thus power and knowledge are 'mutually dependent

51 Foucault, *Society Must be Defended*, 246.
52 Ibid.
53 Ibid., pp. 246–7.
54 Ibid., p. 248.
55 Ibid.
56 Ibid., p. 253.
57 Ibid., p. 245.
58 For a discussion of this combinatorial effect within the context of the project of critical environmental law see e.g. Lange, 'Foucauldian-inspired Discourse Analysis'.

and intersect with one another and are often so interpenetrated as to form a unity'.[59] As Foucault argues,

> we should abandon a whole tradition that allows us to imagine that knowledge can exist only where the power relations are suspended and that knowledge can develop only outside its injunctions, its demands and its interests. Perhaps we should abandon the belief that … the renunciation of power is one of the conditions of knowledge. We should admit rather that power produces knowledge (and not simply by encouraging it because it serves power or by applying it because it is useful); that power and knowledge directly imply one another; that there is no power relation without the correlative constitution of a field of knowledge, nor any knowledge that does not presuppose and constitute at the same time power relations.[60]

The biopolitical control of life which is the goal of biopower, is thus entirely dependent on particular fields of knowledge that first enabled the problematic of life to be altogether entertained, and then enabled its seizure by power, and the development of appropriate technologies for achieving the optimization and regularization of biological processes. This, as we shall see, is a key point to keep in mind in relation to the 'ecosystem approach'.

Continuity and discontinuity between forms of power

An important question to address is whether the different forms of power are discontinuous or continuous. This question is central both in general within the context of the literature on biopower and more specifically in the context of this book; it is also at the core of some of the more important critiques of Foucault, such as Agamben's, as we will see in the third section of this chapter.

Foucault himself is ultimately ambivalent on this question. On the one hand, Foucault considers the emergence of the new technologies of power from the seventeenth century onwards as the 'signal' of a 'historical caesura'.[61] For example, Foucault suggests in some passages that disciplinary power, in particular, stands in a relation of radical discontinuity vis-à-vis sovereign power: disciplinary power, its new mechanisms, strategies and apparatuses, observes Foucault, are '*absolutely incompatible* with relations of sovereignty'.[62]

Yet, in what Esposito characterizes as a sudden 'semantic slippage',[63] Foucault speaks elsewhere of a complementation of the different technologies of power.

59 G. Turkel, 'Michel Foucault: Law, Power, and Knowledge', 17:2 *Journal of Law and Society* 1990, 170, p. 178.
60 Foucault, *Discipline and Punish*, p. 27.
61 Lemke, *Biopolitics*, p. 53.
62 Foucault, *Society Must be Defended*, p. 35.
63 Esposito, *Bíos*, p. 39.

The 'old' sovereign right (to kill), suggests Foucault in an important passage, is not 'replaced', but rather 'complemented', 'penetrated and permeated' by the new forms of power.[64]

Thus, as Esposito notes, Foucault maintains an undecided stance regarding the relationship between sovereignty and biopower (and this is especially true in relation to biopolitics): as Esposito asks, are the two related chronologically or through a conceptual superimposition?[65] If one (sovereignty) is the background for the other (biopower/biopolitics), is this background 'a definitive withdrawal' or 'is it the horizon that embraces and holds what newly emerges within it'?[66] Esposito is not the only commentator on Foucault's work on biopower who recognizes this undecided stance. Foucault's approach to biopower has been variously characterized as 'anything but straightforward',[67] 'rather confusing',[68] full of 'doubts ... second thoughts [and] shifts [in] focus and direction'[69] and even deceptive.[70]

In this respect, and this is the interpretation favoured by the present author, forms of power remain complexly entangled while functioning simultaneously but at different levels or on different planes. This entanglement, in the words of de Sousa Santos, implicates 'complex circulations of meaning' as well as 'possible complicities ... and interpenetration',[71] and is the inevitable result of the fact that 'power is never exercised in a pure form'.[72] Power, rather, operates in terms of a 'power formation', that is, a 'constellation of different forms of power' that combine in multiple and variable ways.[73] Other scholars similarly interpret the entangled relations between forms of power in terms of a cumulative augmentation. Every new form of power, suggests Rutherford, for example, 'invests' the preceding forms with 'new capacities' in relation to different fields of activity and of intervention, through the deployment of new forms of knowledge.[74]

Moreover, and importantly, Foucault's neat demarcation between a political-juridical power (sovereignty) and a technical-scientific power (biopower) seems to reflect a much too narrow concept of law, which supposedly disappears as both a practice and as a strategy of power and government as soon as the power

64 Foucault, *Society Must be Defended*, p. 241.
65 Esposito, *Bios*, p. 33.
66 Ibid.
67 Campbell and Sitze, *Biopolitics*, p. 7.
68 De Sousa Santos, *New Common Sense*, p. 406.
69 Campbell and Sitze, *Biopolitics*, p. 7.
70 Ranciére, quoted in Campbell and Sitze, *Biopolitics*, p. 7.
71 De Sousa Santos, *New Common Sense*, p. 406.
72 Ibid.
73 Ibid. Santos, moreover, considers law and science "isomorphic", particularly in the sense that modern law transforms into an "alter ego" of science, and that the two discourses become fully convergent, ibid., p. 4.
74 P. Rutherford, The Problem of Nature in Contemporary Social Theory, PhD Thesis, The Australian National University, June 2000, p. 111.

over life emphasizes life's enhancement and optimization rather than its destruction. As we will see presently, the operative continuity of law and sovereignty as a crucial dimension of biopower is a central element of critiques of Foucault's work on biopower and biopolitics.

Biopolitics beyond Foucault

Introduction

As already anticipated, while Foucault is usually the starting point for most discussions of biopower and biopolitics, he is by no means the only philosopher who has engaged with the theme.[75] This section will outline some important ways in which the concepts have been further elaborated by subsequent scholarship, with the aim of offering some corrective to, or integration of, Foucault's theory of biopower and biopolitics. Before proceeding however, it is especially important to remind the reader of the fact that the terms biopower and biopolitics will be used interchangeably, in line with the practice of much literature, including Foucault himself.[76] These further elaborations take three key directions. First, sovereignty and biopolitics are shown to be irreducibly linked at the outset, in this sense vindicating the claim of the continuing relevance of modernity even within a postmodern context, where certain of its thresholds and trajectories are in a state of crisis, while others, particularly through their insertion within the structures and operations of biopower, become intensified. Second, the problematic of political conflict and of the transformation of mastery into domination through the subsumption of the entire world within the production process, gains a central place, a centrality somewhat lacking in Foucault. Third, and relatedly, the aporetic logic of biopolitics will be made more explicit. Finally, and this is Esposito's particular contribution, biopolitics will be linked to a trajectory which provides a broad horizon of sense that gives coherence, or at least organizes through relations of congruence, the thresholds and trajectories of modernity as well as their legal articulation, discussed in Chapter 1.

Biopolitics and 'Italian Theory'

Foucault's framework of biopower and biopolitics has had a phenomenal reception in Italy. As Campbell observes, 'modern Italian political philosophy has emerged perhaps as the primary locus of research related to biopolitics'.[77] This

75 See e.g. Esposito, *Bios*, and Campbell and Sitze, *Biopolitics.*

76 Foucault in fact at times almost identifies biopolitics with biopower and refers to disciplinary techniques as disciplinary power, Foucault, *Society Must be Defended.* Rabinow and Rose, among others, underline Foucault's terminological imprecision, Rabinow and Rose, *Biopower*, p. 197.

77 T. Campbell, 'Bios, Immunity, Life: The Thought of Roberto Esposito' in Esposito, *Bios*, 2008, p. xx.

is unsurprising, given the long tradition of Italian political thinkers and philosophers approaching politics in what Campbell calls a 'biopolitical key',[78] a theoretical key that has been recently repeatedly associated with what is denoted as 'Italian Theory' or 'Italian Thought'.[79] A number of the most prominent contemporary Italian philosophers – Agamben, Negri and Esposito, among others – have engaged deeply with the concept of biopolitics and have developed it in a number of radical directions, often in contrast with one another.

Italian philosophy concerned with biopolitics starts its investigations with the idea of responding to

> the underlying question with which Foucault ended: what is the nature and meaning of biopolitics? Are we to understand it as a process that is substantially positive, innovative, and productive, or rather as something negative, as a lethal retreat from life?[80]

Further, asks Esposito with Foucault, '[w]hy does biopolitics', i.e. a politics that affirms life, 'continually threaten[s] to be reversed into thanatopolitics?',[81] that is, a politics of death, more closely reminiscent of the sovereign power to take life than of the affirmative power to foster life?[82]

Esposito speaks of a constitutive dilemma of biopolitics, a 'never-released tension',[83] a 'lacerating effect'[84] that accompanies it, as the two poles of its articulation – life and politics – are engaged in an inevitable and insoluble struggle.[85]

78 Ibid., p. xx. Campbell refers to that Italian tradition of 'living thought' that includes, among others, Machiavellli, Vico, Croce, Gentile and Gramsci and contemporaries such as Negri, Agamben and Esposito, and which Esposito has discussed at length in R. Esposito, *Pensiero Vivente. Origine e Attualità della Filosofia Italiana*, Turin: Einaudi, 2010 (English translation: R. Esposito, *Living Thought. Origin and Actuality of Italian Philosophy*, Stanford: Stanford University Press, 2012).

79 See e.g. the special issue of *Angelaki: Journal of Theoretical Humanities*, 'Italian Thought Today: Bio-economy, Human Nature, Christianity', 2011; R. Esposito, *Living Thought. The Origins and Actuality of Italian Philosophy*, Stanford, CA: Stanford University Press, 2012; D. Gentili, *Italian Theory: Dall'Operaismo alla Biopolitica*, il Mulino, 2012; D. Mariscalco and P. Maltese (eds), *Introduzione a Vita, Politica, Rappresentazione. A Partire dall'Italian Theory*, Ombre Corte; D. Gentili, E. Stimilli and G. Garelli, *Italian Critical Thought: Genealogies and Categories*, Rowman & Littlefield International, 2018.

80 T. Campbell, 'Interview: Roberto Esposito',36:2 *Diacritics* 2006a, 49, p. 50.

81 Esposito, *Bios*, p. 39.

82 Ibid., pp. 38–9.

83 Ibid., p. 32.

84 Ibid.

85 Ibid. As Esposito further observes, '[e]ither life holds politics back, pinning to its impossible natural limit, or, on the contrary, it is life that is captured and prey to a politics that strains to imprison its innovative potential', ibid., p. 32.

Agamben and Negri (the latter together with Hardt), in particular, respond to this question in opposite ways. Such is indeed the difference, that Campbell describes Agamben's reading and development of the concept of biopolitics as dominated by an 'obsessive emphasis on the negative figure of *homo sacer* and the state of exception', while Negri's by contrast is rather described as emphasizing the 'incantations of a vital biopolitics of the multitude'.[86]

Their divaricated articulations will be outlined first. Esposito will provide a third and median line of reflection on biopolitics and will be discussed last. What all three have in common, however, is that each considers biopolitics, despite all the differences, and unlike Foucault, thoroughly modern (if perhaps through a postmodern declension or intensification of modernity), and that each marks the continuity between sovereign power and biopower, so that biopolitics, rather than substituting sovereignty, is linked to it by 'a logic of copresence'.[87] This logic, then, allows two intimately linked interpretive registers, one that understands biopolitics as an 'internal articulation of sovereignty' and the other than understands sovereignty as a 'formal schema of biopolitics'.[88]

Agamben: sovereignty, bare life and thanatopolitics

Giorgio Agamben's approach to biopolitics hinges on the relationship between sovereignty, the exception and bare life, a relationship which for Agamben unfolds inevitably in the form of a thanatopolitics (that is, literally, a politics of death). While Foucault considered different forms of power as 'essentially heterogenous' and 'conceptually distinct'[89] Agamben, by contrast, speaks of a fundamental continuity of power, whose different forms are not distinct, but overlap and are inscribed in the same overall horizon of sense.[90] Foucault (and most modern and contemporary political philosophy) misses for Agamben the 'hidden point of intersection' between sovereign and biopolitical modes of power.[91] Indeed, Agamben argues that 'the two analyses cannot be separated and that the inclusion of bare life in the political realm constitutes the original – if concealed – nucleus of sovereign power'.[92] From this perspective, then, biopolitics and sovereign power are equally old[93] and '*the production of*

86 Campbell, 'Bios, Immunity, Life', p. viii. See also T. Campbell, 'Bios, Immunity, Life: The Thought of Roberto Esposito', 36:2 *Diacritics* 2006b, 2.

87 Esposito, *Bios*, p. 40.

88 Ibid., p. 41.

89 Ibid., p. 49.

90 G. Agamben, *Homo Sacer. Sovereign Power and Bare Life*, Stanford, CT: Stanford University Press, 1998. See also Lemke, *Biopolitics*; M. Coleman and K. Grove, 2009, 'Biopolitics, Biopower, and the Return of Sovereignty', 27:3 *Environment and Planning D: Society and Space* 2009, 489, p. 494; Campbell, 'Bios, Immunity, Life'.

91 Agamben, *Homo Sacer*, p. 11.

92 Ibid.

93 Ibid.

a biopolitical body is the original activity of sovereign power'.[94] Moreover, the classical Aristotelian distinction between the political form of life, endowed with rights (*bíos*) and the bare animal life (*zoe*) collapses. For Agamben the 'decisive fact' of biopolitics is that 'the realm of bare life … gradually begins to coincide with the political realm'.[95]

The concept of sovereignty is what Agamben calls a 'limit concept',[96] the 'blurred'[97] space where right and fact, law and violence, interface. He draws on Carl Schmitt's theory of sovereignty as decision, and on the concept of exception.[98] The latter is a key mechanism for Agamben, as it both suspends and creates law, while simultaneously 'baring' life through the suspension of its legal and political protection (i.e. rights). Like the *home sacer* of Roman origin,[99] bare life can be killed with impunity, as it no longer belongs to the political order, yet it is not entirely 'indifferent to law', since by way of 'its own suspension law encompasses living beings'.[100] This is the paradox of biopolitics for Agamben: bare life is included in the juridical order through its exclusion.[101]

This is, for Agamben, the aporetic core of modernity, the site where inclusion and exclusion, law and violence, life and death, the political form of life endowed with rights and the bare life which can be killed with impunity, enter a 'zone of irreducible indistinction'.[102]

Following Agamben's lead, scholars such as Coleman and Grove theorize then the 'return of sovereignty'.[103] This return, or perhaps rather, re-discovery, given that sovereignty never really disappears, takes the form of a right to kill now grounded and legitimated on biopolitical grounds (through the mechanism that Foucault calls 'racialization', aimed at securing a population through immunization against external, racial, threats).[104] In this account, Foucault's theorization of 'how life is brought within the realm of government' appears 'much more complex' than often acknowledged.[105] Biopolitics functions as the

94 Ibid., emphasis in the original.
95 Ibid., p. 12.
96 Ibid., p. 13.
97 Ibid., p. 12.
98 As Lemke remarks, Agamben's biopolitics 'owes more to Carl Schmitt than it does to Foucault', Lemke, *Biopolitics*, p. 60.
99 A human being that was declared *sacer* in Roman law would be included in the juridical order only in order to be excluded, i.e. to be killed, as anyone declared *homo sacer* could be killed with impunity by anyone.
100 G. Agamben, *State of Exception*, Chicago: University of Chicago Press, 2005, p. 3.
101 A. O'Donoghue, 'Sovereign Exception: Notes on the Thought of Giorgio Agamben, Critical Legal Thinking', HTTP: http://criticallegalthinking.com/2015/07/02/sovereign-exception-notes-on-the-thought-of-giorgio-agamben/ (Accessed 26 September 2018). See also, of course, Agamben, *Homo Sacer*, p. 18.
102 Agamben, *Homo Sacer*, p. 12.
103 E.g. Coleman and Grove, 'Biopolitics, Biopower'.
104 Ibid., p. 494. I discuss the mechanics of biopolitical racialization in relation to nature in depth in V. De Lucia, 'Bare Nature'.
105 Coleman and Grove, 'Biopolitics, Biopower', p. 494.

core horizon for politics and government and is forcefully complemented and corrected by both the disciplining of the individual bodies and the killing (literal or metaphorical) of biopolitical threats inimical to the 'economy of biopower'.[106] This is what Lemke calls the 'bioeconomic imperative',[107] an imperative inscribed within a political economy of life which, however, while inevitably activating thanatopolitical mechanisms such as those Agamben describes, it is not exhausted by them, as Agamben argues.[108]

For a full treatment of some of the important shortcomings in Agamben's rendering of the biopolitical frame, I shall refer the reader to existing literature.[109] Here I offer only two further comments. First, as Lemke rightly highlights, Agamben's analysis appears to conflate the entire history of western civilization within the biopolitical fold, in a theoretical manoeuvre that downplays, if not entirely eliminates, the central complicities that Foucault had enucleated between modernity, capitalism and biopolitics.[110] These complicities and entanglements are crucial elements of the reading of both modernity and biopolitics defended in this book (some of these complicities are addressed in the fourth and fifth sections of this chapter; some have been addressed in Chapter 1).

Second, the exclusive and exhaustive concretization of biopolitics as a thanatopolitics that Agamben outlines, cannot capture the positive, or even 'affirmative'[111] nuances of biopolitics that are potentially at work within the sphere of interest of this book, namely international environmental law, and the 'ecosystem approach' in particular. And while the negative operative mechanisms of biopolitics are always at work, internally, in any biopolitical discourse that aims at the enhancement of life, any exclusive and exhausting reading in either sense would be too narrow and one-sided, considering the complexities revealed by a genealogical method of inquiry.

The main take-away from this succinct (yet hopefully sufficient, at least for our purposes) outline of Agamben's theory of biopolitics, is the intimate, original entanglement of sovereignty and biopower. In light of Foucault's ambivalence, Agamben's more assertive reading will prove useful. However, as Lemke suggests in line with Foucault, sovereign power does not exhaust the spectrum of the practices of power, and, indeed, biopolitics indicates all those sets of operations of power that exceed and operate outside the traditional mechanisms of sovereignty. This reading is borne out by the complex transformations that law itself (and most especially international environmental law), is undergoing, as increasingly noted by legal scholarship,[112] and as will be discussed in the next sections.

106 Ibid.
107 Lemke, *Biopolitics*, p. 60.
108 Ibid.
109 Ibid., pp. 59–64.
110 Ibid.
111 On this distinction between positive and affirmative biopolitics see Esposito, *Bios*, p. 184ff.
112 See e.g. Kuokkanen, *International Law*; De Sadeleer, *Environmental Principles.*; Fisher, 'Maturity and Methodology'.

This more nuanced recognition of the *differentiated* articulations of power must, additionally, remain cognizant of the fact that not even Foucault himself has ever really considered these forms of power entirely or always distinct, but rather, if ambivalently, complementary. As suggested by Coleman and Grove, it is, therefore perhaps more useful to use a language of activation, deactivation and reactivation, where some forms become operatively prevalent at different junctures but never replace one another through a rupturing event.[113] Moreover, 'power's complex hold over life', as Coleman and Grove further suggest, materializes as 'a convergence over time of sovereign-juridical, disciplinary, and biopolitical arts of government'.[114] In this respect, like Gramsci, Foucault considers that coercion and consent are always entangled, and that when the (biopolitical) strategies that elicit consent lose effectivity, the State always resorts to the coercive threat.[115]

Hardt and Negri: biopolitical production and biopotenza

Hardt and Negri's elaboration of the concept of biopolitics takes, in many ways, a direction *opposite* to that of Agamben. The latter in fact, identifies biopolitics with its negative converse, thanatopolitics, into which biopolitics 'necessarily' and 'continually' turns.[116] Hardt and Negri, on the other hand, focus on the role of what they call 'cognitive capitalism' in the increasing exploitation of bodies, intellects and affects; however, they ultimately identify a positive, affirmative potential of biopolitics. And if Agamben makes bare life the protagonist of biopolitical articulations of power, Hardt and Negri find their protagonist in the multitude. They develop their biopolitical analysis in the course of a trilogy comprised of 'Empire', 'Multitude' and 'Commonwealth', respectively discussing the new form of sovereignty in the current advanced and cognitive stage of capitalism; the collective subject at the centre of the biopolitical struggle with Capital; and those cooperative social practices that are the target of current biopolitical forms of capitalist exploitation, but that also represent a new, bio-affirmative emancipatory horizon.

Commonwealth, or the commons, is understood as a production process *and* as the multiple sites where value is produced through social cooperation, and where the production of value is indeed, made possible. Biopower, by contrast, materializes as the capitalist appropriation of the result of these social practices of production of value. Hardt and Negri call this 'biopolitical exploitation' and

113 Coleman and Grove, 'Biopolitics, Biopower', p. 495.

114 Ibid., p. 496.

115 Foucault, *History of Sexuality*, p. 144.

116 Ibid., p. 83 and p. 88 respectively; see also Lemke, *Biopolitics*, p. 59. Agamben considers the Nazi concentration camps as the 'biopolitical paradigm of modernity', a formulation which is indeed the title of Part III of Agamben, *Homo Sacer*.

distinguish at least two forms in which it takes place, both of which are external to the production process itself: the exploitation of nature and the exploitation of labour. This biopolitical exploitation operates mainly in terms of *expropriation* of the commons, both as public and common social wealth and as natural wealth.[117]

The form of expropriation central to Hardt and Negri's analysis hinges on the exploitation of biopolitical labour, a form of social wealth. Biopolitical labour must be understood as the source of the dynamic aspect of the commons, that most directly produced through social practices[118] – and is distinguished from the 'relatively inert'[119] notion of natural commons, i.e. natural resources etc.[120]

Here we must bear in mind Marx's description of the capitalist as the organizer of productive cooperation within the industrial production system. What Hardt and Negri understand as a central specificity of biopolitical capitalism is that cooperation in biopolitical terms is a spontaneous process arising from common social practices no longer located within material sites of industrial production, and it is this *commons* in the form of cooperation, taking place everywhere in the life-world, which capital predates on and expropriates. The biopolitical exploitation of labour assumes then, in economic terms, the form of capitalist rent, rather than profit, and capital becomes increasingly predatory, as it 'seeks to capture and expropriate autonomously produced common wealth' which is external to the capitalist production form.[121]

This is the horizon and objective of Empire: the 'limitless mobilization of individual and collective powers in order to generate surplus value'.[122] In this sense, there is a total identification of economics and politics: '[t]here is nothing, no naked life, no external standpoint, that can be posed outside this field permeated by money; nothing escapes money'.[123] This passage is the central hinge upon which the relevance of Hardt and Negri's articulation of biopolitics for this book emerges. Biopolitics marks the total subsumption of life

117 Marx described this sort of process as primitive accumulation and it has been described and analysed by many critics of neoliberalism such as David Harvey, who refers often to it as 'accumulation by dispossession'. See for example, D. Harvey, *The New Imperialism*, Oxford: Oxford University Press, 2003.

118 Taking the form of languages, social relationships etc.

119 M. Hardt, and A., Negri, *Commonwealth*, Cambridge: Belknap Press of Harvard University Press, 2011.

120 The notion of the inert character of natural commons cannot be shared, though it is not important at this point to critique it. On the contrary, it will be the case that the commons displaces entirely the ontological and political *difference* between natural and social, by re-integrating both under a transversal understanding of ecology, politics and law. I will discuss these themes in Chapter 10.

121 Hardt and Negri, *Commonwealth* p. 141.

122 Lemke, *Biopolitics*, p. 67.

123 Hardt and Negri, *New World Order*, p. 32.

under the gaze of capitalism and economics – with special reference, for the purposes of this book, to the non-human or more-than-human world. This biopolitical subsumption of nature, as I will try to illustrate in this Part of the book, and especially in subsequent sections and chapters, is indeed, *central* to international environmental regimes and a crucial horizon for the 'ecosystem approach'. However, another aspect of Hardt and Negri's biopolitical analysis is important. The biopolitical subsumption of the entire world is, for Hardt and Negri, the hallmark of biopower.[124] In this respect, they distinguish between biopower, which they describe as a 'sovereign authority' which imposes its order on society and the world, and biopolitics, which is, rather, the immanent productive, creative potential of society understood as the entirety of the entangled networks of social life.[125] In this conceptual framework, the realm of the biopolitical (which is in fact dual) marks, most importantly for the purpose of this book, the transformation of nature into capital.

Furthermore, the biopolitical subsumption of nature no longer entails simply the exploitation of nature, but importantly, its re-orientation towards the biopolitical goals of its optimization, of the enhancement of its productive forces.[126] This is the result of what Hardt and Negri call the 'civilization of nature'[127] in the sense of its integration *inside* the production processes so that it must become subject to direct economic calculation. This internalization of nature as a central factor of production,[128] its integration under the (biopolitical) purview of economic calculation, will also prove central.

Esposito: biopolitics and the immunitary paradigm of modernity

Esposito's thinking about biopolitics is complex. Here I will only address two elements that are relevant for my argument: the continuity that Esposito identifies in the relationship between sovereignty and biopower/biopolitics; and the genealogy of biopolitics, which Esposito traces to the double *locus classicus* of modernity (a locus also convincingly visited by de Sousa Santos):[129] Hobbes' philosophy of sovereignty and Locke's philosophy of property. In this respect, Esposito argues for 'the modern origin of biopolitics in the immunizing features of sovereignty, property, and liberty as they emerge in the writings of Hobbes

124 Though they focus on what they call the 'real subsumption of society', rather than of nature, Hardt and Negri, *New World Order*, p. 255.

125 Ibid.

126 Ibid. See also Lemke, *Biopolitics*, p. 68.

127 Hardt and Negri, *New World Order*, p. 187.

128 Traditionally, political economy considers 'land' as a factor of production. Recently however (that is in the last 30 – 40 years) the treatment of the factors of production in most economic textbooks has been reduced to capital and labor, see e.g. H. Daly, *Beyond Growth. The Economics of Sustainable Development*, Boston, MA: Beacon Press, 1997, esp. chapter 2, p. 45ff.

129 De Sousa Santos, *New Common Sense*.

and Locke'.[130] Esposito links the genealogy of modernity with the forceful entry into politics of the question of the conservation of life,[131] thus significantly extending the Foucauldian scope of biopolitics. Modernity and biopolitics are so intertwined for Esposito that he suggests that it is perhaps biopolitics that 'invented' modernity as that complex set of categories 'capable of answering the question of the preservation of life'.[132]

Esposito, like Agamben and Negri, emphasizes decisively the continuity between sovereignty and biopolitics, speaking of a 'logic of copresence' that he sees at work in Foucault's thought already, albeit only on occasion and outside of any coherent or systematic articulation.[133] The central point is that this co-presence suggests that *both* operate as fundamental premises of modernity. Similarly to Agamben's reference to the hidden point of intersection between sovereignty and biopolitics, Esposito speaks of a 'secret and essential' relation between the two modes of power.[134] Sovereignty and biopolitics then are for Esposito two 'opposing vectors superimposed in a threshold of originary indistinction that makes both the ground and the projection … of the other'.[135] In other words, the two modalities, the negative and the positive, sovereignty and biopolitics, can be organized together through an 'internal articulation' and a 'semantic juncture' that makes one always already present in the other. This logic of co-presence then allows two interpretive registers, one that understands biopolitics as an 'internal articulation of sovereignty' and the other that understands sovereignty as a 'formal schema of biopolitics'.[136] Biopolitics then, for Esposito – though he reads this already in some of Foucault's passages where biopower is understood as 'complementing', 'penetrating' and 'permeating' sovereign power[137] – 'accompanies the sovereign regime' and constitutes a 'specific tonality' of it.[138]

130 Campbell, 'Bios, Immunity, Life', p. 3.
131 Esposito, *Bios.*
132 R. Esposito, *Community, Immunity, Biopolitics*, New York: Fordham University Press, 2013a, p.70; Esposito, *Bios*, p. 55.
133 Esposito speaks in this respect of a semantic and conceptual 'slippage' that one can find in Foucault, which maintains a certain linguistic mobility or uncertainty, and speaks of a replacement or substitution as well as a complementation, ibid., pp. 39–40.
134 Ibid., p. 40.
135 Ibid.
136 Ibid., p. 41.
137 Foucault, *Society Must be Defended*, p. 241. But see also, more explicitly, Foucault, *Security, Territory, Population.* Lectures at the Collège de France 1977-1978, New York: Palgrave/Macmillan, 2009, where Foucault clearly states that, and it is worth quoting at length, 'So, there is not a series of successive elements, the appearance of the new causing the earlier ones to disappear. There is not the legal age, the disciplinary age, and then the age of security. Mechanisms of security do not replace disciplinary mechanisms, which would have replaced juridico-legal mechanisms. In reality you have a series of complex edifices in which, of course, the techniques themselves change and are perfected, or anyway become more complicated, but in which what above all changes is the dominant characteristic, or more exactly, the system of correlation between juridico-legal mechanisms, disciplinary mechanisms, and mechanisms of security', p. 22.
138 Esposito, *Bios*, p. 52.

The second key point of Esposito's elaboration of biopolitics, and the pivot of his theoretical construction, is the concept of immunity. To understand Esposito's concept of immunity, in turn, one must start from the concept of community.[139] Esposito carries out a genealogical reconstruction of the concept of community, and emphasizes the *munus*, that is the duty, the burden, that a community poses on individual members. As Esposito puts it, 'to belong entirely to the originary communitas means to give up one's most precious substance, namely, one's individual identity'.[140] From this starting point, where individuals gradually open (and lose) themselves to the communal, Esposito constructs the concept of immunity. If community 'is what binds its members in a commitment of giving from one to the other', immunity is a strategic reversal that enables individuals to 'unburden' themselves from the communal bond and obligation, to 'exonerate' themselves from the communal responsibility.[141] If the community penetrates and deconstructs the 'barriers of individual identity' then, immunity by contrast is a strategy aimed at rebuilding those barriers 'in defensive and offensive forms, against any external element that threatens it'.[142] Moreover, immunity need not be associated only with individuals, but can take also a collective character, where particular communities immunize themselves against external threats.[143]

Immunity is thus, for Esposito, the crucial interpretive key of biopolitics, the fundamental mechanism of modernity, the key hinge in the inextricable relation between life and power. Immunity is, in other words, the 'constitutive nexus' between the negative and positive aspects of biopolitics.[144]

Immunization becomes thus for Esposito the key mechanism through which sovereignty and biopower remain entangled with one another in an intimate relation that ends up producing an aporetic logic from which, apparently, there is no possibility of escape. Each represent one pole – the deadly and the euphoric, the protective and the destructive – of the 'bivalent character of the immune dispositif'.[145] Immunization is thus precisely the biopolitical mechanism through which, paradoxically, life protects itself from its own excesses.[146] Immunization keeps at bay all chaotic, dangerous, exuberantly biological and

139 R. Esposito, 'Community, Immunity, Biopolitics', 18:3 *Angelaki* 2013b, 83, p. 86.

140 Ibid., p. 84.

141 Ibid.

142 Ibid., p. 85.

143 Here there is a clear connection with what Foucault calls biopolitical 'racism', a securitization mechanism that allows the preservation of certain life against the threats posed to it by other lives, see e.g. Foucault, *Society Must be Defended*, and Foucault, *Security, Territory, Population*. I have discussed this aspect in depth, in relation to the international regulation of invasive alien species in De Lucia, 'Bare Nature'.

144 Esposito, 'Community, Immunity', p. 86.

145 Ibid.

146 Ibid.

animal impulses, thus protecting the newly constructed modern subject from all those communitarian obligations that negate, by crossing it, the threshold of the individual autonomy so central to the socio-cultural paradigm of modernity. It is however precisely at this point that the immunitary trajectory shows its aporetic logic, as the protection of life turns out to be also the protection *from* life. This relationship always maintains itself on the brink of a paradox to the extent that in order to protect itself, life must submit to a power that is capable of interdicting life's own development.[147]

This paradox is already evident in the paradigmatic Hobbesian model where men, in order to protect their life from the effects of the war of all against all, must submit to the command of the sovereign, who in turn can justly kill them – and here lies the aporia, the paradox – to protect them (or, rather, kill some to protect others). Esposito indeed underlines the 'necessary link' that exists between the goal of 'the preservation of life with the possibility – always present even if rarely utilized – of the taking of life by the one who is also charged with insuring it'.[148]

Yet sovereignty also always includes a biopolitical dimension, that is, a positive, productive orientation, to the extent that, as Bernini argues, 'the sovereign has to satisfy and modify human biological needs in order to create those docile subjects who submits their will to his'.[149] This emerges from certain careful readings of Hobbes, where security is only the first step in the role of sovereign power. Indeed security, or safety as Hobbes more precisely writes, does not include only 'bare preservation, but also all other contentments of life',[150] in order for life to achieve 'its happiness'[151] and the 'benefit and good'[152] of the people.

Expanding biopolitics to nature

The preceding sections presented, in some detail, the framework of biopolitics. After presenting the concepts of biopower and biopolitics, and their relationship, in the theory of Michel Foucault, I have discussed some key elaborations of biopolitics produced by three Italian political philosophers, Agamben, Negri and Esposito. What emerged is a dual consideration. On the one hand, biopolitics is a novel way to frame power and its operations, one that has become particularly

147 Esposito, 'Community, Immunity', p. 56.
148 Ibid., p. 62.
149 Bernini, 'Sovereignty and Biopolitics', p. 5.
150 T. Hobbes, *Leviathan*, 1651, Chapter: XXX (1), eBooks@Adelaide The University of Adelaide Library
151 T. Hobbes, *De Cive: Philosophical Rudiments Concerning Government and Society*, 1651, XIII (4), http://www.constitution.org/th/decive.htm (Accessed 26 September 2018).
152 T Hobbes, *The Elements of Law Natural and Politic*, 1640, XXVIII (1), http://www.constitution.org/th/elements.htm (Accessed 26 September 2018).

decisive in what Esposito calls a second articulation of modernity, where its mode or tonality becomes prevalent vis-à-vis sovereign power. On the other hand, there is a deep, intimate continuity between sovereignty and biopolitics, so that, with Esposito, it is possible to argue that modernity is *always* biopolitical.

The next step is to articulate an *ecological* declension of biopolitics, that is, to expand the frame of biopolitics to nature, in order to make the frame fully applicable in relation to the 'ecosystem approach'. This specifically ecological declension of biopolitics takes the form of an expansion of the biopolitical framework of rationalization and control of life, life understood as a political and economic resource to be governed, enhanced, optimized and regularized. This expansion invests the entire world, so that, as Lascoumes warns, every living and non-living form on Earth is subjected to extended surveillance and human control.[153] Moreover, it is important to underline how the ecological expansion of biopolitics maintains the aporetic void that inexorably links a positive biopolitics with its negative, thanatopolitical counterpart, and offers a powerful critical framework to analyse the articulations of ecosystem approaches in international environmental law, as will become increasingly clear as the analysis progresses.

Expanding the concept of biopolitics to encompass the natural environment can be approached from two analytically distinct, but ultimately inextricable and complementary perspectives. One entails the expansion of care for the human population through the inclusion, within the purview of the practices of regularization enacted through the biopolitical *dispositif*, of all those environmental processes that affect the well-being and the productivity of human populations. This perspective incorporates all those environmental factors affecting the nutritional, health and psychological well-being of a human population. Foucault considers these environmental factors as one aspect of the biopolitical optimization of populations, and as Rutherford observes, 'the definition and administration of populations simultaneously requires the constitution and management of the environment in which those populations exist and upon which they depend'.[154] As such, they are always already included within the scope of biopower.

The second perspective is more directly focused on natural entities and populations, and on the direct regularization of *their* life. While still ultimately linked to the well-being of human populations, the link may at times

153 P. Lascoumes, *L'éco-pouvoir. Environnements et Politiques*, Paris: La Découverte, 1994; K. Whiteside, *Divided Natures: French Contributions to Political Ecology*, Cambridge, MA: MIT Press, 2002, p. 143.

154 P. Rutherford, 'The Entry of Life into History' in E. Darier (ed.), *Discourses of the Environment*, Oxford: Blackwell Publishers, 1999, p. 45. See also in general E. Darier (ed.), 1999a, *Discourses of the Environment*, Oxford: Blackwell Publishers, 1999b.

appear very tenuous, particularly in the context of some 'ecocentric' approaches to environmental protection and management. In fact, from this particular analytical angle, even the ecological re-calibration of law and politics can be understood ambivalently not only as a critique of the prevalent, increasing instrumental control of the natural world, but also as a new set of normalizing strategies *extending* the scope of biopolitical technologies of power to the entire world.[155]

The 'ecosystem approach' responds simultaneously to both biopolitical perspectives. Yet the *focus* or *angle* from which the different perspectives approach the question of the biopolitical subsumption of *ecological* life processes offers significant scope for meaningful differentiation. In this respect, one may argue that the second perspective is instrumental to the first, which represents the ultimate goal of biopolitics. This relation of instrumentality, whereby the biopolitical subsumption of non-human life is aimed at the ultimate goal of optimizing human life, will be central to the analysis of the relationship between the 'ecosystem approach' and ecosystem services, in yet another example of the continuity between an epistemology of mastery and a (bio-)politics of optimization and control (in the form of its statistical regularization rather than sovereign control) of life, despite the postmodern epistemological re-calibration, or perhaps further enabled by it.

Yet the two articulations also run on a parallel track to the extent that *both* enact those biopolitical mechanisms through which life is subjected and subjugated even as it is fostered and enhanced. This is a consequence of the internal contradiction of biopolitics, as identified and discussed by Agamben and Esposito in particular, which in the same gesture includes and excludes, gives and takes life, creating that irreducible zone of indistinction where the enhancement of life becomes, or always already is, its subjection and subjugation. This double register through which the aporetic void of biopolitics remains operative will be emphasized as relevant throughout the remainder of the analysis, and represents a crucial critical horizon in relation to environmental law in general, and the 'ecosystem approach' more specifically and intensely.

In line with Foucault's insight of the co-implicated relationship between power and forms of knowledge – a relationship captured by the well-known power-knowledge binomial construct[156] – the expansion of biopolitical regimes to the natural environment is historically contingent on the development of the sciences of biology and, especially, ecology. As amply recognized in the literature, and as we have seen in Chapter 5, ecology, in particular, functions in a double and ambiguous epistemic role, and effectively operates as a framework

155 E. Darier, 'Foucault and the Environment: An Introduction' in E. Darier (ed.), *Discourses of the Environment*, Oxford: Blackwell Publishers, p. 23.

156 See e. g. Foucault, *Discipline and Punish*, pp. 27–8.

of ambiguity.[157] Ecology's conceptual framework is, in fact, easily mobilized in defence of highly incompatible projects, especially with respect to the 'ecosystem approach'.[158]

This ambivalence or ambiguity hinges on the fact that, if ecology has, on the one hand, helped to problematize the relationship between the social and the natural world, and its separating threshold, casting doubt on the self-image of the modern subject, it has, on the other, 'provided the political technology for new forms of regulatory intervention in the management of the population and resources'.[159] These new forms of intervention, it has been suggested, combine to 'constitute a form of ecological governmentality',[160] an 'ecopolitics',[161] or an 'ecopower'.[162] One may indeed agree with Rutherford when he suggests that '[m]odern thinking about the natural environment is characterized by the belief that nature can be managed or governed through the application of the scientific principles of ecology',[163] a belief that remains equally operative, if not intensified, in the epistemological context of the Anthropocene.[164]

Biopolitics is, in this respect, in perfect continuity with the epistemological project of modernity and with the trajectory that continues to make that project operative, even if the earlier physical emphasis on pure exploitation has had to be significantly complemented – note, not replaced! – by a biological/biopolitical concern for enhancement and optimization, within a global context of increasing organization of life and biological processes as factors of production and as key economic resources.

A central element of this ecological declension of biopolitics, moreover, is arguably linked to the way ecology expands and develops the regulatory role of biopolitics in relation to populations. Rutherford underlines in this respect the 'regulatory' role of ecology, and points out that 'despite the popularizing of nature in holistic and sometimes organicistic terms, the bio-economic model at its core [which is the prevalent model of nature obtaining within most articulations of scientific ecology] can be said to express an "agronomic attitude toward nature" which seeks to provide the "analytic tools" needed to "intensively farm"

157 D. Worster, 1994, *Nature's Economy: The Roots of Ecology*, 2nd edn, Cambridge: Cambridge University Press, 1994; Darier, *Discourses*; W. Sachs, *Global Ecology. A New Arena of Political Conflict*, London: ZED Books, 1993.

158 Bell, 'Non Human Nature'; De Lucia, 'Competing Narratives'.

159 Rutherford, 'Problem of Nature', p. 4.

160 Ibid.

161 Darier, 'Foucault and the Environment', p. 23.

162 Lascoumes, *L'éco-pouvoir*.

163 Rutherford, 'Entry of Life', p. 37.

164 We have discussed in this respect the contributions of Botkin and Kareiva and Marveir to the interventionist conservation policy (especially Chapter 7); see however also e.g. B. Minter, 'Geoengineering and Ecological Ethics in the Anthropocene', 62:10 *BioScience* 2012, 857; R. Hobbs et al., 'Intervention Ecology: Applying Ecological Science in the Twenty-first Century', 61:6 *BioScience* 2012, 442.

the Earth's resources'.[165] In this respect ecology, suggests Rutherford, operates as a 'new form of political economy'.[166] This 'intensive farming' is crucially enabled, enacted and legitimated through law.

If ecological science's ambivalent role is very explicit, for our purposes the key element is the equally ambiguous, if not outright and thoroughly complicit, role of environmental law. This role is linked primarily to the ways in which environmental law institutionalizes knowledge in particular politico-juridical regimes, thus enabling the biopolitical interventions necessary to optimize life and its productivity and, through the discourse of law, legitimating them. If ecology, as Rutherford argues, can be understood as the 'rationale behind a new, and increasingly influential, form of political economy',[167] then environmental law becomes crucial in facilitating and providing the legal framework necessary to that new mode of capitalism concerned with the construction and maintenance of ecological regimes of accumulation.[168] In this manner, the isomorphic relationship between science and law that, as Santos underlines, is a key element of modernity, is further consolidated.

It is also important to underline how the emergence of ecology as a regulatory science, is deeply connected to the rise of 'big science',[169] that is, industrial and capital-intensive scientific programmes closely linked to state agencies and industrial interests. These interests have played a crucial role in the 'normative constitution of ecological knowledge', particularly in relation to its 'manufacture, negotiation and certification' as knowledge.[170] In other words, ecology is not a descriptive science but is normative and performative to the extent that it 'actively constitute[s] the environment as an object of knowledge and, through various modes of ... intervention, manage[s] and police[s] it'.[171]

Indeed, Foucault underlines how 'biopolitics assumes the dissociation and abstraction of life from its concrete physical bearers. The objects of biopolitics are not singular human beings [vulnerable bodies] but their biological features measured and aggregated at the level of populations'.[172] Because of this, life becomes 'an independent, objective, and measurable factor, as well as a collective reality that can be epistemologically and practically separated from concrete living beings and the singularity of individual experience'.[173] This, as

165 Rutherford, 'Entry of Life', p. 53. See also D. Worster, 'The Vulnerable Earth: Towards a Planetary History', 11:2 *Environmental Review* 1987, 87, and Worster, *Nature's Economy*, on which Rutherford draws.
166 Rutherford, 'Entry of Life', p. 54.
167 Rutherford, 'Problem of Nature', 134.
168 Paterson, 'Legitimation and Accumulation', p. 345.
169 Rutherford, 'Entry of Life', 54. Rutherford, 'Problem of Nature', p. 136.
170 Ibid., p. 136.
171 Rutherford, 'Entry of Life', p. 56.
172 Lemke, *Biopolitics*, p. 5.
173 Ibid.

we will see in Chapter 9, is precisely the conceptual horizon embodied in the concept of biodiversity.

It is through this set of mechanisms, and this is the central claim of Rutherford's analysis, that contemporary forms of biopolitics express a new governmental rationality (an ecological rationality) 'related to the institutionalization of new areas of scientific expertise [such as biology and ecology], which is, in turn, based on a bio-economic understanding of global systems ecology'.[174] It is exactly 'this relatively recent articulation of biopolitics' that Rutherford calls 'ecological governmentality'.[175]

Rutherford also describes the effects of regulatory science in terms of a panopticism subsuming life/nature under a comprehensive 'modality of intervention',[176] an observation corroborated by widespread ecological practices of subjection of nature to a series of intense and comprehensive monitoring programmes such as the International Biological Program,[177] the Global Census of Marine Life[178] or the Global Taxonomy Initiative (we will return on these aspects in Chapter 9).[179] This biopolitical project of panopticist surveillance (entirely attuned to the interventionist horizon of the Anthropocene),[180] is already outlined in the Report of the World Commission on Environment and Development (a key document in the methodological economy of this book for its crucial role in the development of legal meaning, outlining both the normative narrative and the political trajectory of the framework of sustainable development), that, at paragraph 56, envisioned a 'planetary management' enacted through the establishment of surveillance mechanisms aimed at monitoring 'the vital signs of the planet' so as to 'aid humans in protecting its health'.[181] Indeed, the discourse of (ecosystem or ecological) health has become central (along with the intertwined discourse of integrity) in the context of the 'ecosystem approach', and the two are in fact considered by some to be

174 Rutherford, 'Entry of Life', 38.

175 Ibid.

176 Rutherford, 'Problem of Nature', 140.

177 Aimed at 'understanding the biological basis of productivity and human welfare', Rutherford, 'Problem of Nature', 135.

178 'A 10-year international effort undertaken in to assess the diversity (how many different kinds), distribution (where they live), and abundance (how many) of marine life [that] produced the most comprehensive inventory of known marine life ever compiled and cataloged'. Office of Marine Programs (University of Rhode Island), *Census of Marine Life – About the Census*, 2010, http://www.coml.org/about-census (Accessed 26 September 2018).

179 Aimed at removing the so-called 'taxonomic impediment', that is, the lack of taxonomic knowledge, a key to the conservation of biological diversity. Convention on Biologival Diversity, *Global Taxonomy Initiative*, https://www.cbd.int/gti/default.shtml (Accessed 26 September 2018).

180 See in this respect especially V. De Lucia, 'Critical Environmental Law and the Double Register of the Anthropocene: A Biopolitical Reading' in L. Kotzé (ed.), *Environmental Law and Governance in the Anthropocene*, Hart Publishing, 2017b.

181 WCED, 'Our Common Future', para 56.

'inseparable'.[182] One can then agree with Sachs and Darier who characterize ecopolitics – understood 'as an extension of 'biopolitics' … to all life-forms'[183] – as a 'normalizing strategy' attempting 'to extend control ('management') to the entire planet'.[184]

Here we then need to address briefly the relationship between biopolitics, capitalism and nature. As already mentioned, Hardt and Negri suggest that a new relationship is established between nature and culture through the biopolitical declension of capitalism. Life, in all its complex manifestations, for Hardt and Negri had hitherto remained *external* to the production process. Now however it becomes an object of political and technological intervention. Nature becomes capital, or, more precisely, becomes 'subject to capital'.[185] Thus 'biological resources become the object of juridico-political regulation', while nature becomes a central part of economic discourse.[186] Hardt and Negri further suggest that biopolitical capitalism transitions from a discourse of exploitation to one of translation of 'biological and genetic diversity … into economic growth'.[187] In fact, a crucial biopolitical marker can be identified precisely in the transformation of ecology from a 'limit to the economic process to fundamental element of market valorisation'.[188]

These considerations indeed resonate strongly with a number of recent developments in the context of the international biodiversity regime, as will be illustrated in Chapter 9 and further illustrated by a recent OECD report that outlines the concept of 'bioeconomy'.[189] The concept of bioeconomy, the report explains, refers to

> a world where biotechnology contributes to a significant share of economic output. The emerging bioeconomy is likely to involve three elements: the use of advanced knowledge of genes and complex cell processes to develop new processes and products, the use of renewable biomass and efficient bioprocesses to support sustainable production, and the integration of biotechnology knowledge and applications across sectors. There are three main sectors where biotechnology can be applied: agriculture, health, and industry.[190]

182 Siron et al., 'Ecosystem-Based Management', p. 87.
183 Rutherford, 1993 quoted in Darier, 'Foucault and the Environment', 23.
184 Sachs 1993 quoted in ibid.
185 Hardt and Negri, 2000, *New World Order*, p. 32.
186 Lemke, *Biopolitics.*, pp. 69–70.
187 Ibid., p. 70.
188 E. Leonardi, Biopolitics of Climate Change: Carbon Commodities, Environmental Profanations, and the Lost Innocence of Use-Value, PhD Thesis, The School of Graduate and Postdoctoral Studies, The University of Western Ontario, 2012, p. ii.
189 Organisation for Economic Co-operation and Development (OECD), *The Bioeconomy to 2030: Designing a Policy Agenda, Findings and Policy Conclusions*, OECD, 2009.
190 Ibid., p. 8; see also A. Kotsakis, The Biological Diversity Complex: A History of Environmental Government, PhD Thesis, London School of Economics, 2011, which makes the case for the transformation of biological diversity into 'genetic gold', hence offering a case study of what amounts to a 'bioeconomic' regime.

This bioeconomy, according to the report, 'offers technological solutions for many challenges facing the world'.[191] As the report further explains,

> [b]iotechnology offers technological solutions for many of the health and resource-based challenges facing the world. It can increase the supply and environmental sustainability of food, feed and fibre production, improve water quality, provide renewable energy, improve the health of animals and people, and help maintain biodiversity by detecting invasive species.[192]

Bioeconomy and conservation then offer two perspectives on the same biopolitical alignment of nature to the requirements of capitalist biopower. This biopolitical control of life, this ecopower constructed upon the economic productivity of life that it continuously enhances and optimizes, is indeed convincingly situated at the heart of neoliberal conservation by Büscher et al. They understand neoliberal conservation (albeit without explicit reference to the conceptual semantics of biopolitics) as 'an amalgamation of ideology and techniques informed by the premise that nature can only be "saved" through its submission to capital and its subsequent revaluation in capitalist terms'.[193] This is also what Cary Wolfe describes, in Heideggerian terms, as 'biopolitical enframing'.[194] Moreover, this enframing reproduces precisely the aporetic mechanism identified by Esposito whereby life, through the workings of biopower, is at once enhanced and subjugated. This latter is, in a nutshell, the central 'biopolitical' problematics of the 'ecosystem approach'.

Ecopolitics and law: sovereign encoding and technical norm

Before rounding up the discussion on the ecological declension of biopolitics, it is necessary to discuss the role of law within the biopolitical framework presented thus far. The primary aim of the discussion in this section hinges on the problematic equivalence between sovereign power and law that is at work in both Foucault's own work and in the commentaries and elaborations of at least some legal scholarship drawing on Foucault.[195]

Foucault's neat demarcation between a political-juridical power (sovereignty) and a technical-scientific power (biopower) is most often taken to express (a) an equivalence between sovereign power and law and (b) a much too narrow concept of law.

191 OECD, *The Bioeconomy.*

192 Ibid.

193 B. Büscher et al., 2012, 'Towards a Synthesized Critique of Neoliberal Biodiversity Conservation', 23:2 *Capitalism Nature Socialism* 2012, p. 4.

194 Wolfe, *Before the Law*, p. 3.

195 See e.g. A. Hunt, 'Foucault's Expulsion of Law: Toward a Retrieval', 17:1 *Law and Social Inquiry* 1992, 1.; A. Hunt and G. Wickham, *Foucault and Law: Towards a Sociology of Law as Governance*, London: Pluto Press, 1994; and G. Wickham, 'Foucault, Law, and Power: A Reassessment', 33:4 *Journal of Law and Society* 2006, 596; but see also, to a more limited extent, De Sousa Santos, *New Common Sense.*

In this approach, law supposedly disappears as both a practice and as a strategy of power and government as soon as the power over life emphasizes life's enhancement and optimization rather than its destruction (i.e. the deductive mode of operation of sovereign power). The critiques of Foucault's reductive consideration of law (critiques that are, or have been until recently, considered to be a 'widely held opinion'[196] and refer to what is called the 'expulsion thesis'),[197] hinge primarily on the exclusive equivalence that Foucault allegedly makes between law and command. Law is reduced by Foucault, in this reading, to the Austinian sense of a rule-and-sanction complex,[198] that is, law is reduced to those sovereign mechanisms of punishments for breach of a prohibitive rule, aligned with what Michel Villey called the Torah concept of law.[199] It follows, then, in this reading, that once biopower ascends to prominence, law, reduced to this equivalence, clearly wanes in importance.

This reading of Foucault's ideas about law, is for example linked to the introductory differentiation that Foucault makes between sovereignty, discipline and security (i.e. biopolitics) in his lecture series gathered under the name of 'Security, Territory and Population'.[200] There, Foucault indeed illustrates the difference that distinguishes the three forms or articulations of power through the distinct responses that power enacts in relation to a theft.[201] Yet, Foucault speaks of *power* and not of *law*, even though his deployment of the formulation 'juridical' or 'juridico-legal'[202] to denote the operation of sovereign power induces, understandably, some confusion. But as Tadros underlines,

> [t]he term juridical ... refers to the conception of power relations which one might call Austinian. It is neither the case that all law is necessarily 'juridical' in Foucault's understanding of the term nor that the only way in which juridical power manifests itself is legal.[203]

196 C. Gordon, 'Expelled Questions: Foucault, the Left and the Law' in B. Golder (ed.), *Re-reading Foucault on Law, Power and Rights*, New York: Routledge, 2014, p. 13.

197 As formulated, for example in Hunt, 'Foucault's Expulson', 1–38. Munro considers this to be the 'dominant interpretation' of Foucault's conception of law, V. Munro, 'Legal Feminism and Foucault – A Critique of the Expulsion of Law', 28:4 *Journal of Law and Society* 2001, 546, p. 552; and even as critical re-readings of the relation between Foucault and law are emerging (such as Golder, *Re-reading Foucault*), the inevitable starting points seems to remain the same (i.e. the expulsion thesis), Gordon, 'Expelled Questions', 13.

198 A typical instance of this misreading is Hunt and Wickham, *Foucault and the Law*.For a thorough critique of this critique of Foucault, see V. Tadros, 'Between Governance and Discipline: The Law and Michel Foucault', 18:1 *Oxford Journal of Legal Studies* 1998, 75.

199 M. Villey, 'Dikaion-Torah' in *Seize Essais de Philosophie du Droit*, Paris: Dallzo, 1969. Similarly, and drawing on Villey, Douzinas suggests that '[t]he idea of law as commandment or rule accompanied by sanctions originated in Jewish and later Christian concepts of law', C. Douzinas, *The End of Human Rights*, Oxford: Hart Publishing, 2000, p. 39.

200 Foucault, *Security, Territory, Population.*

201 Ibid.

202 Ibid., p. 21.

203 Tadros, 'Between Governance and Discipline', 76.

Indeed, the 'supercession' between modes of operation of power – that is between sovereignty, discipline and biopolitics (even modes of power, as we have seen, entertain a relation of intimate complementarity) – results, for Foucault, *not* in a decisive reduction of the importance of law, but in its transformation. A transformation, moreover, that also translates into its proliferation, given the proliferation of disciplinary and biopolitical apparatuses of power.[204] With Tadros, we can understand that law is a *crucial* interface 'through which governmental decisions can take effect by adjusting the operations and arrangements' of biopower.[205] In fact, what really emerges from a careful reading of Foucault's text, is that he does understand that the operations of law, its role, and the formalities of its construction, expression and enactment 'are at least partially governed by the predominant ways in which power is exercised'.[206] Indeed, one can rhetorically reiterate the question that a recent collection, attempting to 're-read' the role of law in Foucault's work, asks: is it really possible to understand law '*outside* of the rationalities and technologies of power operative in a given social formation'?[207]

However, while Foucault's terminology, as we have seen, has misled a number of legal scholars into believing that law fades with the ascent of biopower (though Ewald recognizes that Foucault never meant to suggest that the emergence of biopower led to a 'decline of law'),[208] the case is arguably the opposite. Law is central to Foucault's concept of biopower. Law's role becomes, indeed, intensified. Its mode of operation and the mechanisms through which it is constructed and enacted change, however, or rather *expand* – significantly. At the most general level, Foucault suggests that there is a significant shift in the central mechanics of law that becomes visible when life becomes the primary concern of the State. This difference hinges, for Foucault, on the fact that within the context of sovereign power '[l]aw cannot help but but [sic] be armed, and its arm, par excellence, is death; to those who transgress it, it replies, at least as a last resort, with that absolute menace. The law always refers to the sword'.[209] On the other hand,

> a power whose task is to take charge of life needs continuous regulatory and corrective mechanisms. It is no longer a matter of bringing death into play in the field of sovereignty, but of distributing the living in the domain

204 Ewald speaks in fact of an 'astonishing proliferation of legislation', Ewald, 'Norms, Discipline' p. 138. This proliferation particularly visible in the field of international environmental law, affected by a 'congestion' determined by the excessive proliferation of treaties (on this respect see B. Hicks, 'Treaty Congestion in International Environmental Law: The Need for Greater International Coordination', 32 *University of Richmond Law Review*, 1998–99); see also Tadros, 'Between Governance and Discipline', p. 77.

205 Ibid., p. 79.

206 Ibid., pp. 79–80.

207 Golder, *Re-reading Foucault*, p. 1.

208 Ewald, 'Norms, Discipline', p. 138. Indeed, Ewald underlines how the normalization of society that follows the ascendancy of biopower 'in no way diminished the power of law'.

209 Foucault, *History of Sexuality*, p. 144.

of value and utility. Such a power has to qualify, measure, appraise, and hierarchize, rather than display itself in its murderous splendor; it does not have to draw the line that separates the enemies of the sovereign from his obedient subjects; it effects distributions around the norm. I do not mean to say that the law fades into the background or that the institutions of justice tend to disappear, but rather that the law operates more and more as a norm, and that the judicial institution is increasingly incorporated into a continuum of apparatuses (medical, administrative, and so on) whose functions are for the most part regulatory.[210]

The biopolitical decision – or its legitimation - does not any longer hinge (not immediately and routinely at least), on a sovereign command, but on the enactment of that complex set of strategies that inserts life within particular matrices of control. This is the double register of law, captured by Esposito for example, that links sovereignty and biopower in that intimate relationship that we find paradigmatically expressed in Hobbes, as we have seen.

For Foucault then, 'juridical' refers to a medieval, monarchical expression of law that finds its central symbolic representation in the politico-theological category of the sovereign.[211] Law within the context of biopower finds, instead, its symbolic representation in the norm, understood, as Ewald elaborates, as that set of 'reference documents' that are crucial 'for the resolution of standard technical' as well as scientific, problems.[212] The norm, then, is a standard against which law is to be measured and modelled. The norm is inserted into a network of power relations which constitute it and reflect the particular regime of truth that enables, and is stabilized through, power. As such, it is detached from a particular sovereign will. The norm is a code without a particular author, a code that regulates. While the sovereign does not disappear, the effect is that of a disappearance. As a related consequence, law does not proceed, substantively, from a legislator: 'parliament', observes Ewald, 'no longer establishes the fundamental principles of law; it can only set forth regulations'.[213]

In this sense, as the prevalent modality in which political power is articulated changes (along the continuum sovereignty – disciplines – regularizations), so law re-articulates itself.[214] Yet law maintains each of the successive characters or modes of operation, overlapping and integrating each other. Moreover, its role of cementing relations of power, albeit re-configured along new regimes of

210 Ibid.

211 Ewald, 'Norms, Discipline',p. 139, suggests that the 'juridical' is considered by Foucault a 'code' that enables the power of the sovereign monarch to 'constitute itself'.

212 Ibid., p. 148.

213 Ibid., p. 155.

214 In similar terms L. Brännström, 2014, 'Law, Objectives of Government, and Regimes of Truth: Foucault's Understanding of Law and the Transformation of the Law of the EU Internal Market', 18 *Foucault Studies* 2014, 173, p. 182.

truths, continues to be operative, importantly, through the objectification of the re-configured network of power relations, an objectification that is encoded in/as/through law.

Ultimately, then, Foucault's argument operates at three different registers. Foucault speaks of law *qua* law, which may and does take a variety of forms; there is law as inserted in the context of networks of power relations, in which case Foucault speaks of the juridical, as distinguished from the disciplinary and biopolitical; there is also 'the code by which power presents itself (which for Foucault is consistently juridical)'.[215] This last point, the encoding of law within a particular symbolic form – what Foucault calls, to reiterate, the politico-juridical form – is important in relation to the shifting of law within a mode of operation or exercise of power that is biopolitical, rather than sovereign. This shift does not amount to a replacement, but, as already suggested, translates into an intensification and expansion of power (given, as we have seen the intimate continuity between forms of power). As Santos suggests, sovereignty and biopower are deeply interpenetrated, an interpenetration that facilitates, or even makes possible the 'isomorphic transformation' of the one into the other.[216]

Law thus plays a central role in relation to the ecological declension of biopolitics insofar as law concretizes the biopolitical discourse and its particular ecological rationality in international legal regimes that legitimize the subsumption of life under the care of power. In this manner, law enables biopolitical modes and practices of intervention aimed at the optimization and regularization of life. Conversely, an ecopolitcal perspective enables a better understanding of the mode of operation of law within the context of international environmental law, as it *both* responds to norms external to the politico-juridical mode of law, *and* is encoded in politico-juridical terms.

This understanding of law, and of its shifts in emphasis as regards its mode of operations, is clearly at work within the field of international environmental law. The trend towards the scientific management of the environment started, as Kuokkanen suggests, in 1972 at the Stockholm Conference on the Human Environment, with the massive 'incorporation of scientific and technical knowledge into policy-making'.[217] The integration of epistemic communities (and their specialized knowledge and regimes of truth) within processes of law-making, entails a shift from 'traditional diplomacy and politics toward expertise'.[218] In the language of Foucault, this shift (which I must remind the reader, is a shift in emphasis, and not a replacement) can be re-phrased as a shift from sovereign power and the juridical form of law, to biopower and the biopolitical form of law that arises not from a sovereign command, but from

215 Tadros, 'Between Governance and Discipline', p. 82.
216 De Sousa Santos, *New Common Sense*, p. 4.
217 Kuokkanen, *International Law*, p. 254.
218 Ibid.

a scientific and technical norm. Indeed, Kuokkanen (albeit without deploying Foucault's language or his theoretical apparatus) suggests that '[t]he new type of policy coordination having recourse to scientific knowledge began to frame a new approach to environmental issues'.[219] This new approach ultimately entails a turn from 'politics to policy-making, from diplomacy to technical management',[220] and from government to governance,[221] as the *locus* of intersection, interaction and mutual structuration of forms of knowledge and modes of power.

However, there is a further aspect that is critically important: biopower reconfigures and re-calibrates the space for critique (of law). The regimes of truth that structure and enable the interventions of biopower on life through law become also, decisively, the 'reference' according to which 'laws can be formulated, justified and criticized'.[222] The 'authority of law' is, within the context of biopower, measured against 'how well it respects "laws of nature"'.[223] This neutralization of politics and epistemic technicalization of law is exactly one of the points subjected to critique in Part II of the book through a genealogical reading of the 'ecosystem approach'.

A final notation. Despite the significant re-calibration of law, in terms of its conformity to scientific and technical norms, the sovereign encoding still remains central in many respects. Indeed, international lawyers still grapple with the central question of sovereign consent as that 'sovereign barrier' that imposes a severe limitation on international environmental law and on effective environmental protection.[224] In addition, while scholarship is increasingly focusing on the role of non-State actors in the processes of law-making and of creation of legal meaning, the point of departure remains the 'bedrock principle' of State sovereignty.[225] A biopolitical reformulation of law within the field of international environmental law, then, entails that law derives its rules and principles from ecological norms and objectifies them, albeit in a fluid way. On the other

219 Ibid.

220 Ibid., p. 237.

221 And in this respect, we will see, the ecosystem approach is a key element in this transformation.

222 Brännström, 'Law, Objectives' p. 182.

223 Ibid.

224 The sovereignty barrier has been described as that set of 'foundational principles and norms of general international law that continue to uphold a state-based system of international politics and law that is often seen as counter-productive to solving global and regional environmental problems,' Call for Papers of the Conference 'The Changing Nature of International Environmental Law: Evolving Approaches of the United States and the European Union', held in Geneva on 23 – 34 November 2013, and organized jointly by the European and American Societies of International Law's International Environmental Law Interest Groups, http://asilesilchangingnature.wordpress.com/conference-programme/ (Accessed 26 September 2018).

225 B. Karkkainen, 2004, 'Post-Sovereign Environmental Governance', 4:1 *Global Environmental Politics* 2004, 72.

hand, given the complexities and genealogical contestations immanent in the science of ecology, law remains linked to a struggle, a political decision, a hegemonic dialectic whose discursive parameters (i.e. the regimes of truth involved in its possible formulations and critique) shift but that remains the same in terms of its mechanics.

However, international environmental law still formally operates according to a language and a grammar of sovereignty. Moreover, biopolitics, which provides an overarching framework that aims at displacing the 'sovereignty barrier' in the name of ecological optimization (and this is a central and recurrent element within the discourse of the 'ecosystem approach'), ultimately also re-asserts the role of sovereignty through the formally and symbolically important role States, and the inter-state practices constituting international law, assume. A biopolitical key allows law to be understood in its multiple and complementary registers of operation. The question of sovereignty, always oscillating between a conceptual obsolescence (empirically registered or normatively advocated by progressive lawyers) and its material re-assertion (underlined by black-letter lawyers or theoretical realists), may thus find a useful re-framing through the analytics of biopolitics. Law finds, in fact, a specific biopolitical articulation in the form of the norm, that is, it produces rules and principles, or rather, regulatory frameworks, based on the incorporation and translation of scientific and technical norms into its grammar. It also simultaneously performs a procedural and symbolic encoding that maintains the legal mode of expression as an absolutely operative and crucial vector of legitimation of biopower. Yet, the sovereign mode of law, in its material operation, remains always latently active, if in the background, but ready to be fully re-activated and enacted through the decision of the sovereign whenever that becomes necessary. The biopolitical decision, however, is made possible and acceptable not in terms of the friend/enemy relation that Schmitt theorized,[226] but in relation to the security and immunitary question of life enhancing/life diminishing. In this respect, this 'biopolitical sovereignty' captures one of the central trajectories of continuity of modernity, particularly in the theorization offered by Esposito of the fundamental immunitary function of biopolitics, as we have seen. Moreover, biopolitical sovereignty is the decisive enabler of the aporetic mechanics of transformation of the politics of life into a politics a death. Yet even death only becomes possible (i.e. legitimate) if its effects are beneficial to life (or at least certain lives).[227]

This mechanism of biopolitical immunization (from the threatening vagaries of biological processes immediately linked to the immunitary logic intrinsic to, and constitutive of modernity, which Esposito explores and exposes) outlines a linkage

226 All political action could be reduced, for Carl Schmitt, to the distinction between friend and enemy, C. Schmitt, *The Concept of the Political*, Chicago: University of Chicago Press,1996, p. 26.

227 De Lucia, 'Bare Nature'.

enacted most specifically in the transformation of nature into an inert extension of resources which can be destroyed in order to expand and optimize human life. This is currently mostly achieved through the declension of the (sovereign) legal subject in the form of the neoliberal corporation, ultimate beneficiary of the biopolitical transformation of life into 'natural capital' susceptible of continuous biopolitical marketization. The corporation captures, conflates and re-assembles the politico-juridical discourse (i.e. the encoding) of sovereign law (i.e. human rights)[228] and re-orients it towards the biopolitical strategy of subsumption and subjugation of life to the imperative of Capital. Here we can also see how the frame of sovereignty is flexibly applicable to a diverse array of structures (the individual, the State, the people, the corporation), and is rather the site of a specific legal epistemology, and of a particular mode of operation of power.

The thanatopolitical consequences of this immunitary logic (whose most spectacular effects are concretized in the Nazi concentration camps)[229] are discussed by Foucault in terms of racism, understood as a process of differentiation, as 'a way of introducing a break into the domain of life that is under power's control: the break between what must live and what must die'.[230] The function of racism then, for Foucault, is 'to fragment, to create caesuras within the biological continuum addressed by biopower'.[231] But there is a second aspect, or function of racism that is of interest for the purposes of this book, provided the term 'racism' is not taken entirely at face value, for it may be misleading, given its usual semantic inflection. This second function relates even more directly to the immunitary logic that Esposito identifies at the genealogical core of modernity. Racism, suggests Foucault, has a 'vital importance'[232] to the exercise of a power that, both conceptually and operatively, conjoins sovereign power and biopower. Racism is situated at their intersection, further consolidating the argument for a fundamental continuity, or even intimacy, between the two modes of power, and hence for the fundamentally biopolitical character of modernity. In this sense, racism is the 'indispensable precondition'[233] for any form of direct and indirect 'killing' as an exercise of sovereign power, to the extent that any killing must be justified in biological terms, that is, in relation to the biopolitical goals of the elimination of biological threats to the (human) species and of the improvement, enhancement and optimization of the species (or race).[234] Here lies the juncture that joins together the conservation of life and its subjugation.

228 As insightfully shown by, e.g. A., Grear, *Rethinking Human Rights. Facing the Challenge of Corporate Legal Humanity*, London: Palgrave Macmillan, 2009.

229 Which for Agamben is the paradigmatic horizon of modernity, Agamben, *Homo Sacer*; the Nazi camp is generally a central theme of the biopolitical literature, see for example, Esposito, *Bios* and Campbell and Sitze, *Biopolitics*.

230 Foucault, *Society Must be Defended*, p. 254.

231 Ibid., p. 255.

232 Ibid., p. 256.

233 Ibid.

234 Ibid.

Conclusions

This chapter performs a decisive function, in relation to both Part III, and more in general in relation to the theoretical and methodological economy of the entire book. Its main goal is that of showing how the biopolitical frame, and the concepts of biopower and biopolitics, may find a specific ecological declension that allows the analytics of biopolitics to find immediate and useful application in relation to the 'ecosystem approach'. What I wanted to emphasize is *a particular set* of interlocking aspects that pertain to biopolitics, and that, we can now add, traverse the theorizations of biopolitics as an *ecopolitics*, rather than any loyalty to Foucault. However, Foucault provides the most immediately applicable framework in relation to the 'ecosystem approach' and to its biopolitical analysis as an advanced strategy for the optimization of life – and ecosystem processes – and for the regularization of its 'alea'. The discourses of sustainability, intergenerational equity and ecological integrity intersect and combine to legitimize biopolitical strategies of control that are objectified in law.

An ecological transposition of biopolitics has a number of key implications for a critique of the 'ecosystem approach'. First, it allows a crucial shift in the critical register, as it envelops both poles of the binary anthropocentrism/ecocentrism. In this respect, it allows the articulation of a critique that is richer and more capable of responding and capturing the genealogical complexities, the competing narratives and the conceptual ambiguities traversing and constituting the 'ecosystem approach'. Moreover, it allows us to overcome the limitations of a binary linear narrative of increasing interpenetration between law and ecology, which constructs critique on an external point of view (ecocentrism), a point of view, or reference framework which remains too un-reflexive and un- or at least severely under-problematized, as we have seen in the previous chapter.

Second, it provides an analytical framework able to theorize how life and all its life processes are subsumed *within* the processes of production of capital accumulation in a way that transforms the problem of ecological limits in its productive opposite.

Third, and relatedly, a biopolitical reading of the 'ecosystem approach' allows to read it *simultaneously* negatively and positively, to the extent that life, situated at the 'moving margins' of intersection and tension between biology and history,[235] is both enhanced and subjugated by power in the same gesture.[236] This is both the underlying aporetic void of biopolitics, its unresolved and undecided – indeed undecidable – tension, and, as I endeavour to show in the concluding chapter, the site where resisting

235 Esposito, *Bios*, p. 31.
236 Ibid., p. 37.

strategies may find a nurturing ground, *dentro e contro* (within and against) biopolitical modernity.

With this theoretical and methodological equipment then, it is now possible to revisit the 'ecosystem approach' more concretely, within the biodiversity legal regime that has nurtured it and erected it as its key conservation strategy.

9 A biopolitical critique of the 'ecosystem approach'

Biodiversity conservation

Introduction

This chapter presents an in-depth case study in order to illustrate more concretely how the analytics of biopolitics allows the articulation of a theoretically nuanced and sophisticated critique of the 'ecosystem approach'. The choice of the case study has fallen on the CBD, for a variety of reasons. The 'ecosystem approach' developed within the ocean cluster, and more specifically within the regime of international fisheries, may be *expected* to operate from a biopolitical perspective, given both the primary focus of fisheries management on resources exploitation[1] and the explicit linkage established by FAO (the central player in the articulation of the 'ecosystem approach' in the ocean cluster) between the 'ecosystem approach' to fisheries and an anthropocentric orientation. By contrast, the articulation of the 'ecosystem approach' in the biodiversity regime offers a more holistic, ecocentric inflection. This chapter argues, however, that the 'ecosystem approach' as it finds expression in the biodiversity regime also responds to a biopolitical logic. This emerges from both its genealogy and its present articulation, notwithstanding the competing narratives at work *within* the CBD. Importantly, however, the CBD offers a nuanced picture of the 'ecosystem approach', and a golden opportunity to unveil, through a biopolitical framework of analysis, competing closures and aspirations entangled within the same field of discursivity.

As the 'ecosystem approach' is primarily articulated in the CBD, and for reasons of space and focus, this chapter will *only* discuss the CBD (and refer to other biodiversity-related international regimes only if and as relevant). The role of the CBD in relation to the development of the 'ecosystem approach' cannot be overstated. The Secretariat, the COP and subsidiary technical bodies have dedicated significant resources to articulating and refining the concept, ever since its endorsement as the primary strategy of the CBD in 1995. Moreover,

1 The 'ecosystem approach' to fisheries is aimed at 'satisfying societal and human needs for food and economic benefits through management actions that focus on the fishing activity and the target resource', FAO, 2003, op. cit., p. 11.

this work has had, and still has, significant influence across international legal regimes, both within and outside the biodiversity cluster.

The CBD articulates the 'ecosystem approach' in holistic terms, hence resonating with 'ecocentric' inflections, aimed at displacing the human subject from the centre and re-defining the relation between man and nature. This holistic orientation transpires already from the definition of the 'ecosystem approach' as a strategy for the integrated management of natural ecosystems. Yet, the 'ecosystem approach' is at the centre of a discursive struggle, with competing narratives trying to impose hegemonic closures in a multiplicity of directions. This chapter explores these themes (and the related contestations), having in mind particularly how the 'ecosystem approach' actively embodies the double epistemic role of ecology, both inspiring an apparent radicalization of legal theory[2] and enabling new biopolitical regimes aimed at the regularization of the provisions of ecosystems goods and services.[3]

The chapter proceeds as follows. The second section briefly locates the CBD and its key elements, and offers a number of preliminary clarifications – conceptual and genealogical. The third section presents the 'ecosystem approach' in the CBD – its role, description, elements and principles. Sections four to six carry out a critical analyis of the concept, focusing respectively on the competing narratives at play, on the relation with the concept of ecoystem servcies, and on the insitutional dimension.

The Convention on Biological Diversity: some preliminary clarifications

Introduction

The Convention on Biological Diversity was adopted in Rio in 1992 and hence emerged from a context dominated by the narrative of sustainable development, an important consideration in relation to the role and orientation of the 'ecosystem approach' within the CBD. The CBD has been described as the 'epitome' of a new 'breed' of multilateral legal regimes combining environment and development.[4] In fact, while all other Conventions relating to biodiversity or wildlife prior to the CBD focus almost exclusively on conservation, the CBD introduces the paired objectives of conservation and sustainable use.[5] Moreover,

2 For a comparative discussion of two such legal philosophical approaches see De Lucia, 'Towards an Ecological'.

3 Darier, *Discourse*.

4 E. Morgera and E. Tsioumani, 2011, 'Yesterday, Today, and Tomorrow: Looking Afresh at the Convention on Biological Diversity', 21:1 *Yearbook of International Environmental Law* 2011, 3.

5 See e.g. L. Kotzé and T. Marauhn (eds), *Transboundary Governance of Biodiversity*, The Hague: Martinus Nijhoff, 2014. One exception is the Ramsar Convention, which focuses on 'wise use'.

the CBD is the first biodiversity-related Convention to operate as a comprehensive regime – most preceding Conventions proceeding in a 'piecemeal fashion', either in terms of subject matter or in terms of geographical scope.[6] The CBD also blazed a new conceptual trail, as it introduced or mainstreamed concepts such as 'biodiversity' and 'ecosystems' into the multilateral environmental legal vocabulary.[7] Since its adoption however, the CBD has been the subject of debate as to whether it is a successful endeavour. Being a process-oriented[8] framework[9] Convention, the success of the CBD cannot, perhaps, be measured at a singular point in time, yet opinions on the matter vary greatly.[10] The CBD has been heavily criticized, for containing too many loopholes;[11] for its language, considered too vague and qualified to impose clear obligations;[12] for having too wide (and widening) a scope;[13] for its deficient implementation, itself a symptom of other underlying causes;[14] and even for its structural failure to offer a true commitment to ecological sustainability.[15] As for the last point, effective implementation has indeed been

6 Birnie et al, *International Law*.

7 As well as other concepts, such as 'genetic resources', 'traditional knowledge' and 'benefit sharing', Morgera and Tsioumani, 'Yesterday', p. 3. It must however be observed that other regional treaties had already deployed the concept and the vocabulary of ecosystems; see e. g. Agreement on the Conservation of Polar Bears, 1973, Article II; and the Convention on the Conservation of Antarctic Marine Living Resources (adopted 20 May 1980, entered into force 7 April 1982), 1329 UNTS 48, throughout. Also UNCLOS contains a generic reference to 'ecosystems' in art. 194 ('The measures taken in accordance with this Part shall include those necessary to protect and preserve rare or fragile ecosystems').

8 Thus V. Koester, 'The Biodiversity Convention Negotiation Process and Some Comments on the Outcome', 27:3 *Environmental Policy and Law*, 1997, 175, p. 187.

9 In the double sense of establishing general principles to be further specified in implementation Protocols, and of leaving implementation decisions and modalities to the individual parties; thus also L. Glowka et al., *A Guide to the Convention on Biological Diversity*, 3rd edn, IUCN, 1994, p. 1. For a partially divergent opinion see A. Johannsdottir, 'The Convention on Biological Diversity. Supporting Ecological Sustainability or Prolonging Denial? ', 1 *Nordic Environmental Law Journal* 2010, 81; which argues that the CBD is a 'mixture of a framework convention and a conventional one, where some of its provisions are frame', ibid., p. 85.

10 For a summary of the different opinions in the five years after the adoption of the CBD (divided in two camps: success or failure) see Koester, 'Biodiversity Convention', pp. 187–8. Koester considers the CBD a success, if seen from the process-oriented perspective, in light of the number of parties ('developed, developing and with economies in transition') to the CBD after two years from its entry into force, ibid., p. 187.

11 M. Chandler, 'The Biodiversity Convention: Selected Issues of Interest to the International Lawyer', 4 *Colorado Journal of International Environmental Law* 1993, 141.

12 Chandler, 'Convention'; Kotzé and Marauhn, *Transboundary Governance*, pp. 117–18. See also Morgera and Tsioumani, 'Yesterday', p. 3. It has been argued however that the 'implementation process' is more important than a 'textual analysis' in order to assess the success (or lack thereof) of the CBD, Birnie et al., *International Law*, p. 617.

13 S. Johnston, 'The Convention on Biological Diversity: The Next Phase', 6:3 *Review of European, Comparative and International Environmental Law* 1997, 219, p. 223

14 A. Vadrot, *The Politics of Knowledge and International Biodiversity*, London: Routledge, 2014, p. 98.

15 Johannsdottir, 'Convention'.

lacking, a fact most notably evident in the failure to meet the global target of significantly reducing biodiversity loss by 2010,[16] a global target that had been agreed in 2002.[17] In addition, while a new Strategic Biodiversity Plan has been agreed for the period 2011–2020, adopting the Aichi Biodiversity Targets,[18] biodiversity is increasingly in jeopardy, despite the encouraging mid-term assessment presented in the 4th Global Biodiversity Outlook.[19] While a full account of the pre-negotiations and of the negotiations themselves exceeds the scope of this chapter,[20] a brief outline of the first steps leading to formal negotiations, as well as the main negotiating issues, which required negotiators to reconcile competing interests in a unitary text, may prove quite useful, as it will provide conceptual anchor points to locate the CBD within a biopolitical horizon.

In 1987, the Governing Council of the United Nations Environment Programme (UNEP) decided to call upon UNEP to convene an Ad Hoc Working Group of Experts on Biological Diversity to explore the 'the desirability and possible form of an umbrella convention to rationalize current activities in [the field of biological diversity], and to address other areas which might fall under such a convention'.[21] The proposal for a comprehensive instrument came from

16 Or as some poignantly put, the target 'did not even come close to being met', 'Bridging the Gaps in Biodiversity Conservation and Use', *Biores*, 8:8. http://www.ictsd.org/bridges-news/biores/news/bridging-the-gaps-in-biodiversity-conservation-and-use (Accessed 26 September 2018). The scientific evidence was presented in the 3rd Global Biodiversity Outlook: 'In 2002, the world's leaders agreed to achieve a significant reduction in the rate of biodiversity loss by 2010. Having reviewed all available evidence, including national reports submitted by Parties, this third edition of the Global Biodiversity Outlook concludes that the target has not been met', Secretariat of the Convention on Biological Diversity, *Global Biodiversity Outlook 3*, Montréal: Secretariat of the Convention on Biological Diversity, 2010, foreword.

17 Decision VI/26, 'Strategic Plan for the Convention on Biological Diversity', COP6, The Hague, 7–19 April 2002, para. 11.

18 Decision XI/2, 'Review of progress in implementation of national biodiversity strategies and action plans and related capacity building support to Parties', COP11, Hyderabad, 8–19 October 2012.

19 The Report underlines the 'significant progress' made towards the meeting of the Aichi Targets, yet acknowledges how 'a range of indicators suggest that based on current trends, pressures on biodiversity will continue to increase at least until 2020, and that the status of biodiversity will continue to decline', Secretariat of the Convention on Biological Diversity, *Global Biodiversity Outlook 4: A Mid-term Assessment of Progress towards the Implementation of the Strategic Plan for Biodiversity 2011–2020*, Montréal: Secretariat of the Convention on Biological Diversity, 2014, Executive Summary.

20 And the reader is referred to the relevant literature, such as, e.g., F. McConnell, *The Biodiversity Convention. A Negotiating History*, London: Kluwer Law International, 1996; Birnie et al., *International Law*, pp. 612–16.

21 Decision 14/26, 'Rationalization of international conventions on biological diversity', UNEP, Report of The Governing Council on the Work of its Fourteenth Session, 8–19 June 1987, General Assembly Official Records: Second Session Supplement No. 25 (A/42/25).

the United States,[22] ironically the only country, together with Andorra and the Holy See, which is not a Party to the CBD to this day.[23]

The three objectives of the CBD reflect the set of competing interests involved. A Report of the Ad Hoc Working Group of Legal and Technical Experts on Biological Diversity on the Work of Its Second Session (7 March 1991) illustrates the complexity of the issues involved and their alignment with the competing interests of industrialized and developing countries in ways that are replicated throughout the international environmental legal landscape. Of the four intertwined issues of central importance for our purposes is the dichotomy between conservation or use that traversed the entire negotiations, and still elicit tensions today. As the report noted,

> the first concerned the fundamental principle that the conservation of biological diversity was a common concern of all people. This principle required the participation of all countries and all peoples in a global partnership. It implied intergenerational equity and fair burden sharing. The common concern called for a balance between the sovereign rights of nations to exploit their natural resources and the interests of the international community in global environmental protection.[24]

These competing interests were carried over into the final text of the CBD and must be borne in mind during the analysis of the 'ecosystem approach' which is to facilitate both the conservation biodiversity and the sustainable use of its component. The relationship between those two concepts (and goals) is precisely the site where the biopolitical logic, and its aporetic core, finds operationalization.

Before proceeding with an analysis of the 'ecosystem approach' in the CBD, it is necessary to further frame the discussion by engaging with some of the key concepts of the CBD, and offer some preliminary conceptual clarifications. In particular, it is important to delineate a brief genealogy of the concepts of 'biological diversity', 'conservation' and 'sustainable use', to the extent that the 'ecosystem approach' is understood within the CBD to be a strategy for the achievement of both.[25] Indeed, the 'ecosystem approach' can only be approached, understood,

22 McConnell, *Biodiversity Convention*, p. 5.

23 The US has signed but not ratified. However, the US participates to the work of the CBD. The main point of contention for the US was the question of intellectual property rights; see in this respect e.g. R. Nayar and D. Ong, 'Developing Countries, "Development" and the Conservation of Biological Diversity' in M. Bowman and C. Redgewell (eds), *International Law and the Conservation of Biological Diversity*, London: Kluwer Law International, 1996.

24 Report of the Ad Hoc Working Group of Legal and Technical Experts on Biological Diversity on the Work of Its Second Session, 7 March 1991, p. 4, para 17.

25 The third objective of the CBD, benefit sharing, is less immediately relevant, and hence will require no specific treatment. However, see e.g. Decision V/6, 2000, op. cit. and Decision VII/11, 2004, op. cit., both of which consider that the 'application of the ecosystem approach will help to reach a balance of the three objectives of the Convention', respectively Section A, para 1 and Annex I, para 1.

interpreted and analysed with the conceptual and normative horizon implicated by the meaning of these concepts firmly in mind.

The objectives of the Convention

In order to explore and fully understand the 'ecosystem approach' within the context of the CBD, it is necessary to understand the primary objectives of the CBD, and for two reasons. First, they perform a crucial function by providing a normative horizon for the 'ecosystem approach', in a manner similar to the imperative of food security within the context of fisheries. The textual formulation and the critical interpretation of the objectives serve to frame the 'ecosystem approach' normatively. Second, and relatedly, these objectives already indicate the biopolitical horizon within which the CBD is located. Given that the CBD has three interlocking objectives that are not, at least prima facie, aligned along the same trajectory, it is useful to spend some time discussing them.

Article 1 establishes three interrelated objectives: The 'conservation of biological diversity'; the 'sustainable use of its components'; and the 'fair and equitable sharing' of benefits deriving from the commercial exploitation of genetic resources.[26] These three goals operationalize, in relation to biodiversity, that 'efficient synthesis'[27] of environmental and developmental concerns that had already emerged as an important element in the 1972 Stockholm Conference, and that would become fully articulated first as a concept in the report of the World Commission of Environmental and Development,[28] and then as a legal principle in the 1992 Rio Declaration[29] and in the so-called 'Rio Conventions', that is, the CBD and the UNFCCC (both adopted during the Rio Conference). Article 2, in turn, defines biological diversity as 'the variability among living organisms from all sources including, inter alia, terrestrial, marine and other aquatic ecosystems and the ecological complexes of which they are part; this includes diversity within species, between species and of ecosystems'.[30]

Recent research discusses the boundary separating 'conservation' and 'sustainable use' of biodiversity within the context of the 'ecosystem approach' to fisheries management.[31] The same delineation is operative in the CBD. The literature recognizes that the CBD is 'not a preservationist convention'; rather, it is aimed at the conservation of the benefit streams that humans may derive

26 CBD, Article 1.

27 D. Leary and B. Pisupati, *The Future and International Environmental Law*, New York: United Nations University, 2010, p. 5.

28 WCED, 1987, op. cit.

29 Rio Declaration on Environment and Development, 31 ILM 874 (1992).

30 UN Convention on Biological Diversity, 1992, op. cit., Article 2.

31 S. Jennings et al., 'The ecosystem approach to fisheries: management at the dynamic interface between biodiversity conservation and sustainable use', 1322 *Annals of the New York Academy of Sciences*, 2014, 48.

from the conservation of biodiversity.[32] Some theorists more explicitly highlight that the CBD establishes a legal framework for the capitalist exploitation of living resources, 'to the detriment of biological diversity and the human, social and cultural diversity that often exists alongside it'.[33] This position resonates with the centrality that the question of access to genetic resources had during the negotiations,[34] where it was repeatedly acknowledged that '[t]he full potential of biological diversity can best be realized when genetic resources remain accessible to all users and if technology and information on their use is transferable to all'.[35] Some commentators however link the conservation of biodiversity and the genetic value of it in the opposite direction, to the extent that as 'new biotechnologies make it possible to utilize the full potential of the world's genetic resources ... the economic incentive to conserve biological diversity increases'.[36] Some recent literature has already established a link between biopower and biopolitics and the genetic bioeconomy that has, in part, developed thanks to the CBD.[37] It can be then noted that the frame of biopower/biopolitics already finds some useful application within the broad context of biological diversity conservation.[38]

However, even the 'conservation' objective is already couched within a biopolitical horizon, insofar as it is fundamentally linked to the imperative of the regularization of the provisions of ecosystem goods and services. The next section will discuss in some detail the meaning of the concept of conservation, and whether and to what extent it is different from the concept of sustainable use.

32 A. Boyle, 'The Convention on Biological Diversity' in L. Campiglio, L. Pineschi, D. Siniscalco and T. Treves (eds), *The Environment after Rio: International Law and Economics*, London/Leiden: Graham & Trotman/ Martinus Nijhoff, 1994, p. 115, quoted in D. Ong, 'International Environmental Law Governing Threats to Biological Diversity', in M. Fitzmaurice, D. Ong and P. Merkouris (eds), *Research Handbook on International Environmental Law*, Cheltenham: Edward Elgar, 2010, p. 534.

33 Nayar and Ong, 'Developing Countries', p. 251.

34 Indeed, Nayar and Ong wonder whether the CBD should be rather called the 'Convention of the Exploitation of Biological Resources', Nayar and Ong, ibid., 251.

35 See e.g. Ad Hoc Working Group of Experts on Biological Diversity, Report of the Ad Hoc Working Group on the Work of its Third Session in Preparation for a Legal Instrument on Biological Diversity of the Planet, Geneva, 9–13 July 1990, Annex IV, para 12(a).

36 K. Rosendal, 'The Convention on Biological Diversity: A Viable Instrument for Conservation and Sustainable Use' in H. Bergesen, G. Parmann, and Ø. Thommessen (eds), *Green Globe Yearbook of International Cooperation on Environment and Development*, Oxford: Oxford University Press, 1995, p. 69.

37 See e.g. Rabinow and Rose, 'Biopower'; Lemke, *Biopolitics.*

38 See e.g. R. Youatt, 'Counting Species: Biopower and the Global Biodiversity Census', 17:3 *Environmental Values* 2008, 393, p. 395; and C. Biermann and B. Mansfield, 'Biodiversity, Purity, and Death: Conservation Biology as Biopolitics', 32:2 *Environment and Planning D: Society and Space* 2014, 252.

Framing biological diversity

Before discussing the CBD, or the role that the 'ecosystem approach' has in it, it is necessary to offer an initial framing of the concept of biological diversity. Detailed treatments are available elsewhere,[39] so I will offer two reflections to illustrate how the concept of biodiversity is: (a) genealogical (in the sense of being inserted in the midst of a 'play of forces') and (b) biopolitical.

The concept of biodiversity is arguably genealogical in at least two senses: in relation to its 'politically instigated' birth;[40] and in relation to how 'it is invoked to further a political agenda'.[41] The very notion of biodiversity emerges from a politically motivated branch of biology, or what Kotsakis calls an 'activist and tactical school of thought'.[42] This political role was claimed for scientists by Michael Soulé and Bruce Wilcox in 1980,[43] opening space for what has been called 'normative science'[44] or 'scientific activism',[45] though the terminological, conceptual and normative referent 'biodiversity' was effectively 'invented' by American conservation biologists at a 1986 Conference entitled 'The National Forum on BioDiversity'.[46] Erasga underlines that the concept of biodiversity was invented, 'as an organizing concept' to function as 'a communicative tool in the broader political arena'.[47]

More important for the purposes of my argument however, is the fact that both the concept of biological diversity, and conservation biology as a branch of science, can be understood biopolitically.[48] The tactics of conservation biology are concretized in a series of prescriptions and interventions aimed at 'the defense of life'.[49] For example, the global biodiversity census proposed by famous conservation biologist Edward Wilson, can be understood as a form of 'panopticism' which, through its processes of 'identification, collection of specimens, and subsequent research ... neatly packages' nonhuman nature into a set of designations which, in turn, facilitate 'conservation *and* commodification'.[50] Youatt argues that from the biopolitical perspective embodied in practices such as the global biodiversity census,

39 See in particular Kotsakis, 'Biological Diversity Complex', especially chapters 2 and 3.
40 D. Erasga, 'Biopolitics: Biodiversity as Discourse of Claims' in D. Ersaga (ed.), *Sociological Landscape - Theories, Realities and Trends*, Rijeka: Intech, 2012, p. 3.
41 Ibid.
42 Kotsakis, 'Biological Diversity Complex', p. 55.
43 M. Soulé and B. Wilcox (eds), *Conservation Biology: An Evolutionary- Ecological Perspective*, Sunderland, MA: Sinauer, 1980, see especially Foreword.
44 Lackey, 'Appropriate Use'.
45 Erasga, 'Biopolitics', p. 4. See also Vadrot, *Politics of Knowledge*.
46 Erasga, 'Biopolitics', p. 4.
47 Ibid.
48 Biermann and Mansfield, 'Biodiversity'; De Lucia, 'Critical Environmental Law'.
49 Biermann and Mansfield, 'Biodiversity', p. 257.
50 Youatt, 'Counting Species', 393.

'nonhumans are regulated and rationalised in matrices of knowledge and science, through which they are readied as productive resources for capitalism and mined as repositories of genetic information'.[51]

Yet, the role of such biopolitical practices, coupled with the uncanny ability of nonhumans to resist and challenge the 'normalising will of biopower'[52] (a point already raised by Foucault),[53] lead Youatt to conclude that '[i]n spite of the fact [that] the biodiversity census may extend the reach of an ecologically unfriendly capitalism ... it will, on balance, reap important ecological goods in hybridising political practice and acknowledging extra-human locations of power'.[54] Youatt acknowledges the potential for productive ambiguities, always present when concepts are immersed in a 'play of forces'. Other theorists, however, point in a different direction. Biermann and Mansfield, for example, discuss the 'racialization' of conservation biology, insofar as 'biopolitical rule must not only arbitrate the classification of life into species and populations', but also decide 'whom to correct and whom to punish, as well as who shall live and who shall die, what life-forms will be promoted and which will be terminated'.[55] Here we face precisely the biopolitical dilemma, the laceration between the life affirming and the life subjugating tendencies. While the motivations behind conservation biology are positive, in the sense of aiming at fostering life and optimizing its potential – i.e. its biodiversity – some of its effects are negative. The latter is especially evident in the regulation of invasive alien species, which hinges on Article 8(h), whereby Parties are obligated to prevent the introduction of alien species and to 'control or eradicate' those species 'which threaten ecosystems, habitats or species'.[56] Importantly, this process shows biopolitics to be fully intertwined with disciplinary and sovereign forms of power and intervention, in what is a seamless matrix of control.

The philosophy behind the global census is integrated into the CBD through the Global Taxonomy Initiative (GTI). The GTI's goal is to provide key taxonomic information necessary for decision makers to achieve conservation and sustainable use of biodiversity resources. GTI aims to provide information on where to establish protected areas; to identify and combat harmful invasive species; and to benefit from the use of

51 Ibid., p. 394.

52 Ibid., p. 395.

53 Foucault indeed suggested that life cannot be 'fully integrated into techniques that govern and administer it; it constantly escapes them', Foucault, *History of Sexuality*, p. 143.

54 Youatt, 'Counting Species', p. 395.

55 Biermann and Mansfield, 'Biodiversity', p. 261, quoting M. Dillon and J. Reid, *The Liberal Way of War: Killing to Make Life Live*, London and New York: Routledge, 2009.

56 CBD, article 8(h), emphasis mine. For an in-depth analysis of the biopolitical logic of the regulation of invasive alien species, see De Lucia, 'Bare Nature'.

biological diversity. All these goals need taxonomic information about components of biodiversity.

Taxonomy is considered 'essential' to the implementation of the CBD as it provides key inputs to the 'management of all types of ecosystems, from marine areas to forests to drylands ... to addressing alien species, access and benefit-sharing, and ... many other cross-cutting issues under the Convention'.[57] As a testimony to the centrality of taxonomic knowledge, the CBD has begun to speak of a 'taxonomic impediment' to biodiversity conservation.[58] To overcome this taxonomic impediment (that is, 'a scarcity of taxonomists, taxonomic collections, and institutional facilities'),[59] the COP, in a series of decisions, has mobilized resources in order to build 'taxonomic capacity', particularly in developing countries.[60] Taxonomic capacity ultimately translates to an adequate capacity to gather biological collections, which form the basis for identification of genetic resources and bioprospecting.[61]

The point that I want to emphasize here is that, 'for conservation biologists, the crisis of biodiversity does not refer to individual organisms, but to gene pools, populations, species, and ecosystems'.[62] And it is at these abstract levels (that is, not at the level of, e.g., the death of individuals, but at the level of mortality rates) that biopower is able to operate through the mechanics and tactics discussed in chapter 8.

The concept of conservation

One of the three overall goals of the CBD is the *conservation* of biological resources. However, the concept is not defined in the CBD, and there is no generally recognized definition in international law. In this respect, it has been suggested that it is best to 'approach it obliquely'.[63] Given that understanding conservation is crucial to understanding the 'ecosystem approach' in the CBD, I will approach it genealogically.

57 Convention on Biological Diversity Website, *GTI Section*, HTTP: http://www.cbd.int/gti/needs.shtml (Accessed 26 September 2018).

58 Decision VI/8, 'Global Taxonomic Initiative', COP6, The Hague, 7 – 19 April 2002, Annex, Section II, para 12; see also Decision X/39, 'Global Taxonomy Initiative', COP10, Nagoya, 18 – 29 October 2010, para 1.

59 Recommendation II/2, SBSTTA 2 Report, 2–6 September 1996, Montreal, 'Agenda Item 3.4: Practical approaches for capacity-building for taxonomy', para 1.

60 See e.g. Recommendation II/2, 1996, op. cit.; Decision III/3, 'Use of languages in the meetings of the Subsidiary Body on Scientific, Technical and Technological Advice', COP3, Buenos Aires, 4–15 November 1996; Decision IV/1, 'Report and recommendations', SBSTTA3, Bratislava, 4–15 May 1998, Section D.

61 See e.g. Recommendation II/2, 1996, op.cit.

62 Biermann and Mansfield, 'Biodiversity', p. 263.

63 Birnie et al., *International Law*, p. 588.

First used in a Treaty adopted in 1781[64] between the King of France and the Prince Bishop of Basel aimed at protecting forests and game fowl,[65] conservation remained close to the idea of preservation at least until 1900, when the idea of conservation slowly started taking a biopolitical inflection,[66] an inflection consolidated in the late 1940s and 1950s. The 1946 Whaling Convention for example associates 'the proper conservation of whale stocks' no longer with preservation (which has a strong protective, 'non-use' orientation),[67] but with 'the orderly development of the whale industry'.[68] The 1958 Geneva Convention on Fishing and the Conservation of the Living Resources of the High Seas associates conservation with 'measures rendering possible the optimum sustainable yield from those resources so as to secure a maximum supply of food and other marine products'.[69] CCAMLR associates conservation with rational use.[70] UNCLOS, finally, associates conservation with optimal utilization and maximum sustainable yield.[71]

All these associations have arguably clear biopolitical resonances, as they highlight the link between conservation and the enhancement and optimization of life for productive and harvesting purposes. De-linked from preservation, conservation enters a horizon where the double gesture of biopolitics, which simultaneously, protects and exploits, enhances and kills, becomes central.

An important passage in this brief genealogical reconstruction is the World Commission on Sustainable Development's definition of conservation, as it unveils the biopolitical tension inhabiting the concept, a tension then carried over into the CBD. The Commission's report considers conservation and

64 P. van Heijnsbergen, *International Protection of Wild Flora and Fauna*, Amsterdam: IOS Press, 1997, p. 45, as referred to in Birnie et al., *International Law*, p. 588.

65 J. Schneider, 'Endangered Species Markets. A Focus for Criminology?' in M. Natarajan (ed.), *International Crime and Justice*, Cambridge: Cambridge University Press, 2011, p. 211.

66 See e.g. Convention for the Protection of Birds Useful to Agriculture (adopted 19 March 1902, entered into force 20 April 1908), IPE Vol. IV, 1615 3, 177, which entailed the establishment of (biopolitically) differentiated lists of useful (protected) and noxious (bare life) birds, art. II.

67 See e.g. V. Nanda and G. Pring, *International Environmental Law and Policy for the 21st Century*, Leiden: Martinus Nijhoff, 2012, p. 247; see also PCA South China Sea, para. 941;

68 International Convention for the Regulation of Whaling (adopted 2 December 1946 and entered into force 10 November 1948), 161 UNTS 72, Preamble, last recital. The full recital states: 'Having decided to conclude a convention to provide for the proper conservation of whale stocks and *thus* make possible the orderly development of the whaling industry', emphasis mine. It must be noted however that the policy orientation of the whaling Convention has undergone a series of significant changes. Currently there is a moratorium (in place since 1981) that prohibits commercial whaling, subject to regular renewals, see International Convention for the Regulation of Whaling, 1946, Schedule as amended by the Commission at the 64th Annual Meeting Panama City, Panama, July 2012.

69 Geneva Convention on Fishing and the Conservation of the Living Resources of the High Seas (adopted 29 April 1958, entered into force 20 March 1966), 559 UNTS 285, art. 2

70 CCAMLR, art. II (2).

71 UNCLOS, art. 61(3) and 62.

sustainable use as a pair,[72] and the maintenance of 'ecosystems and ecological processes essential for the functioning of the biosphere', are linked to 'the principle of optimum sustainable yield in the use of living natural resources and ecosystems',[73] again emphasizing that the overarching goal, to which *both* conservation and sustainable use tend, is the optimum yield of nature.[74] This orientation is also reproduced in the report of the Legal Expert Group of the World Commission on Sustainable Development where conservation is again linked to the 'greatest sustainable' yield,[75] is associated with both use and preservation, and, importantly, with the 'enhancement' of the natural environment, capturing the biopolitical goal of regularization and intervention aimed at the optimization of life.[76]

This biopolitical tension was carried over into the CBD, as evident both by the negotiating documents[77] and by the final text of the Convention.[78] It is also useful to mention that during the negotiations there was significant debate as to whether or not to distinguish, explicitly and terminologically, between conservation and sustainable use.[79] A number of delegations wanted to use conservation broadly, inclusive of sustainable use, following the conceptual approach of the IUCN and WCED. The question of where the emphasis should be placed – protection or utilization – was indeed one of the themes debated throughout the entire negotiations. The final choice of the use of two distinct terminologies reflected the majority view – and more specifically developing countries' – that the emphasis should be placed on the *use* of biodiversity.[80]

72 In the Summary of Proposed Legal Principles, annexed to the WCED Report, Principle I (3) is entitled 'Conservation and Sustainable Use'.

73 WCED, 1987, op. cit. Annex I.

74 The term 'optimum' has been preferred to 'maximum' in order to signify the focus on the optimization of yields dynamically (i.e. over time) and across resource bases (e.g. multiple fisheries), R. Munro and J. Lammers, *Environmental Protection and Sustainable Development: Legal Principle and Recommendations*, London: Graham & Trotman/Martinus Nijhoff, 1987, p. 47.

75 Legal Principles for Environmental Protection and Sustainable Development UN Doc. WCED/86/23/Add. 1, reproduced in Munro and Lammers, *Environmental Protection*, p. 9.

76 Ibid.

77 See e.g. Report of the Ad Hoc Working Group on the Work of Its Second Session in Preparation for a Legal Instrument on Biological Diversity of the Planet, UNEP/Bio.Div.2/3, 23 February 1990, Annex I. See also Report of the Intergovernmental Negotiating Committee for a Convention on Biological Diversity on the Work of its Third Session/Fifth Negotiating Session (UNEP/Bio.Div/N5-INC.3/4, 4 December 1991), draft article 9(b)(v).

78 Article 2 of the CBD defines sustainable use as 'the use of components of biological diversity in a way and at a rate that does not lead to the long-term decline of biological diversity, thereby maintaining its potential to meet the needs and aspirations of present and future generations'.

79 F. Burhenne-Guilmin and S. Casey-Lefkowitz, 'The Convention on Biological Diversity: A Hard Won Global Achievement', 3:1 *Yearbook of International Environmental Law* 1992, 43, p. 49.

80 McConnell, *Biodiversity Convention*.

However, regardless of where the *terminological* emphasis lies, *conceptually* and *operatively*, sustainable use is arguably a 'prerequisite' for the conservation of living resources, in keeping with a modern understanding of the concept.[81] And while some insist on keeping the concepts distinct,[82] it is arguable that they operate as two sides of the same biopolitical *dispositif*, as already underlined.

In fact, while conceptually not identical or equivalent, they are two manifestations of the *same* biopolitical orientation to the management of biodiversity as a problem of life optimization and productivity enhancement. The Addis Ababa Principles and Guidelines for the Sustainable use of Biodiversity further underline how the relationship of instrumentality between the two terms functions in both directions. If conservation is a precondition for sustainable use (i.e. 'sustainable use cannot be achieved without effective conservation measures'),[83] the converse is also true. As explained in the Addis Ababa Principles, when there is a high 'risk of converting natural landscapes to other purposes ... encouraging sustainable use can provide incentives to maintain habitats and ecosystems, the species within them, and the genetic variability of the species'. Indeed, for 'dangerous' species that may 'represent a threat to humans', such as crocodiles, 'sustainable use has provided substantial incentives for conservation'.[84] Understood from a biopolitical perspective, the racialization of species – i.e. the decision about which species must live and which must die – may be avoided through the insertion of 'dangerous' and 'threatening' species into the capitalist processes of exploitation.

The crucial difference between the two concepts rests, then, on their level of abstraction. Conservation addresses biodiversity as an abstract entity that does not occupy a precise ontological space. At this level of abstraction, the central aim is arguably that of regularizing and controlling general properties, quantities, distributions, statistics and population dynamics. This is confirmed, for example, by the consideration that management of biodiversity, as recognized in decision VII/11, 'must focus on maintaining, and where appropriate restoring, the *key structures and ecological processes* (e.g., hydrological systems, pollination systems, habitats and food webs) rather than just individual species'.[85] Sustainable use, on the other hand, addresses *components* of biodiversity, that is, well-defined entities with a precise ontological presence on the ground.

81 Burhenne-Guilmin and Casey-Lefkowitz, 'The Convention', p. 50. Thus also I. Jakobsen, Marine Protected Areas in International Law: A Norwegian Perspective, PhD Dissertation, University of Tromsø, 2009, p. 92, footnote 380.

82 Jing in particular, distinguishes the concepts of conservation and sustainable use on the ground that in the CBD they are 'consistently' distinguished terminologically and that it thus stands to reason that they must be different concepts, Jing, *Preservation of Ecosystems.*, p. 97.

83 Decision VII/12, 'Sustainable Use (Article 10) ', in the Report of the COP7 Ordinary Meeting, Kuala Lumpur, 9–20 February 2004.

84 Ibid., Annex II para 8(c).

85 Ibid., Annex I, Table 1, p. 12.

The concept of ecosystem in the CBD

Finally, it is important to briefly outline the concept of ecosystem as defined and articulated in the CBD (the reader may remember the genealogical nature of the idea of ecosystem discussed in Chapter 5), as that provides the naturalistic underpinning of the 'ecosystem approach'. Article 2 also defines the concept of ecosystem, as 'a dynamic complex of plant, animal and micro-organism communities and their non-living environment interacting as a functional unit'. In this respect, and even if it may appear that ecosystems have, within the context of the CBD a particular material ontology, it is emphasized how misconceptions should be avoided, and that ecosystems are a functional category whose ontological space is to be determined, in both features and scale, by the problem or question to be solved.[86] Decision V/6 further specify that the definition of ecosystems contained in article 2 of the CBD,

> does not specify any particular spatial unit or scale, in contrast to the Convention definition of 'habitat'. Thus, the term 'ecosystem' does not, necessarily, correspond to the terms 'biome' or 'ecological zone', but can refer to any functioning unit at any scale. Indeed, the scale of analysis and action should be determined by the problem being addressed. It could, for example, be a grain of soil, a pond, a forest, a biome or the entire biosphere.[87]

This dovetails with the concept of ecosystem to be found in *certain* ecological literature, particularly the new ecology, which emphasizes a functional understanding of nature and its components, so that ecosystems are not amenable to representation as a map – that is, as an ontological place – but rather as a flowchart, since they focus primarily on processes, energy flows and nutrient cycling. That also contributes to the de-singularization of ecosystems and their insertion as technical and functional entities within a disembodied ontology of nature, a conceptualization that responds perfectly to the biopolitical framework (and its embodiment in the constructions and aims of conservation biology) aimed at the level of populations, processes and rates, rather than at the embodied level of individual entities.

The 'ecosystem approach' as a new strategy for biodiversity conservation

Introduction

All the preceding discussions were a necessary, if lengthy, prelude, whose aim was to *situate* the 'ecosystem approach' within the complex context of the CBD and of its operative narratives.

86 See e.g. Malawi Report, 1998, op. cit., para. 10.
87 Decision V/6, 2000, op. cit., Annex A, para. 3.

The 'fundamental requirement'[88] of the CBD for the conservation of ecosystems and natural habitats is *in-situ* conservation (Article 8). In this respect, one of the primary tools is the establishment of protected areas (Article 8(a)).[89] However, the recognition of the complexity of ecosystems, themselves an integral element of biological diversity (Article 2), has inevitably led to a recognition that 'the conservation and sustainable use of biological diversity and its components should be addressed in a holistic manner, taking into account the three levels of biological diversity and fully considering socio-economic and cultural factors'.[90]

Moreover, traditional conservation has been found to be affected by the so-called 'pathology of natural resources management'.[91] Typically, then, the 'ecosystem approach' is introduced in order to overcome the limitations of classical conservation and management tools such as, e.g., a narrow focus on a particular site, without consideration of the interlinkages with adjacent or otherwise connected sites; too much emphasis on species characteristics or protected areas, considering that biodiversity resides largely outside of protected areas; lack of integration with social and economic dimensions of conservation, including the problematic of market failures or distortion.[92]

During the meeting held in Jakarta in 1995, the 2nd Conference of the Parties (COP) to the CBD adopted the 'ecosystem approach' as 'the primary framework of action to be taken under the Convention',[93] following the recommendation contained in the 1st Report of the Subsidiary Body on Scientific, Technical and Technological Advice (SBSTTA).[94] SBSTTA had been tasked by COP1 with identifying alternative ways, means and actions that could be taken to implement the Convention,[95] and the report had provided that input to the COP.

Another decision adopted during COP2 reiterates the commitment of the CBD to the 'ecosystem approach' with particular regards to marine and coastal

88 CBD, Preamble.

89 Notwithstanding the obligations to also protect biological diversity in areas adjacent to, as well as more broadly outside, protected areas (Article 8 letters c, d, e, f and i), and the complementary objective of *ex-situ* conservation (Article 9).

90 Decision II/8, 'Preliminary Consideration of Components of Biological Diversity Particularly Under Threat and Action Which Could Be Taken Under the Convention', Report of COP2, Jakarta, 6–17 November 1995, Annex II, p. 55.

91 Holling and Meffe, 'Command and Control'.

92 For a comprehensive list see Malawi Report, 1998, op.cit., para 14.

93 Decision II/8, 1995, cit., Annex II, p. 55.

94 Recommendation I/3, 'Alternative ways and means in which the Conference of the Parties could start the process of considering the components of biological diversity particularly those under threat and the identification of action which could be taken under the Convention', SBSTTA 1 Report, Paris, 4–8 September 1995, Annex, p. 25.

95 That particular task had been marked as 'priority item', Decision I/7, 'Subsidiary Body on Scientific, Technical and Technological Advice', Report of the COP1 Ordinary Meeting, Nassau, 28 November–9 December 1994, Annex, where the draft provisional agenda for SBSTTA was set.

biodiversity.[96] On the basis of the SBSTTA's recommendation I/8,[97] a draft Programme for Further Work on Marine and Coastal Biological Diversity was adopted, aimed, among other things, at the identification of 'options for a pragmatic but comprehensive approach in addressing marine and coastal biological diversity on the basis of an ecosystems approach'.[98] This decision, together with decision IV/5, later formed the so-called Jakarta Mandate, an important step considering that there is no explicit and distinct mention of *marine* biodiversity in the CBD. Indeed, the Jakarta Mandate further underlines that the 'ecosystem approach' should be one of the 'basic principles'[99] in the conservation of (marine) biological diversity, and as such, it should be 'promoted at global, regional, national and local levels'.[100]

The COP has developed the 'ecosystem approach' in a number of subsequent decisions. The most salient[101] are decisions V/6 and VII/11. The former endorses a particular description of the 'ecosystem approach', including a set of principles (the so-called Malawi Principles, discussed in the next section) and operational guidance. Decision VII/11 contains a number of significant notations (i.e. the COP 'notes...') in relation to sustainable use, ecosystem services, sustainable forest management, and other compatible frameworks. Moreover, Annex I to decision VII/11 contains 'further guidance on the implementation of the "ecosystem approach" principles'. Annex II to the same decision contains reflections on the relationship between sustainable (forest) management and the 'ecosystem approach' (Section B); and a review of the integration of the 'ecosystem approach' into other programmes of work of the CBD, including the development of further integration strategies (Section B).[102]

96 Decision II/10, 1995, op. cit., Annex II, p. 59.

97 Recommendation I/8, 'Scientific, technical and technological aspects of the conservation and sustainable use of coastal and marine biological diversity', SBSTTA 1 Report, Paris, 4–8 September 1995, Annex, p. 36.

98 Decision II/10, 1995, op. cit., Annex II, para. 2(a).

99 Indeed, it is listed as the first of these principles. The others relate to the precautionary approach.

100 Decision IV/5, 'Conservation and sustainable use of marine and coastal biological diversity, including a programme of work', Report of COP4, Bratislava, 15 June 1998, Annex 'Programme of Work on Marine and Coastal Biological Diversity', Section B, para. 2, p. 85.

101 The ecosystem approach is also mentioned in a number of 'secondary' decisions focusing primarily on other thematic and cross-cutting issues; in these, the ecosystem approach is included as the integrative framework upon which the implementation of these particular areas of biodiversity conservation is to be based; see e.g. Decision IV/4 (Biodiversity of Inland Water Ecosystems); Decision II/10 and IV/5 (on Marine and Coastal Biodiversity: this is the basis for the mentioned Jakarta Mandate); Decision II/16 and III/11 (Agricultural Biological Biversity).

102 Decision VII/11, 2004, op. cit., Annex II Section B explores in particular the integration with marine and coastal biological diversity (Section B (2)), inland water ecosystems biological diversity (Section B (3)), and agricultural biological diversity (Section B (4)).

The Malawi Principles

After the adoption of the 'ecosystem approach' as its primary framework of action, the CBD recognized the need for an appropriate and useful description of the 'ecosystem approach', as well as for the further elaboration of a set of detailed guidelines and principles to facilitate its implementation.[103] The task was again given to SBSTTA, which organized a workshop in Lilongwe, Malawi, from 26 to 28 January 1998. The background for the workshop was the increasing importance that the 'ecosystem approach' had gained since its adoption as the primary framework for analysis and implementation in the CBD. The rationale was that the development of the 'ecosystem approach' was 'still in its infancy', and that there was a 'need for discussion on what it means for the Convention on Biological Diversity'.[104]

An informal discussion, organized by the Secretariat of the CBD, was held in 1997, as a side-event at the third meeting of the SBSTTA.[105] The discussion prompted further agreement on the need to formally – and urgently – discuss the 'ecosystem approach' within the process of the CBD, with the specific aim of addressing problems such as:

> terminology, types of ecosystems ('natural' vs. 'man-modified'), underlying theoretical assumptions, relation between 'ecosystem approach' and ecosystem management, problems of methodology, need for case studies, implications for the implementation of the CBD with special reference to its *modus operandi* and the legal implications.[106]

This discussion was carried over and formalized in the workshop held in Malawi, with the intent of presenting a report to COP4 in Bratislava in 1998. The report was 'taken note'[107] of by COP4, and in turn formed the basis for

103 M. Vierros, 'The Ecosystem Approach of the Convention on Biological Diversity' in G. Bianchi and H. Skjoldal (eds), *The Ecosystem Approach to Fisheries*, Rome: CABI Publishing and FAO, 2008, p. 39.

104 Malawi Report, 1998, op. cit., p. 15.

105 Ibid., para 6.

106 Ibid., para. 6, p. 3.

107 'Taking note' implies an acknowledgement of the material existence of particular document that however resides outside of the set of formal COP documents. Besides this particular case, significant examples of how taking note is a prelude to the integration of the contents of such documents into formal decisions are available within the climate regime (ex to Article 7(2)(c) UNFCCC), where twice the COP of the UNFCCC took note of a document, and both times it led to its eventual transformation into a formal decision of the COP. These two precedents regard the Geneva Ministerial Declaration, which played a crucial role in the adoption of the Kyoto Protocol; and the so-called Copenhagen accord, which played a crucial role in the subsequent re-shaping of the architecture of climate governance (on this particular point see, e.g. V. De Lucia, 2010, 'La Conferenza di Copenhagen tra Fallimento e Nuovi Modelli di Governance Globale', NENS *Nuova Economia Nuova Società*, Occasional Paper, Marzo https://www.nens.it/sites/default/files/cop15-analisi-nuovi-modelli-governance-vdl-25032010.pdf (Accessed 26 September 2018).

further elaboration.[108] The Malawi Principles resulting from this further elaboration were annexed to Decision V/6 and 'endorsed' by the COP.[109]

As I will endeavour to show, the Malawi Principles are traversed by the competing narratives that have been identified as characterizing the discourse of the 'ecosystem approach' in general. A duality of discourses – an anthropocentric discourse aligned with global capitalism and an ecological, holistic discourse aligned with an ecocentric deployment of ecology – can be detected across the various principles. This duality however sustains a tension which, as we shall see, is best understood through an analytics of biopolitics, as biopolitics always encompasses *both* poles of the binary, and the conflict and contestations are not external to the 'ecosystem approach', but fully and inevitably immanent, in genealogical fashion.

The preliminary clarifications contained in the Malawi Report immediately make clear that the combined effects of knowledge and interventions operate according to a biopolitical template. Yet the principles introduce crucial ecological concepts that may, and do, give the impression of an ecological, holistic orientation of the 'ecosystem approach', thus further confirming the complex genealogy of the concept at this finer level of analysis. In particular, the report acknowledges that ecological knowledge – or 'reasoning',[110] as the report puts it – entails non-linearity, interconnectedness, the importance of the human dimension, and a crucial focus on adaptability/resilience (as opposed to stability).[111] In light of this, and of the further consideration that ecosystems are complex and unpredictable, the Malawi Principles reiterate that a central tool for the implementation of the 'ecosystem approach' is adaptive management, that is, a model of management premised on iterative decision-making and learning-by-doing. The report, finally, further underlines the centrality of the 'ecosystem approach' by making it applicable across all thematic areas under consideration of COP and SBSTTA.[112]

Decision V/6, which endorses the description of the 'ecosystem approach' and the Malawi Principles, also *encourages* the 'further conceptual elaboration' as well as 'practical verification' of the principles.[113] To this purpose the COP requested the Executive Secretary to 'collect, analyse, and compare identified case-studies and lessons learned on the ecosystem approach, and to prepare a synthesis'.[114] The effectiveness of the Malawi Principles also needed to be assessed, and that was done through a review of a number of case studies, in order to refine such principles in light of problems encountered in the field to

108 Decision IV/1, 1998, op. cit., paras 1 and 2, respectively.
109 Decision V/6, 2000, op. cit.
110 Malawi Report, 1998, op. cit., para 10(c).
111 Ibid.
112 Ibid., para. 13.
113 Decision V/6, 2000, op. cit., para. 1.
114 Ibid., para 4 and Decision VI/12, 'Ecosystem approach', COP6, The Hague, 7–19 April 2002, para. 2(a).

improve implementation success.[115] To this purpose, a workshop was organized in 2002, with the aim of developing the 'ecosystem approach'.[116] More specifically, the objective of the workshop was to check 'the consistency, redundancy and hierarchy of the Principles',[117] and at the same time to clarify the wording and to develop a more logical sequence in their presentation and concatenation. Indeed, the principles were considered to be 'vaguely worded, poorly structured and overlapping'.[118]

Another problem some lamented about the Malawi Principles was that both the principles and the annexed operational guidance were too 'theoretical and general'[119] and managers required tools and guidance 'relevant at the operational level'.[120] In fact, most of the participants in the pathfinder workshops, a series of workshops that were to form the basis for the review and refinement of the 'ecosystem approach', suggested that the current operational guidance was not sufficient and that there was need for the development of 'specific, problem-oriented guidance'.[121]

An expert meeting was subsequently convened in Montreal in 2003,[122] with the goal of developing 'proposals for the refinement of the principles and operational guidance of the ecosystem approach'.[123] The outcome of this process was formally captured in Annex I of Decision VII/11. Section A of Annex I provides 'further guidance on the implementation of the 'ecosystem approach' principles', while section B offers 'additional explanatory notes on cross-cutting issues related to operational guidance'.[124] The refined principles, however, are the same as the Malawi Principles. What has changed, besides the refinement of their rationale in light of past practice, is that they are grouped differently, and that they follow a different logical sequence. The next section will explore the ways in which the 'ecosystem approach' is articulated in the CBD at a general level, before delving into the competing narratives that are embedded in the Malawi Principles in the fourth section.

Articulating the ecosystem approach

Decision V/6 endorses the description of the 'ecosystem approach' agreed upon in the Malawi Workshop. This description has a number of overarching

115 Decision VI/12, 2002, cit., Section 2(c).

116 Indeed, the workshop was entitled 'Further Development of the Ecosystem Approach', and was held at the Isle of Vim in October 2002; for more details, see Korn et al., *Report.*

117 Ibid., p. 7.

118 Expert Meeting, 2003, op. cit., para 11.

119 R. Smith and E. Maltby, *Using the Ecosystem Approach to implement the CBD: A global synthesis report drawing lessons from three regional pathfinder workshops*, Hunterdale, UK: RHIER, 2003, p. 16.

120 Ibid.

121 Ibid., p. 17.

122 Expert Meeting, 2003, op. cit.

123 As requested in Decision VI/12, 2002, op. cit., para. 2(c).

124 Decision VII/11, 2004, op. cit.

themes that link up with some of the central insights ecology offers to law. Yet they simultaneously link up with some of the themes that are at the core of the Anthropocene, and that are most immediately susceptible to a biopolitical reading. This set of competing narratives that traverse the description of the 'ecosystem approach' endorsed by the CBD deserves further scrutiny.

Decision V/6 offers the best starting point for discussing how the 'ecosystem approach' is articulated in the CBD. The definition given therein describes the 'ecosystem approach' as 'a strategy for the integrated management of land, water and living resources that promotes conservation and sustainable use in an equitable way', and is hence considered to 'help reach a balance' between the three goals of the CBD.[125] Importantly, the 'ecosystem approach … is based on the application of appropriate scientific methodologies focused on levels of biological organization, which encompass the essential structure, processes, functions and interactions among organisms and their environment'.[126] The 'ecosystem approach' thus reflects the definition of ecosystem contained in the CBD, focusing on 'structure, processes, functions and interactions',[127] rather than on a particularly identifiable 'place' with a particular ontology. The 'ecosystem approach' is moreover based on the application of sound science and scientific methodologies. Yet it is recognized that the complexity of ecosystems, the lack of complete knowledge and the non-linear character of natural processes 'require' the adoption of an 'adaptive' model of management.[128] Furthermore, the 'ecosystem approach' – and this is a central point – recognizes that 'humans, with their cultural diversity, are an integral component of many ecosystems'.[129]

The 'ecosystem approach' however is not meant to replace other conservation strategies or approaches such as protected areas, biosphere reserves or single-species management, but rather to integrate them in order better to 'deal with complex situations'.[130] The fact that the 'ecosystem approach' is defined as a 'strategy' points to its policy orientation. Indeed, the COP has explicitly acknowledged that the 'ecosystem approach' does not possess a legally binding character, thus 'allowing for flexibility and experimentation'.[131] This is further explained in the following terms:

125 Decision V/6, 2000, op. cit., Annex A, para 1.
126 Malawi Report, 1998, op. cit., para. 8.
127 Decision V/6, 2000, op. cit., Annex A, para. 3.
128 Ibid., Annex A, para. 4.
129 Ibid., Annex A, para. 2; also reiterated in Review of the Principles of the Ecosystem Approach and Suggestions for Refinement: A Framework for Discussion, Note by the Executive Secretary, Expert Meeting on the Ecosystem Approach, 'Review of the Principles of the Ecosystem Approach and Suggestions for Refinement: A Framework for Discussion', Montreal, 7–11 July 2003, para. 8.
130 Decision V/6, 2000, op. cit., Annex A, para. 5.
131 Decision VII/11, 2004, cit., Annex II, Section A (1)(2). The notation originally referred to both SFM and EA.

> The central concept of the 'ecosystem approach' is integration. The 'ecosystem approach' attempts to facilitate the removal of artificial barriers between economics, social science and ecology, and places humans firmly within the ecosystem model. This is needed to achieve a socially acceptable balance between conservation, resource use and sharing of benefits. The 'ecosystem approach' is not intended to be an alternative or competing concept, but one that unifies and consolidates existing scientific and social wisdom on ecosystem management.[132]

The most salient features contained in this definition, is the concept of integration. The 'ecosystem approach' is to integrate disciplinary barriers; it is to integrate humans *firmly* within the ecosystem model; it is to integrate existing strategies within an overall framework by *unifying and consolidating* other knowledges and wisdoms. Moreover, as noted, the 'ecosystem approach' directly incorporates the central idea of the scientific concept of ecosystem, that is, the interrelation between living and non-living resources, so that they can be considered in an integrated manner.

On the other hand, there is clear emphasis on management. In fact, the 'ecosystem approach' is said to integrate various knowledges 'on ecosystem management', in this respect delineating a distinction between a management practice (i.e. ecosystem management) and a strategic framework (i.e. the 'ecosystem approach'). This distinction however, as we will see, is rather unstable (we have dwelt at length on the terminological and conceptual confusion surrounding the 'ecosystem approach' in Chapter 4).

This articulation of the 'ecosystem approach' can be understood (at least prima facie) as a holistic ('integrated') transposition of Article 8, and reflects the wider paradigm shift that has occurred in conservation philosophy, moving away from a focus on charismatic or key species and towards a broader focus on entire ecosystems.[133] The Expert Meeting on the 'ecosystem approach' states explicitly that '[t]he conservation and … restoration of [ecosystems'] interactions and processes is of greater significance for the long-term maintenance of biological diversity than simply protection of species'.[134] Yet ecosystems are not understood as objectively pre-determined, ontologically given entities, as we have seen. Rather, they are the result of a particular conservation question.[135]

There is also an explicit continuity between the 'ecosystem approach' and earlier management practices. The description of the 'ecosystem approach' reproduced above explicitly acknowledges that the 'ecosystem approach' does not preclude the utilization of other conservation and management tools, such as protected areas or even single-species management. Indeed, the 'ecosystem

132 Expert Meeting, 2003, op. cit., para. 8.
133 This shift is already concretized in the Ramsar Convention.
134 Expert Meeting, 2003, op. cit., Section B, principle 5.
135 Malawi Report, 1998, op. cit., para 10(b).

approach ... is not intended to be an alternative or competing concept, but one that unifies and consolidates existing scientific and social wisdom on ecosystem management'.[136] It does, however, provide the opportunity to integrate them within a more holistic framework. Smith and Maltby observe that the 'ecosystem approach ... represents a codification of previously applied strategies',[137] a perspective not unlike the FAO's in relation to EAF.

The Pathfinder Report identifies some of the 'key distinguishing features' of the 'ecosystem approach' in the following: it is designed to balance the three CBD objectives; puts people at the centre of biodiversity management; extends biodiversity management beyond protected areas while recognizing that protected areas are also vital for delivering the CBD objectives; and engages the widest range of sectoral interests'.[138] One can again detect two intertwined yet incompatible orientations. On the one hand, the 'ecosystem approach' follows the template of sustainable development to the letter, since it 'puts people at the centre of biodiversity protection' and directly espouses biopower in its human-oriented dimension, that is the optimization of human life. Yet on the other hand, it seems to move away from the modern threshold delineating the majestic from the monstrous and urges conservation to take place 'beyond protected areas', i.e. everywhere, fully resonating with the prevailing discourse of the Anthropocene.

Moreover, while the language of management is widely used throughout the CBD, and with particular regards to the 'ecosystem approach' in the Malawi Principles, the 'ecosystem approach' is 'a framework for action based on holistic decision making, not a set of guidelines for managing various ecosystems'.[139] Indeed, the 'ecosystem approach' is a framework for thinking and acting *ecologically*. The 'ecosystem approach', in other words, simultaneously captures competing narratives: a utilitarian perspective aimed at optimal resource use, and an ecological perspective aimed at holistic ecosystem governance, further complicating the ambiguities and tensions traversing the (legal) discourse of the 'ecosystem approach'.

This holistic orientation is, in general, a central strength of the 'ecosystem approach'. In fact, the overall argument behind the adoption of the 'ecosystem approach' rests on the idea that '[c]lassical nature conservation approaches have limitations as the sole tool for management of biological diversity'.[140] In other words, the 'ecosystem approach' allows us – in the vision expressed in the Malawi Report – to overcome a number of the shortcomings of classical conservation strategies, such as: the lack of sufficient recognition that ecosystem functioning is vital; a management model that is too site-specific, missing out on the

136 Expert Meeting, 2003, op. cit., para. 8.
137 Smith and Maltby, *Using the Ecosystem Approach*, p. 6.
138 Ibid.
139 Ibid., p. 13.
140 Malawi Report, 1998, op.cit., para. 14.

importance of cross-site interlinkages; a view that maintains a separation between nature and culture when they should be integrated; too little emphasis on the fact that the vast majority of biodiversity lies outside of protected areas, which conversely receive far too much attention; an approach that is too sectoral; the underestimation of the importance of re-aligning economic incentives (including market so-called failures); and, by converse, makes it possible to appreciate the values of ecosystem goods and services. This list offers an interesting view of the genealogy of the 'ecosystem approach' as it unfolds within the particular context of the CBD.

This genealogy is complex and responds to competing narratives. On the one hand, the articulation of the 'ecosystem approach' in the CBD has a clear ecocentric inflection. Integration, a central ecological insight captured by the expression that 'everything is connected to everything else',[141] is central to the 'ecosystem approach'; the dismantling of the fence between humanity and nature signals the recognition that the central modern threshold between nature and culture is a social, political and epistemological construction, a threshold which *hinders* appropriate strategies for environmental protection; and, finally, the complex epistemological horizon of ecology seems to be incorporated in the framework of the 'ecosystem approach' as evident by the recognition of the need for an adaptive model of management *precisely* in light of the complex and shifting nature of ecosystems, knowledge of which remains inevitably incomplete and dynamic. These are all crucial markers of a significantly different perspective on law, nature and knowledge than traditional conservation practices and modern epistemological trajectories.

Yet this articulation of the 'ecosystem approach' – *even* this articulation – is best understood as the enactment of a biopolitical template, precisely at the point at which ecology as a new form of rationality intersects with the trajectories of modernity: it transforms them and it is in turn operationalized as a crucial element of the biopolitical framework of control. The 'ecosystem approach', from this perspective, can be understood as a mode or strategy of intervention that harbours the central aporia of biopolitics, that is, the fundamental contradiction always at work between the care, enhancement and optimization of life, and its subjugation. The inadequacy of the anthropocentric-ecocentric binary framework is here clearly demonstrated in light of the complexities and ambiguities of the relationship between ecology and law.

The next section reviews the Malawi Principles, their associated operational guidance, and their subsequent review and refinement, in order to further highlight and discuss the competing narratives and the biopolitical tensions that traverse the 'ecosystem approach', and whether these tensions are resolved within the context of the CBD.

141 B. Commoner, *The Closing Circle: Nature, Man and Technology*, New York: Alfred Knopf, 1971, p. 16.

The competing narratives traversing the 'ecosystem approach'

Introduction

Together with the description of the 'ecosystem approach', decision V/6 endorses the so-called Malawi Principles (contained in Annex B), and *recommends* their application, while not formally adopting them.[142] At the outset, the 12 Malawi Principles are described as 'complementary and interlinked',[143] so they each need to be read in light of the others. I choose to present and discuss the principles through particular themes that emphasize the different narratives and genealogies that traverse the ecosystem approach, rather than following a systematic discussion of each principle. The analysis shows how the articulation of the 'ecosystem approach' in the CBD is inserted within a field of competing narratives and complex genealogies. These roughly align with the anthropocentrism/ecocentrism binary, creating what appears to be a significant space of internal tension supporting the view, presented in this book, of the 'ecosystem approach' as irreducibly contested and ambiguous. Yet, as amply explained and anticipated earlier in this book, the anthropocentric/ecocentric binary is not satisfactory. A further analytical register will relocate both narratives within the larger and richer critical frame of biopolitics.

Ecological normativity

The 'ecosystem approach' that emerges from the analysis of the Malawi Principles arguably shows how the central place of the ecological narrative is seemingly to contribute to a more 'ecocentric' orientation (at least to the extent that the articulation of the 'ecosystem approach' is not immediately instrumental). To be sure however, ecology is also a key epistemological framework for the biopolitical subsumption of nature under that complex grid of norms aimed at its care and optimization that is objectified in international environmental legal regimes. Indeed, the added critical value of the analytics of biopolitics rests *precisely* in its ability to envelop both anthropocentric and ecocentric articulations of the 'ecosystem approach'. However, it is important and useful to mark the different ways ecology is deployed within particular regimes. In the CBD then, the ecological narrative underlying and supporting the 'ecosystem approach' takes a specifically normative dimension along the conceptual lines elaborated by Italian legal philosopher Mariachiara Tallacchini.

Tallacchini's concept of 'ecological normativity'[144] captures the relationship between science and law, in the sense of a necessary transformation of law along

142 Decision V/6, 2000, op., cit., para. 1. In this sense the COP implicitly reiterated how the 'ecosystem approach' is not legally binding, but rather is a policy approach.

143 Malawi Report, 1998, op. cit., para 16.

144 See Tallacchini, 'Legal Framework'. and Tallacchini, *Diritto*; for an analysis of Tallacchini's framework see De Lucia, 'Towards an Ecological'.

an ontological and epistemological trajectory that emerges from the science of ecology. Reflecting the key challenges that ecology poses to legal modernity illustrated in Chapter 1 (in part drawing on Tallacchini), ecology introduces a number of insights that are key to the articulation of the 'ecosystem approach' in the CBD, and especially in the Malawi Principles. This normative language, moreover, resonates with the Foucauldian insight of a transformation of law's operative, conceptual and symbolic modality from politico-juridical to biopolitical. This transformation, which is better characterized as the appearance of a new layer in what are relations of complementarity between the different modalities, entails that law is no longer primarily a product of the will of a subject, but is located in the 'objective' norms established in other domains, in this case the norms that ecology draws from the natural laws governing ecosystems.[145] The insights from which norms are derived are the existence of ecological limits; the uncertainty and pluralism of knowledge; the complexity of the temporal and spatial scales involved; the relational ontology of nature, and the interactional effects of processes and structures within and between ecosystems. All of these insights find expression in one or several of the Malawi Principles.

Principle 3 for example, acknowledges the transversal interconnections between ecosystems and/or landscapes, and the potential spillover effects of management actions, which can have 'unknown or unpredictable effects on other ecosystems'. Principle 6 states that 'ecosystems must be managed within the limits of their functioning'. The rationale further explains that particular 'attention should be given to the environmental conditions that limit natural productivity, ecosystem structure, functioning and diversity'. This principle embodies what Tallacchini calls the 'normativity of limits' and more generally the precautionary approach, which is generally considered to be one of the underlying constituent elements of the 'ecosystem approach'. Principle 7 further elaborates the ecological underpinnings of the 'ecosystem approach', with particular reference to the need to calibrate it in relation to the appropriate spatial and temporal scales. While these are contingent on the particular objectives pursued, and hence are defined operationally, the key message of Principle 7 is that 'connectivity between areas should be promoted where necessary'. The 'ecosystem approach', continues the rationale underlying Principle 7, 'is based upon the hierarchical nature of biological diversity characterized by the interaction and integration of genes, species and ecosystems', further consolidating the integrative and relational aspects of the approach. Principle 8, finally, recognizes that '[e]cosystem processes are characterized by varying temporal scales and lag-effects'. Management objectives, then, 'should be set for the long term'. The

145 See in this respect Tallacchini, *Diritto*, who speaks of a delegation of responsibility of law to science, and Wood, 'Sovereign Trust', who similarly observes that US judges hold 'an abiding faith in nonbiased administrative expertise' in light of the perception that 'courts are no match in the scientific and technical realm', p. 60.

rationale underlying Principle 8 further recognizes that these long ecological temporal scales 'inherently [conflict] with the tendency of humans to favour short-term gains and immediate benefits over future ones'.

While these themes appear to be strongly and deeply ecological, and may offer some basis for the characterization of the 'ecosystem approach' as ecocentric, they represent only one of the narratives that traverse the Malawi Principles and the articulation of the 'ecosystem approach' within the CBD. In what is arguably a stark contrast, another narrative organizes nature and natural ecosystems as service providers and as natural capital. Some elements of this narrative are already at work within this ecological (or perhaps ecocentric) narrative, insofar as, for example, the respect of ecological limits is functional to the maintenance of 'natural productivity' of ecosystems (Principle 6), and the focus is on the importance of the abstract properties of biodiversity in relation to the provision of ecosystem goods and services, as evident from the emphasis on the crucial role given to '*functional* biodiversity' in the operational guidance to the 'ecosystem approach' annexed to the Malawi Principles.[146] This approach may lead to conclusions such as that 'the restoration of ecosystem functioning is often best achieved through the addition of non-native species';[147] paradoxically however, these types of conclusions may actually undermine the objectives of the CBD, to the extent that biodiversity, if understood primarily in functional terms, may become disengaged and further abstracted from the *particular* species alive, and, rather, re-formulated in terms of key services provided. This represents a further misalignment with traditional conservation practices hinging on protected areas and endangered species lists, hence moving deeper into the conceptual approach of 'new ecology' and 'new conservation'.

Principle 9, finally, incorporates in the management horizon of the 'ecosystem approach' the idea of change. Principle 9 recognizes that change is inevitable and immanent to all ecosystems and their processes. Because of this inherent dynamics of change, explains the principle's rationale, '[t]he ecosystem approach must utilize adaptive management in order to anticipate and cater' for the multiple complexities and uncertainties involved. This principle is of central importance. Yet the concept of ecological change itself is at the centre of a 'play of forces' which pulls its implications in different and, to a large extent, incompatible directions. On the one hand, as we have seen in Chapter 5, change is the new paradigmatic referent of the New Ecology. Change is used in order to deconstruct what has been called the myth of the balance of nature and to justify interventionist policies that arguably fit rather well the biopolitical template described in Chapter 8. The argument, in brief, goes as follows: since change is

146 Decision V/6, 2000, op.cit., para 8, emphasis mine.

147 E. Hearnshaw, R. Cullen and K. Hughey, *Ecosystem Health Demystified: An Ecological Concept Determined by Economic Means*, Conference paper presented at the Economics and Environment Network Workshop, held at the Australian National University, Canberra, 4–6 May 2005, p. 4.

immanent and inevitable, there is no point in preserving certain particular conditions (the equilibrium point of ecosystems), as they are no more natural than any other set of possible conditions (natural processes lacking teleology or direction). As a consequence, humans may legitimately choose any environmental conditions which best maintains the provisions of particular services and the benefits associated with them. Change, moreover, allows for a focus on the abstract elements of biodiversity (its functions and processes) both in relation to particular species, and in relation to the particular spatial dimensions (hence facilitating policies such as the no-net-loss).

Here we see a double manoeuvre. First, the paradigm of change is used to deconstruct the concept of naturalness (epitomized in the myth of the balance of nature); subsequently, it is used to argue for biopolitical interventions aimed at the regularization of nature, that is, aimed (with paradoxical circularity) at the minimization of change within particularly beneficial parameters. Yet seen from another perspective, change is central to a responsive and responsible ecological law whose epistemological premise rests on a situated, circular (i.e. iterative) and sustained relationship with a particular place (I will return on these points in Chapter 10). For the time being the reader should note exactly how this double perspective further corroborates the conclusions of Part II, that is, that the 'ecosystem approach' is fully inserted in a dynamic and genealogical 'play of forces' and suspended amidst competing narratives.

Governance

Governance is an important aspect of the 'ecosystem approach', and four of the Malawi Principles are exclusively or partially dedicated to it. Principle 1 states that the objectives of management are 'a matter of societal choice'. Importantly, any governance structure should also 'adequately' represent 'the needs of future generations and the natural world'.[148] This seems to open space for a truly ecocentric inclusion of nature as an independent stakeholder, which, in legal terms translates quite naturally into some form of representation grounded in procedural rights, along the lines suggested by Grumbine.[149] And while this is *not* the current case it is important to recognize that there is an opening in this sense that resonates with ecocentric articulations of the 'ecosystem approach'.

Principle 1 states that the objectives of management are a societal choice. This reflects the recognition that the understanding of ecosystems mirrors the particular economic and cultural needs of different social groups. Management operates as a mediating layer, where decision-making implies political negotiations, and, as importantly, 'ought to be participatory'. The rationale attached to Principle 1 emphasizes that 'indigenous peoples and other local communities living on the land are important stakeholders'. Yet, indigenous or local people

148 Decision VII/11, 2004, op. cit., Annex I, at 8, (Annotations to the rational of principle 1).
149 Grumbine, 'What is Ecosystem Management?'

are transformed into biopolitical subjectivities aligned with the double goal of nurturing and subjugating nature, and are transformed into 'environmental managers' that operate schemes such as the Payment for Ecosystem Services, fully integrated into biopolitical mechanisms of control and regularization of nature.[150] There is a second important element to this principle, relating to the political epistemology of nature, which indicates that ecosystems are not ontologically given, but are rather a social construction, or at least are co-produced by social and ecological elements.[151]

Principle 2 further specifies that the governance model should be decentralized to the lowest appropriate level. After participation then, the Malawi Principles include decentralization as a key governance parameter. Decentralization, explains the rationale of Principle 2, 'may lead to greater efficiency, effectiveness and equity'. This principle, in many ways, incorporates, within the framework of the 'ecosystem approach', the principle of subsidiarity. And while this principle has proved difficult to operationalize, it, importantly, carves space for a 'situated' approach that may open space for a shift from a 'law of the centre' to a 'law in the middle' (I will return on this in Chapter 10). Yet, decentralization, as it further specifies the participatory model outlined in Principle 1, may suffer from the same biopolitical closures in relation to the construction of biopolitical subjectivities that have been adumbrated above.

Principle 3 introduces an important ecological truism: ecosystems respect no legal or political boundary. Since management choices may have unpredictable effects on adjacent or other ecosystems, new institutional arrangements may need to be considered. This principle reflects a holistic orientation and acknowledges the transversal interconnections between ecosystems and/or landscapes. The rationale thus outlines the need for re-structuring the institutional arrangements involved in the management of particular ecosystems, in order to match the particular interconnections traversing nature. Indeed, it has long been recognized that State and territorial boundaries often hinder the effective protection of the environment, as they fragment and absorb in multiple jurisdictional spaces what, from an ecological perspective, are best considered as single zones or ecosystems. This was recognized at an early stage as one of the central challenges for the meaningful translation into law – and international environmental law more specifically – of the 'ecosystem approach'.[152] Its importance is further emphasized in the operational guidance annexed to the Malawi Principles, where guideline 5 suggests that the 'ecosystem approach' ought to be integrated across any 'production systems that have an effect on biodiversity',[153] and that

150 Kotsakis, 'Biological Diversity Complex', see more in general Chapter 7, 'Governing in the Biocomplex II: The Participation of Local and Indigenous Communities'.

151 Thus, for example K. Jax, 'On the Different Perceptions of the Ecosystem Concept within the Ecosystem Approach: Problems and Potentials' in Korn, Schliep and Stadler, *Report.*, p. 36.

152 See e.g. Brunnée and Toope, 'Environmental Security'.

153 Decision V/6, 2000, op. cit., para. 12.

inter-sectoral cooperation and communication should increase, taking on board a widespread observation that one of the central problems in any form of holistic ecological management is the lack of alignment between political and ecological boundaries, and the sectoral fragmentation of administrative institutions and competencies.

Principle 12, finally, recognizes the complex nature of ecological systems and reiterates the need for an inclusive and plural epistemological approach, and for participatory decision-making processes based on the principle of subsidiarity. It encourages transversal cooperation, across political and jurisdictional hierarchies, through the involvement of 'the necessary expertise and stakeholders at the local, national, regional and international level'.

Knowledge pluralism

A third theme relates to knowledge pluralism. Principle 11 states that 'the ecosystem approach should consider all forms of relevant information, including scientific and indigenous and local knowledge, innovations and practices'. This is in line with Article 8(j) of the CBD, which emphasizes the importance of traditional indigenous knowledge, for conservation and for the purposes of the equitable sharing of the benefits derived from the utilization of such knowledge.

Principle 11 may be understood as incorporating that epistemological pluralism that is one of the central insights of ecology. And indeed the CBD gives significant space to traditional ecological knowledge, through article 8(j), and through significant rhetorical efforts.[154] However it is often the case that local and indigenous knowledge cannot be brought to bear in the context of large-scale management plans, as it is spatially too local, or it does not fit within the formal structures of mathematical modelling.[155] Wiber offers an example of these issues in relation to Canadian fisheries management in the Bay of Fundy. He argues that 'management regimes increasingly rely on the spatial/temporal expertise emanating from highly technical "epistemic communities", thereby discounting practice-based knowledge',[156] that is, the knowledge of local and indigenous fishers. Natural resources regulation, Wiber argues, occurs in other words at spatial and temporal scales which exceed the capacity of local and traditional knowledges to be effective.[157] Similarly, the incorporation of local or traditional knowledge into large-scale management plans based on the

154 See e.g. 'Message from the Executive Secretary, Ahmed Djoghlaf, on the Occasion of the International Day of the World's Indigenous People' (2007).

155 See e.g. M. Wiber, 'The Spatial and Temporal Role of Law in Natural Resource Management: The Impact of State Regulation of Fishing Space' in F. von Benda - Beckmann and A. Griffths (eds), *Spatializing Law: An Anthropological Geography of Law and Society*, Farnham: Ashgate, 2009.

156 Wiber, 'Spatial and Temporal', p. 7.

157 Ibid.

'ecosystem approach' is difficult if not impossible.[158] Traditional and indigenous knowledge is, moreover, effectively incapacitated in a second way. Since this type of knowledge is not captured in a mathematical, quantitative and formalized manner but rather in qualitative and narrative form, it is as such difficult to fit within highly formalized mathematical models.[159]

Ecosystem services and natural capital

This final theme is increasingly central within the context of the CBD. The theme reflects a biopolitical template whose aim is that of fostering life in order to optimize its productive capacity. Indeed, conservation from this vantage point is seen as a necessary form of management that will ensure the continued (i.e. sustainable) delivery of all those goods and services beneficial to human beings. The converse of this is that, in principle, all those functions or services that are not beneficial to human beings, such that they (a) do not even indirectly fulfil a useful task or (b) are outright dangerous for human beings, may be done away with. From a biopolitical perspective this discernment between the positively, neutrally or negatively useful, operates according to a logic of racialization (as we have seen) – i.e. the decision about which species must live (or function, endure etc.) and which must die (or be done without). Indeed this is the logic underpinning the under-appreciated yet crucial distinction between ecosystem services and ecosystem 'disservices'.[160] Indeed, nature 'often harms and kills us'[161] (in the form of, for example, living beings such as insects, bacteria, or fungi, or of ecosystem processes such as floods, forest fires, hurricanes etc.).[162] This poses a dilemma insightfully described by Redford and Adams thus:

> [N]ot all ecosystem processes sustain and fulfill human life. Processes such as fire, drought, disease, or flood work against this goal, yet they are vital for ecosystem function, structuring landscapes, and providing vital services and regulatory functions to nonhumans. There is a danger that an economically driven

158 Ibid.

159 See e.g. F. Mazzocchi, 2006, 'Western Science and Traditional Knowledge: Despite Their Variations, Different Forms of Knowledge Can Learn from Each Other', *EMBO Reports* 7:5, 463, p. 464, who among other things highlights the problem of 'cognitive mining', a process of knowledge extraction which atomizes traditional and indigenous knowledge threatening its integrity and its very value (which lies in its specific character of being situated, rather than disembedded and de-contextualized). However, there are also examples of successful incorporation of traditional knowledge into modern conservation practices, though they remain the exception, ibid.

160 See e.g. R. Dunn, 'Global Mapping of Ecosystem Disservices: The Unspoken Reality that Nature Sometimes Kills us', 42:5 *Biotropica* 2010, 555.

161 Ibid., p. 555.

162 Ibid.; see also K. Redford and W. Adams, 'Payment for Ecosystem Services and the Challenge of Saving Nature', 23:4 *Conservation Biology* 2009, 785.

> focus on those 'services' that are valuable to humans in their nature, scope, and timing may lead to calls to 'regulate' ecosystem services to times and in flows that match human needs. Such regulation may be highly detrimental to long-term survival of the nonhuman parts of the ecosystems.[163]

It is then here that the immunitary paradigm that Esposito sees as the central mechanism underpinning modern biopolitics, and its inevitable thanatopolitical articulation, becomes most immediately operative.[164]

Yet, as the reader may recall, the consequences of the logic of racialization may be avoided through the valorization of otherwise useless functions or species, and the insertion of 'dangerous' and 'threatening' species into the capitalist processes of exploitation, paradoxically deploying the key insight of ecology (everything is connected to everything else) in order to further subjugate nature through its valorization, in what is perhaps a paradigmatic case of the shift from a sovereign to a biopolitical form of power (one which however simultaneously shows their complementarity). Furthermore, the deconstruction and reduction of nature into isolated (and isolable) goods and services also allows the further abstraction of 'biological diversity'.

Here the same logic of racialization may take a substitutive, calculating form, through market-based instruments that de-localize, abstract and dis-embed biodiversity so that it can be traded across locales through biopolitical policies such as the no-net-loss, which leads to a fungible conceptualization of biodiversity, entirely de-linked from its places, and re-organized at a functional and higher level, and finally re-distributed geographically in the most efficient (i.e. market-driven) manner. Indeed, this in many ways contradicts the central integrative orientation of the 'ecosystem approach' as it emerges from the discussion in the third section of this chapter, further signalling the internal tensions, contestations, instabilities and genealogical complexities at work, as well as the biopolitical resolution of these tensions and contestations. In relation to the latter, we can include Principle 5 within this theme, as it combines precisely the insights of ecology with the biopolitical horizon of regularization and optimization of nature, and Principle 4, which recognizes the tight connection between ecosystems and economy. However, biopolitical references to the productivity of nature are, implicitly or explicitly, at work throughout.

Principle 5 states that 'conservation of ecosystem structure and functioning, in order to maintain ecosystem services, should be a priority target of the ecosystem approach'. The deployment of ecology in a biopolitical direction is further evident from the rationale underpinning this principle: 'Ecosystem functioning and resilience', suggests the rationale, depends 'on a dynamic relationship within species, among species and between species and their abiotic environment, as well as the physical and chemical interactions within the

163 Ibid., p. 786.

164 For a detailed account of this logic in relation to invasive alien species, see De Lucia, 'Bare Nature'.

environment'. Moreover, the 'conservation and, where appropriate, restoration of these interactions and processes is of greater significance for the long-term maintenance of biological diversity than simply protecting species'.[165] This principle offers perhaps the clearest evidence that the 'ecosystem approach' is firmly situated within the general normative narrative of the CBD, and makes it very explicit that the conservation of ecosystems' 'structure and functioning' is instrumental to 'maintain[ing] ecosystem services'. Indeed, this is a 'priority target' of the 'ecosystem approach'. Principle 5 thus embodies exactly what Cary Wolfe describes, in Heideggerian terms, as the 'biopolitical enframing'[166] of nature, a central element of the ecological declension of biopower and biopolitics. Similarly, Principle 4 integrates ecosystems into the human economy,[167] in ways that substantiate Hardt and Negri's claim that biopower brings nature, and life in all its complex manifestations, *inside* the production process as something more than raw materials. Life, and nature, is, through biopower, made into an object of political and technological intervention. Nature thus becomes 'subject to capital'[168] and is entirely re-organized through and by economic discourse.[169] Hardt and Negri suggest that biopolitical capitalism abandons a discourse of exploitation and embraces one that translates 'biological and genetic diversity … into economic growth'.[170] Principle 4 speaks a purely economic language of markets, market incentives, distortions and failures, and of cost internalization. This economic framing however, is only conceivable within a larger context that re-conceptualizes nature as capital. Indeed, the concept of 'natural capital' is crucial to the biopolitical enframing of nature. But the biopolitical declension of ecology as the crucial regime of knowledge for the achievement of the goals of biopower is also at work in principle 6, which deploys the full language of ecology, and the concept of ecological limits, in order to maintain or enhance 'natural productivity'.

The narrative of ecosystem services: a biopolitical resolution?

Introduction

This section discusses the concept of ecosystem services and its increasing influence within the narrative and normative context of the CBD. Ecosystem services, and the related ideas of natural capital and valuation of nature, are

165 Decision V/6, 2000, op. cit., Annex, Section B, principle 5, rationale.

166 Wolfe, *Before the Law*, p. 3.

167 And, crucially, invokes crucial market economic principles: the reduction of market distortions; the re-orientation of market incentives; and the internalization of costs and benefits of biodiversity conservation, Decision V/6, 2000, op. cit., Annex, Section B, principle 4

168 Hardt and Negri, *Commonwealth*, p. 32.

169 Lemke, *Biopolitics*, pp. 69–70.

170 Ibid., p. 70.

arguably resolving the tension between the competing narratives that traverse the 'ecosystem approach' in a decisive biopolitical direction.

First discussed in the late 1970s,[171] the idea of the valuation of the services that nature provides to human beings was progressively refined in the late 1990s.[172] At the most elementary level, ecosystem services (which includes ecosystem goods)[173] are the benefits people obtain from ecosystems.[174] This conceptual framework developed especially within the context of that branch of economics called *ecological* economics.[175] Ecological economics aims at explicitly inscribing the economy within the wider ecology in an attempt at 'greening' the dominant neoclassical economic orthodoxy. The latter, in fact: (a) remains insensitive to the role of ecosystems and natural processes in economic activities, (b) dismisses the idea of limits to resource availability (limits that become a function of technological capabilities, and are subject to infinite substitution) and (c) fails to internalize the environmental costs determined by production processes, costs that consequently are not borne by the producer, but are imposed on society at large, thus creating market distortions and a less efficient allocation of resources.[176]

171 A very early exploration of the idea of measuring the monetary value of ecosystem services appeared in W. Westman, 'How Much Are Nature's Services Worth', 197 *Science* 1977, 960, as mentioned in M. Ntona and E. Morgera, 'Connecting the Dots between SDG 14 and the Other SDGs: The Value Added of the Ecosystem Services Concept and the Integration of Equity Through Marine Spatial Planning', SCELG Working Paper No 6 (2017) 7. For a more detailed history see, e.g., J. Ruhl, and J. Salzman, 'The Law and Policy Beginnings of Ecosystem Services', 22 *Journal of Land Use and Environmental Law* 2007, 157; as well as T. Kaime, 'Symposium Foreword: Framing the Law and Policy for Ecosystem Services', 2 *Transnational Environmental Law* 2013, 211.

172 See in particular R. Costanza et al., 'The Value of the World's Ecosystem Services and Natural Capital', 387 *Nature* 1997, 253.

173 Ibid.

174 Costanza et al., 'The Value', p. 253; UNEP, *Ecosystems and Human Well-being: A Framework for Assessment*, Island Press 2003, 53 (UNEP 2003); UNEP, *Ecosystems and Human Well-being: Current State and Trends*, Island Press 2005, 26:1 (UNEP 2005).

175 There are various currents within ecological economics, some more radical and oriented towards structural modifications of the economic system (see, e.g., H. Daly, *Economics, Ecology, Ethics: Essays Toward a Steady-State Economy*, New York: W.H. Freeman and Company, 1980; and J Martinez-Alier, *Ecological Economics: Energy, Environment and Society* Oxford and New York: Basil Blackwell, 1990), rather than towards what can be called ecosystem services and 'natural capitalism' (see, e.g., G. Daily, 'Introduction: What Are Ecosystem Services?' in G Daily (ed.), *Nature's Services: Societal Dependence on Natural Ecosystems*, Washington DC: Island Press, 2008; and P Hawken, A Lovins and H Lovins, *Natural Capitalism: Creating the Next Industrial Revolution*, Columbus, OH: Back Bay Books, 2008). I will mostly refer to the more mainstream current. For the trajectory followed by ecological economics from its beginnings in the early 1980s to its 'fall' once absorbed back into the mainstream economic framework of utility theory, M. Sagoff, 'The Rise and Fall of Ecological Economics a Cautionary Tale', *The Breakthrough Journal*, 2012, https://thebreakthrough.org/index.php/journal/past-issues/issue-2/the-rise-and-fall-of-ecological-economics (Accessed 26 September 2018).

176 See, e.g., Daly, *Economics, Ecology.*

Robert Costanza, one of the pioneers of ecological economics, is the main author of a seminal study that attempted to value the global flow of ecosystem goods and services.[177] Such services, the study argued, 'represent part of the total economic value of the planet'.[178] The study put the economic value of ecosystem services at US$16 – 54 trillion per year.[179] By comparison, the global gross national product, at the time of the study, was US$18 trillion. However, the value of ecosystem services remains invisible and largely outside the market, hence their deterioration cannot be factored in economic decisions. Gretchen Daily, another prominent ecological economist, described ecosystem services as those

> conditions and processes through which natural ecosystems, and the species that make them up, sustain and fulfil human life. They maintain biodiversity and the production of ecosystem goods, such as seafood, forage timber, biomass fuels, natural fiber, and many pharmaceuticals, industrial products, and their precursors.[180]

The key and intertwined ideas then are that, on the one hand, ecosystem services must be made visible to the market, and that can only happen if they receive appropriate dollar valuation. On the other hand, a transparent and efficient market becomes the key for the conservation of biodiversity.

The conceptual framework of ecosystem services was mainstreamed more recently, in two subsequent reports published in 2003[181] and 2005[182] under the auspices of the United Nations Environment Programme (UNEP), and together forming the Millennium Ecosystem Assessment (MA) series. Before exploring the concept of ecosystem services in some detail, it is important to show their overall conceptual orientation articulated in the MA. While it is 'recognized' in the preamble of the CBD that biodiversity and ecosystems (included indeed under the definition of biodiversity) 'have intrinsic value', the conceptual framework presented in the MA 'places human well-being as the central focus for assessment'.[183] Indeed, both reports are titled 'Ecosystems and Human Well-being', thus immediately establishing a key functional linkage between the framework of ecosystem services and the concern with the well-being of humans.

177 Ibid.
178 Ibid.
179 Costanza and his team published an updated valuation in 2014, where the value of ecosystem services was estimated at between US$125 and US$145 trillion; see R Costanza et al., 'Changes in the Global Value of Ecosystem Services', 26 *Global Environmental Change*, 2014, 152.
180 Daily, *Economics, Ecology*, p. 3
181 UNEP, 2003, op. cit.
182 UNEP, 2005, op. cit.
183 Ibid., p. 28.

The MA series takes as its starting point the ecological economics literature,[184] and emphasizes the fact that '[h]uman demands for ecosystem services are growing rapidly'. Simultaneously, it observes how 'humans are altering the capability of ecosystems to continue to provide many of these services'. The solution lies, according to the MA, in ensuring the '[m]anagement of this relationship', so as to 'enhance the contribution of ecosystems to human well-being without affecting their long-term capacity to provide services'.[185] Yet while ecosystem services underpin and sustain a significant part of the global economy,[186] their importance, value and 'pivotal role' is under-recognized (at best).[187] This, according to the MA, is a crucial element in the continuing deterioration of the capacity of ecosystems to deliver their services. This view is echoed in other contexts. The Global Biodiversity Outlook 3 identifies for example some of the central causes for the failure to achieve biodiversity conservation targets in the 'insufficient integration of biodiversity issues into broader policies, strategies and programmes'.[188] In this respect, '[t]he real benefits of biodiversity, and the costs of its loss, need to be reflected within economic systems and markets'.[189]

In the MA framework, ecosystem services are organized into four categories: provisioning services; regulating services; cultural services; and supporting services. These categories, however, overlap significantly and, as the MA underlines, 'the purpose is not to establish a taxonomy but rather to ensure that the analysis addresses the entire range of services'.[190] *Provisioning services* supply goods of *direct* benefit to people. These goods are usually already commodities with a clear monetary value (timber, medicinal plants, wild fish, etc.). *Regulating services* are all those regulatory functions performed by ecosystems (e.g. climate regulation, water and air filtering, protection against landslides, pollution removal etc.). These functions, unlike provisioning services, generally lack a monetary value in conventional markets. *Cultural services* usually do not provide direct material benefits and are not exchanged in formal markets, and need to be estimated by so-called contingent valuation methods. In other words, their value can be measured only indirectly (rather than through direct market transactions).[191] These services include spiritual and sacred places, landscapes of aesthetic value, and tourist attractions. Finally, *supporting services* operate 'underneath' all other ecosystem services and enable them to deliver their direct or indirect benefits to human beings. These include all large-scale (in both

184 This is made explicit in UNEP 2003, op. cit., pp. 55–6.

185 Ibid., p. 27.

186 'The production and manufacture of industrial wood products in the early 1990s contributed on the order of $400 billion to the global economy. The world's fisheries contributed $55 billion in export value in 2000' (references omitted); UNEP, 2003, op. cit., p. 27.

187 Ibid., p.28.

188 Secretariat of the Convention on Biological Diversity, 2010, op. cit., p. 9.

189 Ibid., p. 12.

190 UNEP, 2003, op. cit., p. 38.

191 See, e.g., R Costanza et al., *An Introduction to Ecological Economics*, Boca Raton, FL: St Lucie Press, 2010, especially Chapter 3.

temporal and spatial terms) biogeochemical processes supporting life, plant growth, soil formation, etc.[192]

Even though there remain many inconsistencies, uncertainties and even contestations as to the appropriate delineation of the concept of ecosystem services,[193] the primary and dominating orientation is arguably linked to benefits humans can extract from nature, regardless of the typology of benefits.[194] The concept of ecosystem services, in other words, while it can be considered an 'organising principle [that] consider multi-scale and cross-sectoral synergies and tradeoffs',[195] remains primarily a tool aimed at a particular calculating, disaggregating and abstracting perspective, where natural ecosystems are deconstructed into a set of properties or streams of services that can be separately addressed, organized and reorganized in order to optimize and enhance their productivity and foster the positive streams (rather than the negative ones, also called ecosystem disservices).[196]

Ecosystem services and the 'ecosystem approach'

The relevance of the conceptual framework of ecosystem services for the further development of the 'ecosystem approach' within the context of the CBD is immediately clear. Decision VIII/9, entitled 'Implications of the findings of the Millennium Ecosystem Assessment', requested SBSTTA and invited Parties, 'to make use as appropriate of its [MA's] conceptual framework and methodologies in further developing work on … the ecosystem approach'.[197] Moreover, the pathfinder workshops mentioned above recognized that a central impediment to the full implementation of the 'ecosystem approach' was the '[l]ack of scientific assessment and quantification of the services provided by most ecosystems', and '[t]he absence of generally agreed mechanisms and procedures for the valuation of ecosystem services'.[198] Similarly, the then Norwegian Minister of

192 The description of these ecosystem services is taken from UNEP, 2003, op. cit.

193 Ntona and Morgera, 'Connecting the Dots', p. 9. For an overview of the criticisms and contestations over the concept see M. Schröter et al., 'Ecosystem Services as a Contested Concept: A Synthesis of Critique and Counter-Arguments', 7 *Conservation Letters* 2014, 514.

194 A full assessment of the concept of ecosystem services is outside the scope of this article. Some commentators are hopeful in relation to their potential role (e.g. Ntona and Morgera, 'Connecting the Dots'), while others are decisively critical (e.g., S. Sullivan, 'Banking Nature? The Spectacular Financialisation of Environmental Conservation', 45 *Antipode* 2013, 198; S. Sullivan, 'Green Capitalism, and the Cultural Poverty of Constructing Nature as Service Provider', 3 *Radical Anthropology* 2009, 18.

195 M. van den Belt et al., 'Scientific Understanding of Ecosystem Services', in United Nations, *The First Global Integrated Marine Assessment: World Ocean Assessment 1*, Cambridge: Cambridge University Press, 2016, p. 6.

196 On ecosystem disservices see, e.g., R. Dunn, 'Global Mapping of Ecosystem Disservices: The Unspoken Reality that Nature Sometimes Kills Us', 42 *Biotropica* 2010, 555.

197 Decision VIII/9, 'Implications of the findings of the Millennium Ecosystem Assessment', COP8, Curitiba, 20–31 March 2006.

198 Smith and Maltby, *Using the Ecosystem Approach*, p. 26.

International Development and Human Rights, when summarizing the 1999 Norway/UN Conference on the 'ecosystem approach' for Sustainable Use of Biological Diversity, observed that '[t]o make rational choices, we need to know both which ecosystem goods and services are provided by the environment and what those services are worth to us'.[199]

The relation between the framework of ecosystem services, and the 'ecosystem approach' has also been explored by scholarship, and from multiple perspectives. Morgera and Tsioumani, for example, have suggested that while 'the concept of ecosystem services as developed within the CBD framework has attempted to reconcile a rights-based and an economic approach to biodiversity policy, tensions persist'.[200] They note that the concept of ecosystem services has increased the focus on the 'ecosystem approach'.[201] However, they also recognize how the narrative of ecosystem services emphasizes economic valuation 'as a key instrument for mainstreaming biodiversity … and for tackling effectively all the drivers of biodiversity loss'.[202] Platjouw offers a similar assessment of the potential benefits and tensions inherent in utilizing the concept of ecosystem services in order to 'facilitate the application of the ecosystem approach'.[203]

In a recent article, McIntyre suggests that the general adoption of a 'meaningful ecosystem approach' would have 'truly far-reaching' implications.[204] However, even as he imagines the possible dissolution of international water law (the thematic context of McIntyre's article) as a discrete body of rules, and its subsumption 'into a broader corpus of international ecosystems law', he frames such possibility in terms of its capacity to 'facilitate the integrated sustainable management' of the relevant ecosystem (corresponding to, or even exceeding, the drainage basin).[205] McIntyre further acknowledges that the 'parameters and practical implications of a meaningful ecosystem approach to the protection of [international] water resources are only now becoming clearer'. However, he observes, 'the 2005 Millennium Ecosystems Assessment now provides a detailed elaboration of the concept of "ecosystem

199 Minister of International Development and Human Rights Hilde F. Johnson, Norway/UN Conference on the Ecosystem Approach for Sustainable Use of Biological Diversity Trondheim, 6–10 September 1999, https://www.regjeringen.no/nb/aktuelt/norway-un_conference_on_the_ecosystem/id263361/ (Accessed 26 September 2018).

200 Morgera and Tsioumani, 'Yesterday, Today, and Tomorrow: Looking Afresh at the Convention on Biological Diversity', *Yearbook of International Environmental Law*, 21:1, 2011, 3.

201 See e.g. Decision X/4, 'The Third Edition of the Global Biodiversity Outlook: Implications for the Future Implementation of the Convention', COP10, Nagoya, Aichi Prefecture, 18–29 October 2010, and especially para. 5(d) and 5(f).

202 Morgera and Tsioumani, 'Yesterday', p. 13.

203 F. Platjouw, *Environmental Law and the Ecosystem Approach: Maintaining Ecological Integrity through Consistency in Law*, London: Routledge, 2016, p. 79, but see the entire Chapter 4.

204 McIntyre, 'Ramsar Convention', p. 88

205 Ibid., p. 88.

services"',[206] a framework that, he further suggests, is '[c]losely linked to the continuing evolution of the ecosystem approach'.[207] This clearly resonates with the official position of the CBD.[208]

There is, then, a widespread perception that the two conceptual frameworks of 'ecosystem approach' and ecosystem services, unified by the common reference to the notion of ecosystem, are mutually supportive and synergic.[209] This synergy however, from the point of view of the analysis carried out in this book, arguably embodies a biopolitical closure of the 'ecosystem approach'.

However, there are also different assessments of the relationship between ecosystem services and the 'ecosystem approach'. John Ruhl offers one such example. Ruhl considers that

> [e]cosystem management and ecosystem services management share many methodological traits. Both take a multiscalar, cross-boundary perspective. Both will demand robust data, performance monitoring, and adaptive decision-making capacity. Both depend on managing ecosystem integrity to sustain their goals.[210]

At the same time though, he also underlines a key difference. It is at the level of *goals*, Ruhl observes, that the two frameworks differ significantly, and that tensions emerge. Ecosystem management in fact takes, according to Ruhl, a 'biocentric ... perspective'.[211] Ecosystem services on the other hand take a decisive anthropocentric perspective.[212] And indeed, part of the initial and growing interest in ecosystem services is according to some commentators linked to the 'search for new revenue streams for landowners to support sustainable practices'.[213]

206 McIntyre, 'Protection', p. 88.
207 Ibid., p. 92.
208 See e.g. Decision VII/11, 2004, op. cit., para. 6, where the COP '*Notes* the relevance of the conceptual framework of the Millennium Ecosystem Assessment in supporting the implementation of the ecosystem approach'.
209 See, again, McIntyre, 'Protection', esp. p. 89, where he links the Guidelines on the 'ecosystem approach' in Water Management *and* the 2007 Recommendations on Payments for Ecosystems Services in Integrated Water Resources Management but also others. See also S. Brels, D. Coates and F. Loures, *Transboundary Water Resources Management: The Role of International Watercourse Agreements in Implementation of the CBD*, CBD Technical Series No. 40, Montréal: Secretariat of the Convention on Biological Diversity, 2008.
210 Ruhl, 'Ecosystem Services', p. 116.
211 Ibid. It must however be mentioned that Ruhl deploys Grumbine's framework as the reference point against which he carries out his analysis, itself only one of many articulations of the 'ecosystem approach'.
212 Ibid., p. 116.
213 L. Scarlett and J. Boyd, 'Landscape-wide Conservation and Ecosystem Services. Leveraging Federal Policies' in K. Robbins (ed.), *The Laws of Nature: Reflections on the Evolution of Ecosystem Management Law and Policy*, Akron, OH: University of Akron Press, 2013, p. 218.

Unlike McIntyre, and more clearly than Morgera and Tsioumani or Platjouw, Ruhl thus emphasizes the different narratives within which the 'ecosystem approach' and ecosystem services operate. Ruhl's assessment however, doesn't go far enough. In fact, Ruhl's analysis remains firmly aligned with the binary framework that I have shown to be inadequate and insufficient to account for the genealogical complexities of the 'ecosystem approach', and its biopolitical orientation.

This biopolitical orientation of the 'ecosystem approach' is to a significant extent contingent on the fact that the framework of ecosystem services increasingly provides the overarching narrative within which the 'ecosystem approach' is articulated. The combined logic of ecosystem services and of the 'ecosystem approach' in fact ultimately reflects the biopolitical *dispositif* that simultaneously aims at, and circularly oscillates between, the conservation and subjugation of nature. It is useful in this respect to recall Malawi Principle 5, which is key to the interpretation of the relationship between the 'ecosystem approach' and ecosystem services. Principle 5 states that 'conservation of ecosystem structure and functioning, in order to maintain ecosystem services, should be a priority target of the ecosystem approach'. It is evident here that the ecological framework informing the 'ecosystem approach' is to be put to use in order to ensure, as a priority, the maintenance of ecosystem services, which are the key metric for the measurement of the usefulness of biodiversity, and the central framework for the achievement of human well-being. This orientation becomes even clearer with the proposal for refinement of the Malawi Principles and for the further elaboration of 'ecosystem approach',[214] which organizes the original Malawi Principles under seven different headings, one being 'provision of environmental goods and services'.[215]

This also brings Principle 4 into the equation as there is a further linkage emerging between the 'ecosystem approach' and PES. While McIntyre has traced the connection between the two to the extent that PES is an instrument aimed at operationalizing *both* Malawi Principle 4 *and* the conceptual framework of ecosystem services,[216] there is also a deeper linkage. Deborah McGrath and Travis Greenwalt, for example, describe PES (and the underlying valuation process that underpins it and indeed makes PES possible) as a 'tool to improve ecosystem management'.[217] The authors build their argument on the fact that a significant number of services provided by ecosystems are not captured in financial markets and are thus invisible, taken for granted and under-protected.[218] These invisible services, which, the authors note, tend to underlie and support the direct provisioning services, accrue

214 Decision VII/11, 2004, op. cit., Annex I: 'Refinement and Elaboration of the Ecosystem Approach, Based on Assessment of Experience of Parties in Implementation'.

215 Ibid.

216 McIntyre, 'Protection'.

217 D. McGrath and T. Greenwalt, 'Valuation and Payment for Ecosystem Services as Tools to Improve Ecosystem Management' in Robbins, *Laws of Nature*.

218 Ibid. These services are also 'undervalued' (p. 284) and 'underpriced' (p. 284).

'directly to humans without passing through a market exchange'.[219] The strategy they outline is premised on two steps. First, ecosystem services must be mapped and valued; second, markets or market incentives must be created in order to protect these services. A full discussion of McGrath and Greenwalt's paper is not necessary. It is, however, important to underline that the framework of ecosystem services carves out significant space for the further commodification of nature, under the claim that without markets biodiversity and ecosystem services will further deteriorate. This orientation has been fully taken on board by the CBD, and will be further discussed in the section dedicated to TEEB.

Another example taken from the context of the Ramsar Convention, highlights the increasing entanglement of ecosystem services and the 'ecosystem approach' within the broader biodiversity discourse, and further illustrates how the narrative of ecosystem services is imposing its biopolitical closure by transforming the 'ecosystem approach' into an instrument for the achievement of its goals (this risk had already been flagged by Anne Bell in 1994[220] and in 2000[221] in relation to the deployment of the 'ecosystem approach' in the context of the 1978 Great Lakes Water Quality Agreement). The Ramsar Convention has updated its vocabulary (and the concept of wise use) in relation to the emerging narrative of ecosystem services, again integrating the 'ecosystem approach' and ecosystem services in a unitary framework where wise use must be understood within the context of sustainable development.[222] Finlayson et al. also comment on how, in the Ramsar Convention, discourses and narratives are intertwined. They observe that the Ramsar Convention has since the beginning 'fully recognised and addressed the importance of wetlands to people and biodiversity and that maintaining the ecological character of wetlands through an 'ecosystem approach' is critical to the continued provision of ecosystem services'.[223]

The institutional dimension

One final aspect to discuss relates to the institutional dimension, and the central role of institutions to the biopolitical articulation of the 'ecosystem approach'. The CBD has, over time, set up comprehensive institutional arrangements, comprising law-making institutions and scientific and technical advice bodies. From

219 Ibid., p. 283.

220 Bell, 'Non Human Nature'.

221 A. Bell, 'Frog Reflections: The Ecosystem Approach to Conservation' in G. Beck and B. Littlejohn (eds), *Voices from the Watershed: Environmental Issues in the Great Lakes-St Lawrence Drainage Basin*, Montreal: McGill-Queen's University Press, 2000.

222 Ramsar Secretariat, 2010, op. cit., p. 27 – 8 and Resolution IX/1, 'A Conceptual Framework for the wise use of wetlands and the maintenance of their ecological character', adopted at 9th Meeting of the Conference of the Parties to the Convention on Wetlands, Kampala, Uganda, 8 – 15 November 2005, Annex A, throughout and para. 24.

223 Finlayson et al., 'Ramsar Convention', p. 193.

this point of view, the objectives of the CBD are to be achieved through a comprehensive biopolitical apparatus that corresponds to one of the *de minimis* elements of biopower identified by Rabinow and Rose, and more specifically a 'truth discourse' (or several) and an 'array of authorities considered competent to speak that truth'.[224] Such biopolitical apparatus takes control over every living and non-living form on Earth through the application of rational, scientific, technical expertise enacted and legitimated through political power.

The Conference of the Parties to the CBD, that is the central governing institution of the CBD, guides, directs and enacts, in policy and legal terms, the biopolitical strategies aimed at the optimization of life[225] that are identified by three central technical and scientific bodies (one, the first, internal and two of formally external to the CBD) namely SBSTTA, the Intergovernmental Platform on Biodiversity and Ecosystem Services (IPBES) and the Economics of Ecosystems and Biodiversity (TEEB).[226] The next four sections will discuss each in turn.

Conference of the Parties

A Conference of the Parties (COP) is the decision-making organ of modern Multilateral Environmental Agreements (MEAs), tasked generally with the implementation of their provisions,[227] as well as with the coordination of the activities of other bodies and institutions (usually scientific, technical or financial) created pursuant to, and in support of, the objective of the particular agreement whose implementation the COP supervises and guides. The role of the COP however is not univocal and in large measure depends on the particular substantive and procedural rules agreed within the context of each individual MEA. Indeed, it has been noted that this role may range 'from the adoption of texts that are subsequently ratified by MEA parties to what appear to be more autonomous forms of law-making'.[228]

Within the context of the CBD, particularly considering that it operates largely as a framework Convention, the COP takes a role of active decision making.[229] A detailed discussion of the role of the COP, which is regulated by

224 Rabinow and Rose, 'Biopower Today', p. 197.

225 Lascoumes, *L'éco-pouvoir*; Whiteside, *Divided Natures*.

226 I will not discuss the role of the Secretariat, nor of the formal working groups established under the CBD, as tangential to the central objective of this section, albeit they could be usefully integrated in the institutional set-up as further contribution to consolidation of the operative biopolitical apparatus.

227 MEAs consists in general of two stages: the diplomatic conferences aimed at the adoption of the formal text of the agreement; and the implementation stage, which is where the COP enters into stage, J. Werksman, 'The Conference of Parties to Environmental Treaties' in J. Werksman (ed.), *Greening International Institutions*, London: Earthscan, 1996, p. 57.

228 J. Brunnée, 'COPing with Consent: Law Making Under Multilateral Environmental Agreements', 15:1 *Leiden Journal of International Law* 2002, 5.

229 A. Johannsdottir, 'Convention', p. 85.

Article 23 of the CBD, is not important for our purposes.[230] However, the important role the COP has played in relation to the introduction, development and articulation of the 'ecosystem approach' must be emphasized. The COP has led the coordination of scientific and institutional activities leading to its introduction, development, review and refinement. It has also formally endorsed the 'ecosystem approach' as the central strategy for the achievement of the objectives set out in the CBD. The COP is in this respect central to the enactment of biopolitical strategies identified by the technical and scientific bodies, and sanctions their truth discourse by transforming it into the formally authoritative language of law. The COP in this respect is an excellent example of a biopolitical institution that however still harnesses legitimacy through the sovereign encoding (i.e. the authoritative language of law) of its activities.[231]

Subsidiary Body on Scientific, Technical and Technological Advice

The Subsidiary Body on Scientific, Technical and Technological Advice (SBSTTA) was established as a permanent body by the CBD (Article 25) with the aim of providing the COP with 'timely advice' in relation to the implementation of the Convention. More specifically, SBSTTA is to '[p]rovide scientific and technical assessments of the status of biological diversity'; '[p]repare scientific and technical assessments' in order to optimize the effects of measures taken pursuant to the CBD; '[i]dentify innovative, efficient and state-of-the-art technologies and know-how' relevant for the conservation and sustainable use of biodiversity; offer advice in relation to the development and/or transfer of relevant technologies; '[p]rovide advice on scientific programmes and international cooperation in research and development'; finally, 'respond to scientific, technical, technological and methodological questions' that the COP may have. In general SBSTTA carries out its 'mandate under the authority of, and in accordance with, guidance laid down by the Conference of the Parties, and upon its request'.[232]

Decision VIII/10 endorsed a consolidated *modus operandi* of SBSTTA,[233] that is a set of more detailed rules of procedure, pursuant to Article 25(3) of the CBD, which establishes that '[t]he functions, terms of reference, organization and operation of this body may be further elaborated by the Conference of the Parties'. Moreover, the decision (among other things) adds a new function to the ones listed in Article 25, namely the identification of 'new and emerging

230 And I will refer to Article 23 itself, and to existing literature, such as, e.g. Johannsdottir, 'Convention'.

231 Regardless of the formal status of COP decisions, that can be debated, the encoding is thoroughly legal.

232 CBD 'Modus Operandi of the Subsidiary Body on Scientific, Technical and Technological Advice' (2–6 September 1996) (SBSTTA 2nd Meeting) UN Doc UNEP/CBD/SBSTTA/2/16', p. 1.

233 Ibid.

issues relating to the conservation and sustainable use of biodiversity' (Appendix A),[234] and lays out the process for the conduct of scientific and technical assessments that are initiated by SBSTTA.[235] As we have seen, SBSTTA has played a central role in relation to the development of the 'ecosystem approach' within the context of the CBD, in particular through the organization of expert meetings and workshops, and the presentation of recommendation to the COP.

Intergovernmental Platform on Biodiversity and Ecosystem Services

The Intergovernmental Platform on Biodiversity and Ecosystem Services (IPBES) was formally established in Panama City on 21 April 2012,[236] pursuant to UNGA Resolution 65/162,[237] as an independent intergovernmental platform.[238] The establishment ended a specialized intergovernmental process, which culminated in the adoption of the so-called Busan Outcome during the third ad hoc intergovernmental and multi-stakeholder meeting on an intergovernmental science-policy platform on biodiversity and ecosystem services.[239] While IPBES is not a body constituted under the CBD or the COP,[240] and it is, as mentioned, independent, IPBES's role is of central relevance for the implementation of the CBD. In fact, the primary task of IPBES is 'to strengthen the science-policy interface for biodiversity and ecosystem services for the conservation and sustainable use of biodiversity, long-term human well-being and sustainable development'.[241] In particular, IPBES 'should respond to requests from Governments, including those conveyed to it by multilateral environmental agreements related to biodiversity and ecosystem services'.[242]

Thus IPBES responds to the need of decision-makers for 'scientifically credible and independent information that takes into account the complex relationships

234 Decision VIII/10, 2006, op. cit., Appendix A, para. d.

235 Ibid., Appendix C.

236 Resolution 'Intergovernmental Science-Policy Platform on Biodiversity and Ecosystem Services' in Report of the second session of the plenary meeting to determine modalities and institutional arrangements for an intergovernmental science-policy platform on biodiversity and ecosystem services, 18 May 2012, Annex I.

237 Resolution 65/162, Report of the Governing Council of the United Nations Environment Programme, Governing Council 11, 11 March 2011, para. 17.

238 UNEP Decision 26/4, 'Intergovernmental science-policy platform on biodiversity and ecosystem services', United Nations Environment Programme Governing Council/Global Ministerial Environment Forum 25, Nairobi 2011d, para 1.

239 Busan Outcome, Report of the Third ad hoc intergovernmental and multi-stakeholder meeting on an intergovernmental science-policy platform on biodiversity and ecosystem services, 2010, Annex.

240 Decision XII/25, "Intergovernmental Science-Policy Platform on Biodiversity and Ecosystem Services", COP12, Pyeongchang, 6–17 October 2014.

241 Busan Outcome, 2010, op.cit., para. 6; see also UNEP Decision 26/4, 2011d, op. cit.

242 Busan Outcome, 2010, op.cit., para. 6(a).

between biodiversity, ecosystems services and people',[243] as well as their need for 'effective methods to interpret this scientific information in order to make informed decisions'.[244] This is a pervasive, inevitable problem in relation to any environmental question, which is generally dependent on science for its operationalization into rules, principles, thresholds etc. In this regards McIntyre notes that

> the technical complexity of the issues underlying international environmental law dictates that scientific expertise plays an absolutely central role in its implementation and requires intense interaction between environmental scientists and environmental lawyers at every stage in its development and application.[245]

However, from a Foucauldian point of view, this technical complexity operates to establish particular configurations of power and knowledge, that is, a particular mode and institutionalization of power (a biopower in this case) that is sustained through a particular regime of truth.

In this respect, the main functions of IPBES are knowledge generation, assessments, policy support (through the preparation of tools and methodologies) and capacity-building.[246] One of the objectives under IPBES' work programme relates to the preparation of global and regional assessments on biodiversity and ecosystem services. Such assessments, as requested by the COP, provide an evaluation of 'the status and trends' with regard to such services, as well as 'the impact of biodiversity and ecosystem services on human well-being and the effectiveness of responses, including the Strategic Plan and its Aichi Biodiversity Targets'.[247]

Given the centrality of the framework of ecosystem services for the articulation of the 'ecosystem approach' and, especially, for its biopolitical closure, IPBES, with its mandate,[248] plays an increasingly decisive role in the subsumption of life (i.e. biodiversity) under the biopolitical matrix of optimization, control and subjugation, keeping in mind the aporetic lacerations that biopolitics always contains and enacts, lacerations which, transforming every biopolitical gesture into thanatopolitical effects, capture the central contradictions, the central aporetic core of the Anthropocene.

243 http://www.ipbes.net/index.php/about-ipbes.html (Accessed 26 September 2018).

244 Ibid.

245 McIntyre, 'Protection', p. 90.

246 Busan Outcome, 2010, op. cit.

247 Decision IPBES-2/5, 'Work programme for the period 2014–2018', IPBES-2, Antalya, 9 – 14 December 2013, para 9(b); The request of the COP is in Decision XI/2, 2012, op. cit., Section E, para 28.

248 For a detailed list of IPBES objectives and the associated deliverables, see COP 12, Item 30 provisional agenda, Summary Report on Progress in Implementing the Work Programme of the Intergovernmental Science-Policy Platform on Biodiversity and Ecosystem Services, Twelfth Meeting, Pyeongchang, 6–7 October 2014.

The Economics of Ecosystems and Biodiversity

The Economics of Ecosystems and Biodiversity is another central institution in relation to the 'ecosystem approach', particularly in its increasingly entwined relationship with the framework of ecosystem services. If the MA has made a 'path-breaking contribution' in relation to both the conceptualization and popularization of the concept of ecosystem services, 'the genuinely economic aspects of [ecosystem services] were subsequently carried forward by the more recent studies on the Economics of Ecosystems and Biodiversity (TEEB)'.[249] TEEB's mandate, very simply, is to value nature. TEEB's website welcomes visitors with the heading 'making nature's value visible'.[250]

TEEB is a process and a programme that was initiated in 2007 during the G8 + 5 summit held in Potsdam, and in particular by the Environment Ministers Meeting. The outcome was the Potsdam Initiative. 'Inspired … by the momentum … created by the Stern Review of the Economics of Climate Change',[251] the Potsdam Initiative set out to carry out a global study aimed at 'analysing the global economic benefit of biological diversity, the costs of the loss of biodiversity and the failure to take protective measures versus the costs of effective conservation'.[252] The programme, in other words, aims to end the 'struggle to find the value of nature'.[253] Here lies the key of the entire project: '[n]ature is the source of much value to us every day, and yet it mostly bypasses markets, escapes pricing and defies valuation. This lack of valuation is, we are discovering, an underlying cause for the observed degradation of ecosystems and the loss of biodiversity'.[254]

The first output of the programme was an interim report published under the auspices of the European Commission, DG Environment and of the German Federal Ministry of Environment.[255] After this interim report, a new more comprehensive study was carried out and published by TEEB, which outlines the economic and ecological foundations of TEEB.[256] In the introduction, the study taps into a long tradition of scholarship that identifies the cause of

249 Statement by Mr. Braulio F. De Souza Dias, Executive Secretary Convention on Biological Diversity to the International Seminar 'Towards Linking Ecosystems and Ecosystem Services to Economic and Human Activity', New York, 27 – 29 November 2012, p. 2.

250 The Economics of Ecosystems & Biodiversity, HTTP: http://www.teebweb.org (Accessed on 26 September 2018).

251 TEEB, The Economics of Ecosystems and Biodiversity: An Interim Report, European Communities, 2008, p. 4.

252 Potsdam Initiative – Biological Diversity 2010, Environment Ministers Meeting, Potsdam, 15–17 March 2007, http://www.g-8.de/Content/EN/__Anlagen/2007-03-18-potsdamer-erklaerung-en,property=publicationFile.pdf (Accessed 26 September 2018)

253 TEEB, 2008, op. cit., p. 4.

254 Ibid.

255 Ibid.

256 P. Kumar (ed.), TEEB, *The Economics of Ecosystems and Biodiversity Ecological and Economic Foundations*, London and Washington: Earthscan, 2010.

environmental problems in the so-called tragedy of the commons.[257] A long quote is worth reproducing here, as it gives a very clear picture of the underlying orientation of TEEB:

> Fundamental changes are needed in the way biodiversity, ecosystems and their services are viewed and valued by society. A major difficulty is that many ecosystem services are (mixed) public goods, and use levels are therefore difficult to regulate, even when they are at or near the point of exhaustion. Although many people benefit from ecosystem services, individuals or groups usually have insufficient incentives to maintain ecosystems for continued provisioning of services. For example, open access fisheries provide valuable harvests but often suffer from over-exploitation that leads to declines in fish populations and lowered future harvests.[258]

This is essentially the same conceptual approach that we saw in McGrath and Greenwalt, and that finds its original articulation in Garret Hardin's 1968 essay.[259] The central argument is that in a regime of open access, people will not have conservation incentives, and ecosystem services will continue to deteriorate. TEEB maintains that '[w]ithout changes in institutions and incentives, further declines in natural capital are likely',[260] or indeed, inevitable. TEEB's diagnosis emphasizes that lacking market signals, 'those who gain from actions that deplete natural capital will continue to avoid paying the full costs of their actions'.[261] TEEB's aim is to provide knowledge and data, with the emphasis falling on 'the economic, notably monetary, effects of the loss of ecosystem services'.[262]

Interaction between TEEB and IPBES

While there has not been, to date, any formal interaction between the two programmes, it is useful to frame the ways in which the two institutions indirectly contribute to the same biopolitical organization and conservation of nature. While TEEB elaborates the economics of biodiversity conservation in terms of how ecology and economy relate and interact (thus fully aligned to the ecological economics re-orientation of economics that is also embraced by the MA), IPBES has the double role of producing science and facilitating the interface between science and policy, that is, facilitating the legal articulation of scientific knowledge. Their significance however is best appreciated together. Together they function as central biopolitical institutions and they enact

257 See the seminal paper of Garret Hardin, 'The Tragedy of the Commons', 162:3859 *Science* 1968, 1243.
258 TEEB, 2010, op. cit., p. 4, footnotes omitted.
259 Which however was focused on the problem of human population, see Hardin, 'Tragedy'.
260 TEEB, 2010, op. cit. p. 5.
261 Ibid.
262 Ibid.

a biopolitical programme that first seeks knowledge in order to regularize nature's service provisions, and then (or simultaneously) seeks to assess the value of each service in order to enact the efficient optimization of conservation measures in relation to their highest value (or, in economic terms, return on investment). Together, they arguably contribute to the construction of a specifically biopolitical regime of truth which the CBD, especially through its COP, articulates and consolidates in policy and legal language.

Conclusions

This chapter has explored the 'ecosystem approach' within the context of the CBD. However, this case study is of extreme importance for the articulation of the 'ecosystem approach' more generally in international law for at least three interrelated reasons. The first is the comprehensive work done within the CBD to develop and refine the 'ecosystem approach'; the second is the increasing influence that this work has on the development of the 'ecosystem approach' in other regimes (both within and without the larger biodiversity cluster); and the third is the central role that the framework of ecosystem services has acquired, within the CBD, in relation to the 'ecosystem approach', in a way that anticipates the centrality of ecosystem services in other regimes.

Unlike the case of other, sector articulations of the 'ecosystem approach', such as the case of fisheries, where there is an explicit anthropocentric orientation, in the CBD the competing narratives and complex genealogies of the 'ecosystem approach' are more visible and the tensions more palpable. Yet, the distinction between an ecocentric and an anthropocentric articulation or orientation of the 'ecosystem approach' is not decisive. Indeed, *both* can be folded within a biopolitical horizon. As the reader may recall, the ecological extension of biopower comprises two complementary, adjacent, successive modes of biopolitical intervention which, first, focus on the nutritional, health and psychological well-being of a human population, as it depends on its environment, and then focus on natural entities and populations – ecosystems and biodiversity – and on the direct protection, conservation and regularization of *their* life. The focus on the protection of structure, function, health and integrity of ecosystems falls squarely within this second perspective, and can be read, at least to an extent, together with the preambular recognition of the intrinsic value of nature.

However, and here we could see in practice a point articulated theoretically earlier, rather than offering an ecocentric turn, a radical challenge to existing modes of law and governance, this second mode enacts the extension of the insertion of nature within the biopolitical framework of control. This is exactly the centre of ecopower: the insertion of every living and non-living form on Earth to extended surveillance and human control.[263] This system of surveillance, already discussed in the previous chapter, is equally central in the context

263 Lascoumes, *L'éco-pouvoir*.; Whiteside, *Divided Nation*, p. 143.

of the CBD. We have seen how one of the decisive impediments to the successful conservation and sustainable use of biodiversity rests in fact on the so-called 'taxonomic impediment', discussed in the second section of this chapter. Programmes such as TEEB and IPBES are equally oriented to the subsumption of life under various grids of control that require the careful and comprehensive subjection of nature to various systems of surveillance – economic, ecological, regulatory. What the analysis shows then, is that the 'ecosystem approach', while inserted visibly within competing narratives and genealogical complexities – what we have identified as the theme of ecological normativity and the theme of capitalization of nature, particularly through the framework of ecosystem services – is increasingly subjected to a biopolitical closure that continuously enfolds *both* themes.

But, we must ask, is the biopolitical closure, the subsumption of life in biopolitical apparatuses of control and optimization, *entirely* without remainder, that is, thorough, comprehensive and suffocating? Or is it possible to identify productive ambiguities? Is the genealogical approach not able to reveal any opening from which the critical environmental theorist (who incidentally is perhaps 'not some irrelevant interloper, some high-minded abstractionist' but the most pragmatic problem-solver of all)[264] can 'change the conversation',[265] and perhaps even locate some fracture points in the surface of biopolitical modernity? I end this part of the book leaving these questions suspended, as it will be the primary task of the last substantive chapter to provide a tentative answer to these questions (and to the main research question: whether the 'ecosystem approach' represent a paradigm shift in relation to legal modernity), and to outline where possible the contours of any available productive ambiguities. The analytics of biopolitics, after all, has, as already mentioned, a specific ethicopolitical orientation, and its critique aims at being 'productive and transformative'.[266]

264 M'Gonigle and Takeda, 'Liberal Limits', p. 1112.
265 Ibid.
266 Lemke, *Biopolitics*, p. 122.

10 Inside and against biopolitics

Towards a productive reading of the 'ecosystem approach'?

Introduction

The last chapter ends with a set of questions for which there seem to be no answer. It shows that in fact that there is no inexorable movement from anthropocentrism to ecocentrism in international environmental law; that there is no ecocentric horizon towards which international law might tend; that there is only biopower at work, with its ambiguities and circularities, and with its constant transformations and genealogical conflicts. The biopolitical *dispositif* embodied in the 'ecosystem approach' facilitates the establishment of a regulatory system of surveillance and protection aimed at the care, enhancement and optimization of the health and integrity of ecosystems. Yet, in an irreducible and unresolvable circularity, this same *dispositif* always ultimately transforms into its opposite, as the optimization of living biological resources also enframes them as 'standing reserves' whose ultimate horizon is their instrumentalization. The protection of nature, through the biopolitical aporia, is continuously transformed into its subjugation. Indeed, protection and subjugation are always simultaneously enacted. Herein lies, I have endeavoured to show, the biopolitical aporia. What then? Is it possible to conceive of a resolution of this aporetic logic that continuously reproduces itself? Is the biopolitical closure, the subsumption of life in biopolitical apparatuses of control and optimization, *entirely* without remainder, or is it possible to identify productive ambiguities?

The aim of this chapter is to show that the 'ecosystem approach' is susceptible of a number of productive readings enabled by its productive ambiguities, even as the tensions and the competing narratives that traverse the concept, and the genealogical 'play of forces' within which it is inserted, tend to be subjected to a decisive biopolitical resolution. Yet it is precisely at the aporetic core of the biopolitical resolution that one may find some open, unstable, ambiguous terrain where the critical legal scholar can operate counter- or, better, non-hegemonically. While apparently without remainder in terms of its conceptual and material reach, in fact, the biopolitical closure can never actually entirely fill the genealogical field of discursivity where the 'ecosystem approach' is situated. It can only provisionally, and to various degrees of stability, *hegemonize* it. Biopolitics then, as a prevalent framework of control and mode of operation of power, cannot but leave *some* space for the articulation of counter-trajectories, for the utterance of counter-readings, for the drawing out of antagonist interpretive registers that are available through the productive ambiguities that inhabit the conceptual field of the ecosystem approach.

Genealogy, biopolitics and the immanence of critique

If biopower enfolds and enframes the entire world,[1] there is no remainder *outside* of it.[2] Moreover, the genealogical method suggests that the development of legal concepts (or conceptual trajectories) such as the 'ecosystem approach' is not the inevitable result of objective processes, of an inevitable progression 'towards perfection'. It is rather, in ways that are not deterministically 'prefigurable',[3] the contingent and contested result of the combined actions of the subjectivities involved in the process itself. These two considerations lead to two methodological consequences. First, critique is inevitably *inside* biopower; it is immanent. Second, while inside, critique is immersed in the genealogical 'play of forces'. This means that critique is performed from a particular perspective, a perspective which in the context of this book is aligned with the urgent calls for a critical and fundamental reconstruction of law in the Anthropocene.[4] Critique, then, is not only *inside* biopower, but it is also *against* it.[5]

1 I have outlined the inextricable relation between modernity, capitalism and biopower/biopolitics at multiple junctions in this book, see in particular Chapter 8.

2 Hardt and Negri, *New World Order*, p. 186; see in general p. 186ff. Indeed, consistently with Foucault's theorization of power, biopower exists in and through a relation, ibid..

3 A. Negri, *Cinque Lezioni di Metodo su Moltitudine e Impero*, Soveria Mannelli: Rubbettino Editore, 2003, p. 20.

4 It may be useful to mention how critical legal theory falls for example under the headings of 'fundamental research' according to H. Arthurs, *Law and Learning: Report to the Social Sciences and Humanities Research Council of Canada by the Consultative Group on Research and Education in Law*, Information Division, Social Sciences and Humanities Research Council of Canada, Ottawa, 1983, pp. 63–71; see also P. Chynoweth, 'Legal Research', in L. Ruddock and Knight, A. (eds), *Advanced Research Methods in the Built Environment*, Oxford: Wiley-Blackwell, 2008, esp. pp. 28–9,

5 This consideration is neatly captured by the expression '*dentro e contro*'. The expression – which literally means 'inside and against' – is borrowed from the political language of Italian operaism, and in particular Mario Tronti (M. Tronti, *Operai e Capitale*, Rome: Derive Approdi, 2006) and Antonio Negri (e.g. Negri, *Cinque Lezioni*; A. Negri, *Dentro/contro il Diritto Sovrano: Dallo Stato dei Partiti ai Movimenti della Governance*, Verona: Ombre Corte, 2009, albeit already Pashukanis moved his critique of law from a perspective of '*dentro e contro*' (see e.g. E. Pashukanis, *Law and Marxism: A General Theory*, London: Pluto Press, 1989 and B. Cava, 'Pashukanis e Negri: do antidireito ao direito do comum', 4 *Revista Direito e Praxis* 2013, 2). The precise reconstruction of the historical emergence of the expression exceeds the scope and interest of this book. What is important to note however, is the present significance (and expediency) of the expression as a methodological approach to critique that is fully attuned to both genealogy and the analytics of biopolitics. *Dentro e contro*, in its basic connotation, refers to the historical movements that occur inside and yet against capitalism, as the result of social conflict and struggles (Negri, *Cinque Lezioni*, esp. p. 17). Indeed, it is the

If critique in a biopolitical register can only be developed *within* the genealogical 'play of forces', the productive ambiguities of the 'ecosystem approach' must also be explored from within. The 'ecosystem approach' can thus be read precisely as an ambivalent trajectory, caught in the middle of the biopolitical aporia. Yet if critique remains inevitably *inside* biopower, is it even possible to think of law in novel ways? If biopolitics engulfs even ecocentric articulations of legal concepts such as the 'ecosystem approach', if it maintains within its grip even radical theories of ecological legal personhood that are a key to ecocentric readings of the 'ecosystem approach', how is it possible to think law beyond law, that is, to read and theorize the 'ecosystem approach' outside of the aporetic mechanisms of biopolitics?

This, ultimately, is the central question of this book, as it emerges after the recalibration prompted by the shift in critical register, whose challenge, originally implicit in the main research question, becomes visible now. As the reader may recall, the original research question is whether and to what extent the 'ecosystem approach' represents a paradigm shift. So, how is it possible to speak of a strong paradigm shift, which, I remind the reader, entails a fundamental re-elaboration of the categories of (legal) thought *away* from the closures of modernity and of its legal articulation? Is it possible to protect life without hindering it, without subjugating and inserting it within a matrix of biopolitical control and surveillance, without always re-activating a thanatological return?[6] And if it is possible, through which legal construct, principle, theory? Is legal theory at all useful as a critical practice of thought?

These questions linger. And yet, recognizing the immanence of critique also implies the recognition that there is always a space, within a relation of power, that remains available to resistance against hegemonic closures.[7] If indeed power is a relation, if it is 'a complex strategical situation in a particular [situation]',[8] as Foucault[9] (and Hardt and Negri with him)[10] suggest, there is always a space, however minimal, that exceeds the hegemonic grip of the biopolitical closure, and exceeds even the biopolitical aporia. With particular reference to the 'ecosystem approach', this is the space where it is possible to explore its productive

subjectivities (the workers, in Tronti's and Negri's analysis, the multitude, in Hardt and Negri's updating of this conceptual approach, within the novel logic of Empire) that operate within the conflict that are the engine of historical change. This is a genealogical approach to historical change that points to the inevitable immanence of critique.

6 The circular production and re-production of the thanatological exclusion that always emerge as the seemingly inevitable result of biopolitics rests, as we have seen, on the intimate relation between modernity, biopower – and its specifically biopolitical declension – and capitalism.

7 This is indeed one of the central insights of Foucault's work, see especially Foucault, *History of Sexuality*. In this sense, drawing on Foucault, also Hardt and Negri, 2009, op. cit.

8 Foucault, *History of Sexuality*, p. 93. Foucault uses the term 'society'. Situation however, is arguably a more flexible and immanent referent.

9 Ibid.

10 Hardt and Negri, *Commonwealth*.

potential – or, better, the productive potential that opens up at the empty juncture of its multiple articulations.

Importantly, however, the recognition that critique operates always within the genealogical 'play of forces', inevitably leads to the recognition that critique always takes a partial perspective in both epistemological and political terms. Epistemologically, since knowledge, being always situated, is also entirely produced from a particular point of view;[11] politically, since theory is always already immersed in conflict, and as such it is a practice that, resonating with a particular horizon of sense, inevitably takes sides.

This space, then – and here we find the crucial link to the third pillar, or element, of the 'analytics of biopolitics' articulated by Lemke, and adopted in this book – is the space where subjectivities are produced, reproduced, resisted and where they actively resist. Being *inside* and *against* the 'ecosystem approach' then entails that there is no pure, innocent perspective. Being *inside* and *against* means then that from a methodological perspective, a crucial role and significance is assigned to critical legal scholarship, understood as a form of subjectivity that is always and inevitably epistemologically situated, and politically involved.[12] Indeed, as I have already noted, with M'Gonigle and Takeda, the critical legal theorist has an important, perhaps even crucial, role to play. Rather than being dismissed as 'some irrelevant interloper, some high-minded abstractionist',[13] M'Gonigle and Takeda suggest, critique is pragmatic, and it allows us to solve problems.[14] It does so, however, *against* the existing frame of reference; *inside* and *against*. This resonates with the idea that theory *is* practice,[15] and inexorably points to the *responsibility* of critique. Indeed, law is not the result of particular acts that posit it, but of continuous '*processes of positivization*'[16] to which there exist a multiplicity of participants. This

11 This inevitable and multiple constitutive relation between knowledge, politics, life and critique (the latter as both thought and practice), is a crucial element of what is increasingly referred to as 'Italian Theory'; see e.g. R. Esposito, *Pensiero Vivente. Origine e Attualità della Filosofia Italiana*, Turin: Einaudi, 2010, and Esposito, *Living Thought*, as well as D. Gentili and E. Stimilli, *Differenze Italiane. Politica e Filosofia: Mappe e Sconfinamenti*, Roma: Derive Approdi, 2015.

12 In this sense see for example the lucid reflections of Marrti Koskenniemi, *From Apology to Utopia. The Structure of International Legal Argument. Reissue with a new epilogue*, Cambridge: Cambridge University Press, 2005, esp. p. 533ff., though Koskenniemi pairs up epistemology with ethics, rather than with politics.

13 M'Gonigle and Takeda, 'Liberal Limits', 1112.

14 M'Gonigle and Takeda suggest in fact that the critical theorist may be the most pragmatic problem solver of all; ibid.

15 See e.g. Esposito, *Pensiero Vivente*. Particularly important in this respect is Antonio Gramsci, who for example organized his political philosophical project under the headings of 'philosophy of praxis', which he understands as a 'fusion' of 'philosophy and politics, of thinking and acting', A. Gramsci, *Selections from the Prison Notebooks*, Q. Hoare and G. Nowell-Smith, (eds), London: Lawrence and Wishart, 2007, notebook 7, §35; see also, more comprehensively nos. 10 and 11.

16 L. Lombardi Vallauri, 1978, *Corso di Filosofia del Diritto*, Paduaa: CEDAM, pp. 186–7.

participation and the processes of emergence of law indicate a clear *responsibility* on the part of each participant.[17]

With these considerations in mind, we can now move to the exploration of the productive ambiguities available to a critique situated *inside* and *against* the biopolitical closure of the 'ecosystem approach'.

The productive ambiguities of the 'ecosystem approach'

Introduction

As mentioned, and as the reader may recall, we ended Chapter 9 with important questions: is the biopolitical closure, the subsumption of life in biopolitical apparatuses of control and optimization, *entirely* without remainder? Or is it possible to identify productive ambiguities the critical environmental lawyer can seize to 'change the conversation',[18] and perhaps even locate some fracture points in the surface of biopolitical modernity?

The aim of this section is to show that the 'ecosystem approach' is susceptible of a number of productive readings enabled by its productive ambiguities, even as the tensions and the competing narratives that traverse the concept, and the genealogical 'play of forces' within which it is inserted, are entirely seized by the biopolitical *dispositif*. Yet it is precisely at the aporetic core of biopolitics that one may find some open, unstable, ambiguous terrain where the critical legal scholar can operate counter- or, better, non-hegemonically.[19] While apparently without remainder in terms of its conceptual and material reach, in fact, the biopolitical closure can never actually entirely fill the genealogical field of discursivity where the 'ecosystem approach' is situated. It can only provisionally, and to various degrees of stability, *hegemonize* it. Biopolitics then, as a prevalent framework of control and mode of operation of power, cannot but leave *some* space for the articulation of counter-trajectories,[20] for the articulation of counter-readings,[21] for the drawing out of antagonistic interpretive registers that are available through the productive ambiguities that inhabit the conceptual field of

17 This is for example the delineation of the role of the legal scholar that Italian legal philosopher Luigi Lombardi-Vallauri developed in his 1978 course on legal philosophy, Lombardi Vallauri, *Corso di Filosofia*.

18 M'Gonigle and Takeda, 'Liberal Limits', 1112.

19 As Barabanova underlines in fact, the expression 'counter-hegemonic' refers to a typical Gramscian political vocabulary, and indicates the substitution of one dominant discourse with another. A non-hegemonic perspective on the other hand, remains open to plurality, and does not aim at the consolidation of its own particular hegemony, Y. Barabanova, Emerging Climate Justice Discourse: Perspectives of Grassroots Networks in the UK, PhD Thesis, Budapest: Central European University, 2013, pp. 60–61.

20 Here the adjective 'counter' plays a different role than it does in the expression counter-hegemonic, as it merely signals the inevitable immanence of resistance, rather than the direction, or goal, of its political horizon.

21 Ditto.

the 'ecosystem approach'. Indeed the very aporetic core underlying biopolitics makes sure that any attempted biopolitical closure remains unstable, always caught in its own circular movement, in its own internal articulations and transformations.

The available space is, however, narrow, as biopower is able to subsume within the same horizon of sense *both* anthropocentric and ecocentric orientations (indeed I suggested that this is precisely one of the central advantages of the analytics of biopolitics vis-à-vis the anthropocentrism/ecocentrism binary framework). As the reader may recall, the 'ecosystem approach', particularly in the context of the case studies reviewed, aims at managing ecosystems *both* 'for their intrinsic values' *and* 'for the tangible or intangible benefits for humans'.[22] As biopower continuously scrambles and re-encrypts every ecocentric signal into a biopolitical outcome, and even (re-)deploys ecocentrism towards its own legitimation, drawing out the productive ambiguities available in the 'ecosystem approach' will require us to be constantly alert. Moreover, the ability of biopower to subsume both anthropocentrism and ecocentrism within the same horizon of sense, is continuously pushed further and intensified by the incessant movement spurred by the internal contradictions of environmental law. The latter, as M'Gonigle and Takeda suggest, is locked in fact in an inescapable vicious circle of reforms that only further reinforces the status quo,[23] and constantly folds and re-folds the attempted ecological (re-)imagination of law within its own biopolitical parameters.[24] It is this logic that renders linear readings of the ecological transformation of law inadequate, as a result of their inability to capture this central insight. Yet, as anticipated, there are arguably a number of productive ambiguities. I briefly allude to some of them, and explore and draw out some of their implications for thinking law beyond law and outside of the biopolitical closure, in the next sections. This remains a limited, provisional treatment, and a tentative map of future trajectories of research and critique.

Teasing out some productive ambiguities of the 'ecosystem approach'

Introduction: critique inside the biopolitical closure

The biopolitical closure is at work visibly, albeit in different ways, in many international legal regimes. In the international fisheries regime, for example, there is a significant overlap between biopolitics and anthropocentrism, particularly in the avowedly anthropocentric articulation of the ecosystem approach to fisheries developed and promoted by the FAO. However, demands for an ecocentric turn in international fisheries widen the scope of the biopolitical subsumption.

22 Malawi Principles, 2010, op. cit., Principle 1.
23 M'Gonigle and Takeda, 'Liberal Limits'.
24 See on this De Lucia, 'Beyond Anthropocentrism'.

That ultimately encompasses a multiplicity of articulations of the 'ecosystem approach' that exist in the broad space of fisheries discourse. The internal biopolitical articulation between conservation and use that is at work within the international fisheries regime (and which points to the alignment of both to the underlying biopolitical goal of the optimization of life and the enhancement of its productivity under the narrative of sustainable development), is however even more visible and intense within the context of the CBD. This transpires already from the objectives of the CBD, conservation of biodiversity and sustainable use of its components: that is, protecting and fostering life in order to optimize and enhance its productive capacity. It is precisely here that the aporia of biopolitics lies, as it always inevitably transforms the positive into its negative, the politics of life into a politics of subjugation and control. Life, in fact, is fostered, enhanced, optimized, protected and conserved (and now I am not referring to the practical question of implementation, nor to its failures to achieve even the objectives of the CBD, but rather to its philosophical and legal objective), while simultaneously integrated into a regularized regime of productivity that ultimately responds to the requirements of human well-being.

This biopolitical closure, which exceeds the anthropocentric/ecocentric binary to the extent that it simultaneously optimizes and subjugates life through the insights of an ecological and holistic care for nature, can be confirmed by reading Malawi Principles 1, 4, 5 and 6 in combination. Considering that, as we have seen in Chapter 9, according to the Malawi Principles 'management objectives are a matter of societal choice',[25] and that ecosystems must be managed in an economic context, '[c]onservation of ecosystem structure and functioning, *in order to maintain ecosystem services*, should be a priority target of the ecosystem approach'.[26] Moreover, with a view to achieve the objective of a sustainable development, '[e]cosystems must be managed within the limits of their functioning',[27] i.e. maintained within their current state in order to dispel the risk of a state change, and hence ensure a predictable (i.e. sustainable) provision of goods and services valuable to humans. To this purpose, 'attention should be given to the environmental conditions that limit natural productivity'.[28] This is a widely shared understanding of the purpose of the 'ecosystem approach', whose core goal has been concisely captured as being 'aimed at conserving and sustaining ecosystem services to benefit current and future human generations'.[29] Here the 'ecosystem approach' seems entirely deployed as a biopolitical strategy.[30]

25 Decision V/6, 2000, op. cit., Section B, para 6, Principle 1.

26 Ibid., Principle 4, emphasis mine.

27 Ibid., Principle 5.

28 Ibid., Principle 6, Rationale.

29 Michael Sissenwine, former Chief Science Advisor, National Marine Fisheries Service, USA, as quoted in UNEP, 2011d, op. cit., p. 10.

30 The distinct uses of the concept of ecosystem – theory, object, method – were discussed in Chapter 7, to which I refer the reader.

Moreover, in this biopolitical 'regime of life' there is a passage, a slippage, which is decisive for biopower. Life is abstractly protected (and in representative terms)[31] as biodiversity, yet concretely destroyed as its material components are harvested and inserted in the global flows of capitalist accumulation. It is abstractly conserved, while simultaneously and completely, that is without remainder or excess, enframed as standing reserve, completely harnessed to the economic process in the terms insightfully expressed by Hardt and Negri. Biodiversity itself, as an abstract category, is ultimately ordained to the satisfaction of the requirements of human- wellbeing, as even unknown species are cared for (through the conservation of ecosystems that may accommodate them) in relation to their productive potential (genetic resources that may be transformed into medicine, for example). This passage, crucially, is immediately linked to the 'ecosystem approach' as it purports to overcome the fence between humanity and nature, subsuming under its broader strategic framework the protected areas model (an operation that closely aligns with the biopolitical closure enfolding conservation biology and ecology under the Anthropocene). However, here the first productive ambiguity emerges: the 'ecosystem approach' provides traction for dismantling the 'fence' between humanity and nature.

Deconstructing the 'fence' between humanity and nature

The first productive ambiguity available to a critical reading relates to the 'fence' or threshold that in the ideo-ontology of modernity separates humanity and nature. As we have seen, the 'ecosystem approach' explicitly recognizes the need to see humans as immersed in nature. The 'ecosystem approach' may be mobilized to help deconstruct the modern bifurcated vision of nature as either an object of control – subjected to property rights – or as 'wilderness to be preserved apart from human society'[32] (a bifurcation that Hasley, as we have seen, characterizes in terms of majesty and monstrosity).[33] However, rather than a radical new ecological concept, this immersion of humanity in nature is a trivial truism, masked by the thresholds that modernity has endeavoured to construct and that it continuously guards and enacts through its political epistemology.[34] In and of itself, then, it is not enough. In fact, itself inserted within a 'play of forces', this (re)framing of humanity as inevitably immersed in nature, and thus inextricably part of it, constantly runs the risk of being subjected to particular biopolitical closures. The narrative of humanity immersed in

31 For the problematics involved in relation to the concept of representativeness, which is central for the selection and designation of particular areas as protected under the CBD, see, e.g. T. Stevens, 'Rigor and Representativeness in Marine Protected Area Design', 30:2 *Coastal Management* 2002, 237.

32 Godden, *Nature as Other.*, p. 2.

33 Hasley, 'Majesty and Monstrosity'.

34 On the notion of nature as a political epistemological category of modernity, see B. Latour, 'An Attempt at a "Compositionist Manifesto"', 41 *New Literary History* 2010, 471.

nature risks in fact being immediately re-configured and re-deployed as an opportunity to dismantle the protection of nature that protected areas, with all their problems and limitations, may afford.[35]

The threshold of majestic and monstrous is not necessarily removed, rather, it is shifted and made flexible, mobile, adaptive. In other words, it is de-territorialized (in the sense of not being demarcated on the terrain, in sovereign fashion). All nature, including that part of nature enclosed under sovereign strategies of prohibition (i.e. the protected areas), may be submitted, not to sovereign command, but to a biopolitical regime of optimization, in line with the immanentization of biopower described by Hardt and Negri in terms of the introjection of Empire (i.e. Capital) within the folds (*every* fold) of the world.

At a conceptual level, the immersion of man in nature is precisely what Argyrou identifies as the crucial *continuity* of environmentalism with modernity, despite its radical point of departure and the challenges it poses to the paradigm of modernity.[36] Argyrou argues that environmentalism reproduces the logic of modernity, which he calls the logic of the same, through effacing 'the divisions of the Whole and maintain[ing] its imaginary unity'.[37] Indeed, it also intensifies it – and here Argyrou's suggestion overlaps, albeit using a different language, with the argument proposed in this book in relation to the 'ecosystem approach' and its intensification of the biopolitical regime – by taking it to its 'logical and ontological extreme'.[38] Environmentalism, suggests Argyrou, 'strives to efface ... the last and grandest of all modernist divides of the Whole – the division between humanity and nature'.[39]

Locke and Dearden also capture the mechanics of the biopolitical closure of the man-in-nature notion, though from a more on-the-ground perspective. They warn that the deployment of the conceptual framework of 'humans in nature' may provide space for the introduction of pervasive management of nature legitimated by a rhetorics of conservation, ecosystem management and sustainable development. They point out that 'under the new categories and supported by the 'new paradigm' (i.e. humans in nature), protected areas are

35 Besides their limited effects deriving from the insufficient magnitude of the areas protected, there are the long- standing questions of how protected areas have been used as an instrument of colonial penetration. As Kareiva reminds us in fact, 'for 30 years, the global conservation movement has been racked with controversy arising from its role in expelling indigenous people from their lands in order to create parks and reserves', P. Kareiva et al., 'Conservation in the Anthropocene: Beyond Solitude and Fragility', *Breakthrough Journal*, Winter 2011. See, for a more detailed account, A. Agrawal and K. Redford, 'Conservation and Displacement: An Overview', 7:1 *Conservation and Society* 2009, 1, which without hesitation argues that conservation programmes impose misery on people, p. 1.

36 V. Argyrou, *The Logic of Environmentalism: Anthropology, Ecology and Postcoloniality*, New York and Oxford: Berghahn Books, 2005.

37 Ibid., p. 74.

38 Ibid.

39 Ibid.

being re-cast as tools for social planning and income generation'.[40] Inserted within an overall 'ecosystem approach' strategy, protected areas become essentially biopolitical tools. Both strategies, then, especially once conceptually nested into one another, are arrayed along a continuum of possibilities available to the biopolitical subject for the achievement of its objectives. Philippopoulos-Mihalopoulos warns in this respect that while considering man immersed in nature moves us closer to the dismantling of the fence separating human culture and nature, 'merely focusing on ecosystemic processes and their management' keeps man at the centre, regardless of how 'included' man may be 'in the ecosystem itself'.[41] The iterative shifting between these complementary strategies, between sovereign control and biopolitical optimization, contains the dynamic and dialectic of conservation within an artificial space of biopolitical closure, within an interventionist paradigm that admits no alternative, in close alignment with the particular narrative of the Anthropocene favoured by the new ecology and new conservation.

There appears to be then little room for a productive reading here. Indeed, this productive ambiguity is the one most thickly overlaid with biopolitical closures. However, the question of the immersion of humanity in nature can be posed in entirely different terms; that is, in terms of *how* humanity operates within the context of its *inevitable* immersion; and in terms of *where* this immersion takes place. The 'ecosystem approach' may thus present itself as a vehicle for the articulation of a specific way of being immersed in nature: ontologically, through the concept of the '*middle*';[42] epistemologically, through the concept of 'epistemic location'[43] – both of these are further explored and explained later, in the fifth and sixth sections of this chapter, respectively.

Thinking and acting ecologically

Useful space for drawing out the productive ambiguity of the 'ecosystem approach' can be further identified by looking at the aims of the 'ecosystem approach' as articulated within the context of the CBD. The 'ecosystem approach' is generally aligned with the achievement of the overall objectives of the Convention. Its biopolitical resolution, moreover, is evident, particularly through the gravitational pull that the narrative of ecosystem services exerts on it, thus absorbing it into its biopolitical orbit of sense. Yet, a productive reading can be attempted in relation to the more specific goal that the 'ecosystem approach' is assigned, a goal that can be read *against* albeit *inside* its biopolitical resolution.

40 H. Locke and P. Dearden, 'Rethinking Protected Area Categories and the New Paradigm', 32:1 *Environmental Conservation* 2010, 1.

41 Philippopoulos-Mihalopoulos, 'Actors or Spectators?', 863.

42 See ibid.

43 L. Code, *Ecological Thinking: The Politics of Epistemic Location*, New York and Oxford: Oxford University Press, 2006.

The 'ecosystem approach' is not only understood as 'a set of guidelines for the management of various ecosystems',[44] but also, and more comprehensively, as 'a framework for *thinking and acting ecologically*'.[45] Indeed, the 'ecosystem approach' 'is a framework for action that links biological, social and economic information'.[46] It is true that the Malawi Principles repeatedly refer to the 'ecosystem approach' as a *management* framework,[47] and that the biopolitical horizon of management traverses most articulations of the concept, hence reducing the scope of its implications and the productive space that I am trying to delineate and tease out. Yet it is precisely in the sense just alluded to, as a framework for 'thinking and acting ecologically', that the 'ecosystem approach' appears capable of being read productively. Perhaps the space thus opened up allows for a 'transversal' reading that, as we will see in the fifth section of this chapter, is embraced by the present author as the appropriate way to think and critique ecology and law together, in many ways further complicating the ambiguities and tensions traversing the (legal) discourse of the 'ecosystem approach'. Here, perhaps, lies one of the decisive sites where the ambiguities harboured in the biopolitical resolution of the 'ecosystem approach' can be made productive.

Law and ecological limits

The question of ecological limits may offer further opportunities for a productive reading of the 'ecosystem approach', one that escapes biopolitical closures. Here the productive ambiguity rests on the frame of limits invoked in the context of the 'ecosystem approach' (e.g. Malawi Principle 6). Decision VII/12 on sustainable use, for example, recognizes that,

> [t]he supply of biological products and ecological services available for use is limited by intrinsic biological characteristics of both species and ecosystems, including productivity, resilience, and stability. Biological systems, which are dependent on cycling of finite resources, have limits on the goods they can provide and services they can render. Although certain limits can be extended to some degree through technological breakthroughs, there are still limits, and constraints, imposed by the availability and accessibility of endogenous and exogenous resources.[48]

However, within the broader narrative frame of sustainable development (which provides a central normative horizon for environmental protection in general

44 Smith and Maltby, *Using the Ecosystem Approach*, p. 12.
45 Ibid., emphasis mine.
46 Ibid.
47 See e.g. Principles 1, 2, 3, 4, 6 and 9.
48 Decision VII/12, 2004, op. cit., Annex II, para 8(e).

and for the 'ecosystem approach' in particular,[49] as seen), these limits are considered contingent, rather than insuperable.[50] From a sustainable development perspective, moreover, limits are also re-distributable through the flows and predatory practices of global trade.[51] Additionally, as we have seen, sustainable development is discussed as a strategy aimed at securing a 'new era of economic growth'.[52] Finally, even if the notion of limits *is* taken seriously, that does not mean that a biopolitical regularization of the provisions of ecosystem services is excluded. Indeed, in many ways, taking ecological limits seriously *facilitates* the biopolitical optimization of life. This re-construction of limits into opportunities is implicit in the framework of ecosystem services and is the central thrust of initiatives such as TEEB, and, as we have seen, represent a crucial biopolitical marker.[53]

Yet the question of ecological limits has potentially important implications for a non-hegemonic reading of the 'ecosystem approach', particularly if limits are understood as a moment of engagement with the 'vulnerable living world',[54] rather than an economic category that suggests re-negotiation of the relationship between expectations and the productive capacity of ecosystems as mediated through current technological capabilities. From this perspective, limits speak of vulnerability,[55] rather than of economic calculation or biopolitical valorization. As Philippopoulos-Mihalopoulos suggests, vulnerability entails an entirely new way of thinking about law.[56] Thinking about limits through the lens of vulnerability shatters the immunitary armour (to refer to Esposito's language) that modernity has constructed in order to protect its fears, an armour most explicitly manifest in the concept of the legal subject (both sovereign and owner), paradigmatically in the centre. Here Philippopoulos-Mihalopoulos observes how the subject's central positioning is both shield and misdirection: 'the centre hides behind the blinding light and conceals its all-controlling

49 As also Bell, 'Non Human Nature', suggests, albeit within the particular context of the Great Lakes.

50 WCED, 1987, op. cit., p. 24.

51 See in this respect Hornborg, *Power of the Machine*.

52 Ibid., p. 28 (but the reference is used repeatedly throughout the report).

53 Leonardi, 'Biopolitics', p. ii.

54 Grear, 'Vulnerable Living Order'.

55 On the relation between immunity and fear (a fear that transforms in the subsumption of life under politics with the goal of conserving life), see obviously, see Esposito, *Living Thought*; see also, however M. Horkheimer and T. Adorno, *Dialectic of Enlightenment*, London: Verso Classics, 1997, p. 3, where they observe how: 'In the most general sense of progressive thought, the Enlightenment has always aimed at liberating men from fear and establishing their sovereignty' (and they also recognize the thanatological effects of an excess of immunization: 'the fully enlightened earth radiates disaster triumphant', ibid., p. 3); see also Philippopoulos-Mihalopoulos, 'Actors or Spectators?', p. 857, where he also suggests, although in passing (through Peter Sloterdijk), an immunological interpretation, resulting from fear, of the modern ontological and epistemological thresholds.

56 Ibid. on vulnerability, and its role in relation to human rights and the environment; see also Grear, 'Vulnerable Living Order'.

position with a thick blanket of hinterlands, purely in the service of confirming centrality while avoiding exposure'.[57] Thinking law through vulnerability, on the other hand, makes visible the 'fundamentally precarious position' of life – all life, including human life. '[A]cknowledging one's personal vulnerability', suggests Philippopoulos-Mihalopoulos, implies accepting one's 'limits and therefore limitations. It is precisely the awareness of the latter that is lacking in environmental legal discussions.'[58] By making explicit the vulnerability of the living order, the very biopolitical paradigm (and the immunitary horizon of sense that is its most fundamental interpretive key) is itself exposed. Perhaps, then, its overturn can be attempted by imagining the intersection between life, politics and law differently.

Plural epistemologies and the epistemological role of ignorance

Another productive ambiguity is linked to the recognition of a plurality of epistemologies, and of the epistemological role of ignorance, that ecology (despite the ambiguities and genealogical complexities that itself exhibits), underlying as it does the 'ecosystem approach', brings fully into view. The epistemological role of ignorance was introduced into the discourse and practice of international environmental law through the precautionary principle, and indeed precaution is a central element of the 'ecosystem approach'.[59] Precaution is arguably the 'established' international environmental principle with the most far-reaching potential (putting aside for the moment the question of its formal legal status in international environmental law[60] and the fact that it has been the object of 'much controversy'[61]). Yet it remains largely framed as a risk management tool, organized around the central concepts of 'lack', 'serious or irreversible damage' and 'cost-effectiveness'.[62] It operates moreover, arguably within a clear biopolitical horizon, in multiple ways, and is continually exposed to a biopolitical closure. With particular respect to damages, it privileges those that are out of the ordinary, hence upholding the threshold between monstrous and majestic (albeit along a different demarcation line, that between ordinary and extraordinary).[63] It is conditioned on analyses of costs and benefits, that, while in some cases framed in wider terms than those strictly premised on

57 Philippopoulos-Mihalopoulos, 'Actors or Spectators?' p. 857.

58 Ibid., p. 858.

59 See e.g. Trouwborst, 'Precautionary Principle'.

60 See e.g. J. Peel, 'Precaution – a Matter of Principle, Approach or Process?', 5 *Melbourne Journal of International Law* 2004.

61 M. Ahteensuu, 'Defending the Precautionary Principle Against Three Criticisms', 11:4 *TRAMES A Journal of the Humanities and Social Sciences* 2007, 366, p. 367.

62 Rio Declaration, 1992, op. cit., Principle 15.

63 Precautionary action is to be triggered only when faced with potentially 'serious or irreversible damage' (Rio Declaration, Principle 15).

economic calculation,[64] remain within the immunitary paradigm that sustains biopolitics.[65] It embodies, finally, a structurally provisional stance (i.e. the 'lack' of knowledge is only temporary in a manner similar to the contingency of limits in the vision of sustainable development presented in the Brundtland report) rather than being internalized as a permanent modality expressing an inherent – and epistemologically insuperable – state of uncertainty.[66] The latter is a particularly problematic stance under conditions of uncertainty, ignorance and epistemological pluralism.

A second important aspect that is susceptible of being read productively relates more immediately to the epistemological pluralism that the 'ecosystem approach' purportedly embraces. We have seen that Malawi Principle 11 states that 'the 'ecosystem approach' should consider all relevant (scientific, indigenous and local) knowledges and practices. Indeed Article 8(j) of the CBD emphasizes the importance of traditional indigenous knowledge in relation to conservation. Again, the full significance of this productive ambiguity (beyond the institutional opportunities through which indigenous and local knowledges are already included within the formal, hegemonic international legal regimes), are drawn out later, through the concept of 'epistemic location'.[67] Unlike current inclusions of marginal knowledges, which are injected into the hegemonic regime of truth with the view of instrumentalizing them biopolitically (and from a position of hierarchical superiority), epistemic location shifts the perspective in an interesting way, as we shall see.[68]

64 'Examination of the pros and cons cannot be reduced to an economic cost-benefit analysis. It is wider in scope and includes non-economic considerations. However, examination of the pros and cons should include an economic cost-benefit analysis where this is appropriate and possible. Besides, other analysis methods, such as those concerning the efficacy of possible options and their acceptability to the public may also have to be taken into account. A society may be willing to pay a higher cost to protect an interest, such as the environment or health, to which it attaches priority', thus Communication from the Commission on the Precautionary Principle, COM, 2000, para 6.3.4.

65 As can be evinced from the emphasis on *health* as the overarching reference for the precautionary principle, see ibid.; the stringency of precaution is, moreover, more evident in cases regarding human health and safety, rather than the environment, see e.g. N. de Sadeleer, 'The Precautionary Principle in EU Law', *AV&S*, 2010, p. 173.

66 One of the crucial elements of the precautionary principle is the 'lack of full scientific certainty' (Rio Principle 15). Such lack is in most cases considered a temporary condition, to be progressively eliminated: 'measures based on the precautionary principle should be maintained *so long as scientific information is incomplete or inconclusive*', Communication, 2000, op. cit., para 6.3.5, emphasis mine.

67 Code, *Ecological Thinking*.

68 See e.g. F. Berkes, J. Colding and C. Folke, 2000, 'Rediscovery of Traditional Ecological Knowledge as Adaptive Management', 10:5 *Ecological Applications* 2000, 1251.

Sovereignty and governance

DISPLACING SOVEREIGNTY

Another productive ambiguity relates to the principle of sovereignty. This is perhaps the theme that has been interrogated most frequently in the international environmental legal literature, given the obvious mismatch between the legal concept of territorial sovereignty (denoted as a 'barrier' to effective environmental protection),[69] and the transversal concept of ecosystem.

The centrality of sovereignty, however – for all the claims of its erosion, or decreasing relevance, and for all the puzzled interrogations about whether it is an anachronistic concept – remains in many ways rock solid.[70] Authoritative textbooks of international law make the case unequivocally. Brownlie states, for example, that the 'sovereignty ... of states represent[s] the basic constitutional doctrine of the law of nations, which governs a community consisting primarily of states having a uniform legal personality'.[71] Crawford similarly considers states as the 'basic unit of international law'.[72]

The same picture emerges by looking at general international legal scholarship. Günter Handl candidly affirms that the 'sovereign equality of states [is] the most axiomatic premise of the international legal order'.[73] Two essays on

69 The sovereignty barrier has been described as that set of 'foundational principles and norms of general international law that continue to uphold a state-based system of international politics and law that is often seen as counter-productive to solving global and regional environmental problems,' Call for Papers of the Conference 'The Changing Nature of International Environmental Law: Evolving Approaches of the United States and the European Union', held in Geneva on 22–23 November 2013, and organized jointly by the European and American Societies of International Law's International Environmental Law Interest Groups, http://asilesilchangingnature.wordpress.com/conference-programme/, (Accessed 26 September 2018).

70 See K. Liftin (ed.), 1998, *The Greening of Sovereignty in World Politics*, Cambridge, MA and London: MIT Press, 1998, which emphasizes that the geological reference in the metaphor of erosion (should one submit to the erosion of sovereignty thesis) resonates with a process which operates according to very slow geological time. Moreover, Sol Picciotto describes how 'stateness', through the role that States take in international legislation, is bleeding across national boundaries, hence counteracting what Miéville calls the 'widespread and simplistic assertion that globalization is eroding the nation-state', C. Miéville, *Between Equal Rights. A Marxist Theory of International Law*, Chicago, IL: Haymarket Books, 2006, p. 63; see also, in a similar vein, albeit from a broader socio-historical perspective, S. Sassen, *Territory. Authority. Rights: From Medieval to Global Assemblages*, updated edition, Princeton, NJ and Oxford: Princeton University Press, 2006.

71 I. Brownlie, *Principles of Public International Law*, Oxford: Oxford University Press, 2008, 7th edn, p. 289.

72 J. Crawford, *Brownlie's Principles of Public International Law*, 8th edn, Oxford: Oxford University Press, 2012a, p. 216.

73 G. Handl, 'Trail Smelter in Contemporary International Environmental Law', in R. Bratspies and A. Miller (eds), *Transboundary Harm in International Law: Lessons from The Trail Smelter Arbitration*, New York: Cambridge University Press, 2006, p. 128.

sovereignty (both included in an edited collection on the philosophy of International Law),[74] from two distinct perspectives, conclude that sovereignty remains a central and relevant concept. Sovereignty, they observe, is not only compatible with the contemporary international legal system, but also with the 'constitutionalization' processes that have been at play from 1945 to the present. Indeed, as one of the two essays puts it, 'the core of [the contemporary world order] remains the segmentally differentiated international society of sovereign states, creating consent-based international law'.[75] Crawford observes that, despite the increasing normative challenges to the notion of sovereignty, '[r]eports of [its] death are much exaggerated'.[76] Similarly, Antonio Cassese concedes in a recent essay that, while States today must 'much more than in the past ... take into account pressures, incentives, and exhortations of other subjects', still continue *qua* sovereign 'to pursue [their] own interest[s]'.[77]

Other scholarship mounts a normative challenge to sovereignty, while recognizing its central role. Nollkaemper, for example, while arguing *for* a reduced liberty of States in matters relating to their own environment, admits to the fact that State sovereignty is still central in matters of International (Environmental) Law, and that the basic principle remains the no-harm rule.[78] Similarly, Ebbesson argues, from the perspective of environmental justice, against the 'veil' of state sovereignty, considered to impede the effective satisfaction of environmental justice; yet Ebbesson also recognizes that 'the current paradigm of international environmental law ... is primarily focused on inter-state concerns'.[79]

Granted, the monolithic character of State sovereignty is no longer unscathed. There are 'multiple outlines'[80] and 'multiple meanings'[81] regarding the State,

74 T. Endicott, *The Logic of Freedom of Power* and J. Cohen, *Sovereignty in the Context of Globalization: A Constitutional Pluralist Perspective*, both in S. Besson, and J. Tasioulas (eds), *The Philosophy of International Law*, Oxford: Oxford University Press, 2010.

75 Cohen, *Sovereignty*, p. 262.

76 J. Crawford, 'Sovereignty as Legal Value', in J. Crawford and M. Koskenniemi (eds), *The Cambridge Companion to International Law*, Cambridge: Cambridge University Press, 2012b, p. 132.

77 A. Cassese, *States: Rise and Decline of the Primary Subject of International Law* in B. Fassbender et al. (eds), *The Oxford Handbook of the History of International Law*, Oxford: Oxford University Press, 2013, p. 69.

78 A. Nollkaemper, 'Sovereignty and Environmental Justice in International Law' in J. Ebbesson and P. Okowa (eds), *Environmental Law and Justice in Context*, Cambridge: Cambridge University Press, 2009.

79 J. Ebbesson, 'Piercing the State Veil in Pursuit of Environmental Justice' in J. Ebbesson and P. Okowa (eds), *Environmental Law and Justice in Context*, Cambridge: Cambridge University Press, 2009, p. 270. Moreover, Ebbesson argues for taking the perspective of *individuals*, rather than States, hence still moving within the oscillation between private and public subject.

80 K. Knop, 'Statehood: Territory, People, Government' in J. Crawford and M. Koskenniemi (eds), *The Cambridge Companion to International Law*, Cambridge: Cambridge University Press, 2012, p. 97.

81 Ibid., p. 101.

and the internal relationships of its constitutive elements of territory-population-government. Moreover, international human rights law provides some evidence in this direction. Yet the dynamic is always articulated in terms of subjects (individuals vs. States), consequently operating well within the hegemonic boundaries of legal modernity. As Crawford observes, while the human rights regime may 'qualify' State sovereignty, it does not 'displace' it, and may actually, 'in subtle ways ... reinforce it'.[82] Restrictions and obligations on sovereign rights, then, do not alter its intrinsic quality, as observed by Crawford,[83] but also recognized elsewhere.[84]

Yet the 'ecosystem approach' may offer a useful perspective to deconstruct and disarticulate sovereignty, and perhaps the very essence and usefulness of modern *international* law along with it. As the concept of ecosystem, setting aside for a moment the contestations reviewed in Chapter 5, materially traverses and conceptually lacerates the territorial integrity of sovereign States, it also opens up productive space for a legal theoretical re-thinking of the principle of sovereignty, and of the State as the only (or at any rate primary) agent in international relations.

Twenty years ago, in a seminal article,[85] Brunnée and Toope explored the ways in which an 'ecosystem approach' might affect and re-orient sovereignty. However, the authors dismiss the most potentially far-reaching implications, in light of the deep reluctance of States to relinquish their formal privileged status as central units of reference and ultimate sources of law on the international plane. Brunnée, in particular, identifies early on the central problem in relation to what they call the question of environmental security, namely, sovereignty:

82 Crawford, 'Sovereignty', p. 122.

83 Sovereignty is in fact for Crawford a *quality* 'inhering' in the subject (i.e. The State), independently of the concrete *de facto* situation, or the concrete ways in which it becomes manifested on the 'plane of immanence', to use a Deleuzian expression indicating the world: A State is sovereign 'no matter how fragile its condition, how diminutive its resources'; indeed, such sovereign quality is 'irrespective of ... size, structure, population, resources and other potential', Crawford, 'Sovereignty', pp. 117, 118 and 119, respectively.

84 See the Montevideo Convention on the Rights and Duties of States, at Articles 3, 4 and 5, where a clear picture emerges of States as intrinsically sovereign, regardless of 'power which it possesses to assure its exercise' (Article 4). Moreover, the only limitation to the exercise of a State's sovereign rights is 'the exercise of the rights of other states according to international law' (Article 3). Finally, Statehood is entirely the consequence of the will of a sovereign ('The political existence of the state is independent of recognition by the other states', Article 3), and this entails that the 'fundamental rights of states are not susceptible of being affected in any manner whatsoever' (Article 5). Montevideo Convention on the Rights and Duties of States (signed 26 December 1933, entered into force 26 December 1934), League of Nations Treaty Series, vol. 165, pp. 20–43.

85 Brunnée and Toope, 'Environmental Secutity'; see also J. Brunnée and S. Toope, 'Environmental Security and Freshwater Resources: Ecosystem Regime Building', 91:1 *The American Journal of International Law* 1997, 26.

> [t]he first challenge is rooted in the very foundations of public international law. As a legal order built upon the sovereignty of states, international law does not readily accommodate the development of rules and regimes premised upon ecological unity.[86]

In their joint article Brunnée and Toope identify four ways through which the 'ecosystem approach' offers, or promises to challenge, this 'very foundation', each increasingly 'invasive' with respect to the principle of sovereignty.[87] These four ways in which an ecosystem-orientation may affect international environmental law are, however, mostly low-intensity transformations already ongoing (already ongoing, indeed, in 1994, when Brunnée and Toope wrote the article) that pose, effectively, no real challenge to the principle of sovereignty, but rather smooth its sharpest edges in relation to the increasingly paramount environmental question. In fact, Brunnée and Toope first outline a procedural (the 'least sovereignty-invasive')[88] and a substantive way in which the 'ecosystem approach' may contribute to the re-articulation of the principle of sovereignty. These two ways induce only marginal change, however, and follow a non-confrontational approach (procedural) or offer incremental improvements of existing rules (substantive).[89] The third way the authors consider is an 'ecosystem orientation in content'.[90] Indeed a more 'explicit ecosystem orientation'[91] would be 'more onerous from the perspective of state sovereignty'.[92] Sovereignty still finds its primary limits in the territorial rights of other States (or in the obligation not to cause harm to the global commons). Any duty to utilize their own domestic resources sustainably is linked to the 'responsibility to ensure that activities within their jurisdiction or control do not cause damage to the environment of other States or of areas beyond the limits of national jurisdiction'.[93] The crucial difference hinges on whether these limits refer to transboundary harm and an injured party, or descend from exclusively

86 J. Brunnée, 'Environmental Security in the Twenty-First Century: New Momentum for the Development of International Environmental Law?', 18 *Fordham Journal of International Law* 1995, 1742, p. 1744.

87 Brunnée and Toope 'Environmental Security', p. 57.

88 Ibid.

89 In both cases the reference example used by the authors was the then International Law Commission, Draft articles on the law of the non-navigational uses of international watercourses and commentaries thereto and resolution on transboundary confined groundwater, in Yearbook of the International Law Commission, 1994, vol. II, Part Two (hereinafter ILC Draft Article 1994), see ibid., pp. 57–65.

90 Ibid., p. 65.

91 Ibid.

92 Ibid.

93 Rio Declaration on Environment and Development, 1992, op. cit., Principle 2. See also N. Schrijver, *Sovereignty over Natural Resources. Balancing Rights and Duties*, Cambridge: Cambridge University Press, 1997.

ecological criteria.[94] The authors identify three principles that represent, they argue, 'tentative yet important steps towards a conceptual framework regulating human activity from the perspective of its ecological context':[95] sustainable development (which the authors suggest ought to be reformulated as *ecologically* sustainable development),[96] intergenerational equity and precaution. But it appears to the present writer that neither of the three examples is capable of challenging legal modernity in ways 'that really count', to recall Anna Grear's expression. We have seen for example in some detail that sustainable development is not able to offer any significant horizon of change, firmly embedded as it is in the conceptual and ideological fabric of legal modernity, and operating as both an anthropocentric frame of action, and a biopolitical *dispositif*, exactly at the point of their most intense intersection.

However, there is another, more radical way to imagine an ecosystem orientation, according to Brunnée and Toope, one that could respond to the need to re-orient the legal system towards ecosystems rather than state boundaries.[97] This ecosystem orientation 'in scope'[98] is the 'most invasive' articulation of an 'ecosystem approach' from the perspective of sovereignty.[99] Brunnée and Toope suggest in this respect that water (they focus on international freshwater law, but we can substitute fisheries, wildlife, wetlands etc.), from this ecosystem orientation would be considered as *one* part of the wider set of processes and elements included in an ecosystem. The perspective is then turned, and the focus would be not on water (or fisheries/wildlife/wetlands etc.) management 'in light of the broader ecosystem of which it is part, but on the ecosystem itself, with water being but one component'.[100] However, Brunnée and Toope suggest that time 'may not be ripe for [the] international implementation' of this type of 'ecosystem approach'.[101] Examples of a timid ecosystem orientation of this type can already be found, according to the authors, in the UNECE Convention on International Lakes (which includes both riparian and non-riparian parties, and links harm with the broader environmental effects that activities may have on non-riparian states).[102] They also point to the concept of common concern as a principle that has begun to impinge on the use rights of States in relation to their own territories and resources, to the extent that the exercise that said use right may have detrimental effects on global commons.[103]

94 Ibid., p. 65.
95 Ibid., p. 70.
96 Ibid.
97 Ibid., p. 71.
98 Ibid.
99 Ibid.
100 Ibid., p. 72.
101 Ibid.
102 Ibid., p. 74.
103 Ibid.

Thinking the 'ecosystem approach' *inside* and *against* biopolitics, however, requires us to *unthink* sovereignty, to epistemologically disarticulate and politically re-articulate the relationship between law and natural ecosystems *outside* of the sovereign frame. As sovereignty is a crucial structural category of international law – indeed, of legal modernity altogether, both in its specific public law inflection, and in its representation of the fundamental category of the subject – it is arguably difficult to remove the sovereignty barrier in practice and yet retain sovereignty as a category. Thinking of the entire conceptual trajectory of the 'ecosystem approach' as a productive ambiguity may suggest that sovereignty is *no longer a useful concept.* While much work is necessary to *unthink* sovereignty, work that cannot even be attempted here, some empirical evidence suggests that, in fact, the 'ecosystem approach' has already prompted novel modes of transversal governance.[104]

However, there is a second layer of discourse that is implicated when talking about sovereignty from within a biopolitical framework of analysis, a layer that reflects the relationship between sovereign power and biopower, sovereignty and biopolitics. Sovereignty in fact, is not only located in the State. It is rather a mobile *dispositif*, which traverses and exceeds the debates about the role of the legal principle of sovereignty, or of the State as the sovereign subject of international law.[105]

The emerging modes of governance discussed, while re-articulating the role of sovereign States in ecosystem-oriented governance arrangements, do not necessarily move beyond biopower or the sovereign *dispositif.* From the first perspective, as we have seen, sovereign power is *already* no longer crucial, even though the sovereign encoding remains central for the production of legitimacy of international legal regimes premised on the rule of law as a central (liberal) strategy for the neutralization of political conflict and for its technicalization.[106] From the second, sovereign power, with its intimate relationship with biopower,

104 See e.g. Karkkainen,'Post-Sovereign Environmental Governance'; B. Karkkainen, 'Marine Ecosystem Management and A "Post-Sovereign" Transboundary Governance', 6 *San Diego International Law Journal* 2004 – 5, 113; and B. Karkkainen, 'Towards Ecologically Sustainable Democracy?' in A. Fung and E. Wright (eds), *Deepening Democracy Institutional Innovations in Empowered Participatory Governance*, London, New York: Verso Books, 2003.

105 This is, for example, the lesson to be learned from Hardt and Negri's reconstruction of sovereignty in terms of Empire and Biopower.

106 Indeed, some critical scholars have emphasized how the *dispositif* of rule of law has been deployed for colonial purposes (U. Mattei and L. Nader, Plunder: When the Rule of Law is Illegal, Malden, MA: Wiley-Blackwell, 2008); others consider this mechanism as a form of 'denial of democracy' that operates precisely through the abstraction of law, recast as a technical domain subject to objective rules rather than subjective passion, from political confrontation (G. Hincapíe and R. Restrepo, 'La Constitución Encriptada. Nuevas formas de emancipación del poder global', 4:8 *Revista de Derechos Humanos y Estudios Sociales* 2012, 97).

remains the underlying violent mechanism that, while perhaps discursively dormant, is always available for lethal re-activations.[107]

Some examples may elucidate what I mean. Brunnèe and Toope recognize that one of the most potentially far reaching implications of an 'ecosystem approach' is the ability to re-articulate spaces and places transversally with respect to sovereign boundaries and delimitations. More recently, McIntyre also insists on this same far reaching potential, one adumbrating the possibility of a shift from a law of States to a law of ecosystems. Bradley Karkkainen, even more concretely, speaks of emerging 'post-sovereign environmental governance',[108] that is 'new governance arrangements' whose characteristic is that of being 'polycentric'.[109] This mode of post-sovereign governance, suggests Karkkainen, is characterized by its being 'non-exclusive', 'non-hierachical,' and 'post-territorial'.[110] It is non-exclusive in that States do not exercise exclusive authority in relation to the formation of environmental policy, but governance is rather the result of 'multi-party collaborations'.[111] However, Karkkainen acknowledges that States remain 'highly influential'[112] actors in these governance arrangements. Thus, rather than substantively losing their significance, States may simply be reconsidering their role in the global flows of accumulation, so that sovereignty becomes porous and perhaps shifts its locations, especially considering the biopolitical emphasis on the norm as the paradigmatic reference for law, rather than the sovereign command.

It is 'non-hierarchical', to the extent that formal legal regimes emanating from the sovereign authority of States are replaced by 'on-going, open ended commitments by multiple parties to do what it takes'. Yet, sovereignty and its formal legal structures, as well as its 'hierarchically binding legal rules', concedes Karkkainen, 'may continue to play some role in establishing overarching institutional frameworks and implementing major substantive commitments'.[113] The 'richer mixed strategy' of governance that Karkkainen describes may more readily be accommodated within the conceptual horizon of biopower and its restructuring of the modes through which law operates in order to optimize life through conceptual tools and management arrangements such as, indeed, the 'ecosystem approach'. It is, finally 'post-territorial', insofar as governance

107 For an illustration of the double layered relation between law as violence and law as ideology (and their dialectic articulated in relation to the notions of coercion and consent), which is captured in the Gramscian concept of 'hegemony', see, besides Gramsci's own work, D. Litowitz, 'Gramsci, Hegemony and the Law', 2 *Brigham Young University Law Review*, 2000, 515.

108 Karkkainen, 'Post-Sovereign Environmental Governance', and Karkkainen, 'Marine Ecosystem Management'.

109 Karkkainen 'Post-Sovereign Environmental Governance', p. 74.

110 Ibid., p. 75.

111 Ibid.

112 Ibid.

113 Ibid., p. 76.

becomes problem-driven, and its spatial and conceptual boundaries are defined not by reference to fixed territorially delimited jurisdictional lines, but by reference to shared understanding of the nature, scale, and causes of the problem to be addressed.[114]

However, the mechanisms of biopower and the relevant ecopolitical strategies of governing space from an ecological security perspective (like Brunnée and Toope's), shift the problematic in terms of its formal arrangement and the specific *dispositifs* and apparatuses and institutions of power that are deployed and arranged, but not the substance of the biopolitical control and surveillance of nature. The shift invests the vectors and sites of power, but not its overall mechanics.[115] The problem of environmental harm from a sovereignty perspective is *territorial* integrity (and the no-harm rule). From a biopolitical perspective, which understands the complex ecosystem ramifications of human activities, it is, however, no longer the integrity of the territory as a political-legal concept that matters, but the integrity of the ecosystems, understood primarily as that set of living processes whose productivity must be protected, fostered, regularized, enhanced and optimized. This is exactly the site where it is possible to think the 'ecosystem approach' *inside* and yet *against* biopolitics.

FROM BIOPOLITICAL GOVERNANCE TO GOVERNANCE AGAINST BIOPOLITICS

As anticipated in the previous section, governance *does offer* the possibility of a productive reading of the 'ecosystem approach'. The 'ecosystem approach' in this respect may open up space for a political and legal articulation of alternative forms of governance that are as epistemically located as they are operatively autonomous. There is arguably an open space of articulation of a law beyond law. One such alternative law 'resides' and is produced in the transversal practices of communities. This law exceeds the restricted space to which it is now confined in international environmental law, where participation is reduced to an enabling and legitimating element of the subsumption of life under biopower, for example through the deployment of mechanisms such as payment for ecosystem services.[116] Indeed, the theme of governance emphasizes a central element of an analytics of biopolitics, and precisely the dynamic that ties together the active and passive constitution of subjectivities. If the governance architecture of the CBD, and of the 'ecosystem approach' more specifically, constructs local and indigenous peoples as 'environmental managers',[117] the

114 Ibid., p. 77.

115 See in this respect, Foucault, *Security, Territory, Population*, esp. p.25ff.

116 With specific regards to the context of the CBD, see in this regard Kotsakis, 'Biological Diversity Complex', and especially chapter 7 'Governing in the Biocomplex II: The Participation of Local and Indigenous Communities'.

117 'Message from the Executive Secretary, Ahmed Djoghlaf, on the Occasion of the International Day of the World's Indigenous People' (2007); see Kotsakis, 'Biological Diversity Complex' for a comprehensive analysis.

understanding of power that underlies the biopolitical (re-)construction of indigenous subjectivities always entails a space of freedom where resistance is possible.[118] Critical legal scholarship has the opportunity – and the obligation – to divaricate this productive ambiguity.

Thinking law beyond law, and biopolitics beyond biopolitics

In the previous section I have outlined some of the productive ambiguities available to a critical reading of the 'ecosystem approach', and in so doing I have concretized that part of critique that is situated *inside* and *against* biopower. Now it is possible to explore in more general, theoretical terms the possibility of thinking law beyond law, and biopolitics beyond biopolitics. What does it mean, however, to think law beyond law? Critical environmental law is *both* inside and against legal modernity, and fulfils the goal of opening up, problematizing and showing the contingencies and contestations involved in the emergence and descent of environmental legal concepts (and of environmental law more broadly). Yet it needs a further manoeuvre in order to avoid remaining itself caught in the biopolitical aporia that condemns legal modernity, and environmental law, to its tragic circular oscillation. Critique is, however, only one element of a more complex project.[119] The question is: how can one circumvent, displace or otherwise extricate oneself from biopower, so that one may think law beyond law, and biopolitics beyond biopolitics?

The (tentative) answer is still to be found in the immanence of critique, in an ulterior, more subtle sense that resonates and draws inspiration (that is, takes its starting point, to then recalibrate), on the one hand, from Negri's re-articulation of Pashukanis' critique of law,[120] and on the other, from Esposito's mapping of an affirmative biopolitics. Thinking law beyond law means, I argue,

118 Kotsakis for example highlights this double register thus: 'these newly included environmental subjects are taking advantage of this role attributed to them, by providing sites of resistance within the biocomplex. Positioning themselves as the concrete counterpoint to the forces of global markets, they delineate in themselves a real locality to counteract the abstract globality', ibid., p. 218.

119 Some discuss this further complexity of the critical project in terms of a coupling of epistemology and ethics (thus e.g. S. Singh, 'The Critic (-al) Subject', in W. Werner, A. de Hoon and M. Galán (eds), *The Law of International Lawyers Reading Martti Koskenniemi*, Cambridge: Cambridge University Press 2017. Koskenniemi indeed argues that critique needs a normative element, and that 'criticism without an ideal of community is without direction and degenerates into cynism', Koskenniemi, *Apology to Utopia*, p. 547. However, I speak rather of an epistemological and a political coupling, as we have seen in multiple occasions throughout this book.

120 There is no space for engaging in any significant way with this re-articulation, and it will only be sketched at this point. I refer the reader to Negri's own work (e.g. A. Negri, *Dentro/contro il Diritto Sovrano. Dallo Stato dei Partiti ai Movimenti della Governance*, Verona: Ombre Corte, 2009) and to the excellent summary of Negri's re-articulation available in Cava, 'Pashukanis e Negri'.

to think law beyond itself, that is, beyond the restricted space of legal modernity, and of modern environmental law in particular. Restricted space, in turn, refers to, on the one hand, the formal space and hegemonic delimitations of legal modernity – its method, its constitutive link with sovereignty, its sources, its methodological thresholds, its biopolitical horizon of sense. On the other hand, it refers to the increasing framing of law as a technical command, or as a sovereign norm[121] (formulations that equally capture the combination of the sovereign command with the technical, biopolitical dimension of the norm). It refers, in other words, to thinking law beyond the biopolitical aporia that has accompanied and shaped legal modernity at least since Hobbes and Locke, as we have seen in Chapter 8.

In order to find some indication of how is it possible to think law beyond law we might interrogate Hardt and Negri. Yet their articulation of both biopolitics and of the commons (through the key category of the multitude) risks remaining trapped in a binary dialectic where Biopower (Imperial Capital) is opposed to Biopolitics (the social labour embodied by/in the multitude). They remain thus within a paradigmatic modern oppositional thinking framework, amply deconstructed by Derrida (through and with Nietzsche).[122] However, while within this dialectic, they do imagine a 'line of flight'[123] which, starting inside and against, becomes a trajectory of 'exodus',[124] a 'process of *subtraction*',[125] a - 'refusal',[126] a withdrawal from the very relation of power organized following the biopolitical *dispositif*.[127] This 'exodus' is precisely non-hegemonic and exploits the opportunity available in what they call the contemporary 'rupture' of the veil that capitalist biopower lays over the social practices of production.[128] These are all those practices that produce value in common, and that capitalism exploits even as they exceed capitalism's capacity for enfolding them within its operative and organization boundaries.[129]

121 Both expressions aim at consolidating the complementarity of sovereign power and biopower, particularly in its legal manifestation.
122 See e.g. A. Schrift, *Nietzsche's French Legacy. A Genealogy of Poststructuralism*, New York/London: Routledge, 1995, esp. p. 15ff.
123 To use a Deleuzian term also used by Hardt and Negri, for which Deleuze is an important theoretical referent, see Hardt and Negri, *Commonwealth*, p. 152.
124 Ibid.
125 Ibid., emphasis in the original.
126 Ibid.
127 This refusal is a transposition of the operaist refusal of work which demarcated theoretically and politically operaism from orthodox Marxism in the 1960s, see e.g. F. Tomasello, 'Le stagioni dell'Operaismo Italiano', IV:1 *Cosmopolis* 2011, http://www.cosmopolisonline.it/20110715/tomasello.php (Accessed 26 September 2018).
128 Hardt and Negri, *Commonwealth*, p. 150ff.
129 Under conditions of Empire, suggest Hardt and Negri, the exploitation follows the logic of rent, rather than of profit and capital becomes increasingly predatory, as it 'seeks to capture and expropriate autonomously produced common wealth' which is external to the capitalist production form, Hardt and Negri, *Commonwealth*, p. 141.

This entails, again, an epistemological and a political commitment. Yet this operation is also immanent. Indeed, exodus – and this is a crucial point, as it eliminates the possibility of a transcendental retreat – does not mean an actual spatial movement. Exodus implies transformation from within, a transformation that is also a return, given the ontological priority that Hardt and Negri give to the social practices they call biopolitical production. Similarly, thinking law beyond law does not imply an attempt to find law beyond law in the sense of *somewhere else* (in a transcendental plane, in a time to come etc.), as much as it does try to think law, from within, differently, disengaged from its capitalist form and from its biopolitical closure.[130] Here we see the first glimpse of that 'jurisprudence of the middle' that I will discuss further in the sixth section.

We may also interrogate Esposito. He has explicitly sought to outmanoeuvre the biopolitical aporia that he has identified and brought to light in such a lucid manner, by articulating the initial contours of what he calls an 'affirmative' biopolitics.[131] Esposito tries to do this by 'overturning' the logic of the thanatopolitical *dispositif*, whose most 'terrible' and yet paradigmatic (as Agamben suggested) manifestation was the Nazi camp. If Nazism, through the camp, entirely subsumed life in all its bareness under a biopolitical norm that turned into a tragic thanatopolitics of death, Esposito suggests that an affirmative biopolitics can be imagined through a reversal, that is, through a vitalization of the norm.[132] This is not the place to explore, in any depth or detail, Esposito's attempt to articulate an affirmative biopolitics. What can be briefly mentioned, however, is that this overturning entails a transformation *inside* biopolitics. It entails a de-activation of that immunitary declension of the intersection of life and politics that has dominated modernity, and that has constitutively kept biopolitics within the cage of its aporetic core, always denying life in the very gesture that purports to protect it, through mechanisms of immunitary exclusion operating through the legal categories of modernity – chiefly sovereignty and property. Esposito, then, intends to transform biopolitics from a politics *on* life to a politics *of* life,[133] by 'penetrating within [the biopolitical black box], and overturning one by one its bio-thanatological principles'.[134] Esposito intends to think biopolitics beyond biopolitics, cognizant of the fact that the intersection of politics and life is inevitable – so that the biopolitical horizon remains necessarily central to contemporary thought. Yet its *dispositif* can be turned *inside out*; it can be transformed from a self-negating immunitary declension to an affirmative articulation that, for Esposito, re-opens the question of *communitas*. This is a key passage for the purposes of this book, as we will see.

130 Indeed, Negri's horizon of radical change is not located outside of capitalist society, but immanent to it, see Cava, 'Pashukanis e Negri', esp. p. 22–3.

131 Esposito, *Bíos*.

132 Ibid., p. 184ff.

133 Ibid., entire last chapter; see also G. Pezzano, 'Comunità, Immunità, Alterità: una Biopolitica Affermativa e Oltre-umana?' 2 *Tropos* 2011, 167.

134 Esposito, *Bíos*, p.157.

Esposito's attempt at articulating an affirmative biopolitics, however, is perhaps not entirely convincing. Carey Wolfe suggests that Esposito remains ultimately trapped within a vitalistic frame, which leads back to well-known biocentric (and ecocentric) problematics that already arise with deep ecology, a flat biocentrism that posits the equivalence of all life forms.[135] Such flat biocentrism, where 'every life is a form of life',[136] that is, every living being, to the extent that it is an individualized manifestation of Life, cannot be axiologically distinguished from another in any meaningful way, is vulnerable to Curry's critique, which Wolfe reiterates, of being 'intellectually and metaphysically implausible',[137] and a 'hopelessly impractical' strategy.[138] It certainly does not seem to offer any practical way to guide action in the face of difficult questions: does a lethal virus, asks Luke, for example, have equal value and equal right to its own self-realization, a self-realization which, as he notes, may consist in wiping out a sheep herd or a human community, as any human being?[139] (I shall add that the question, importantly, can only be asked from an inevitably immanent human perspective, immediately located inside and against life, producing the fracture that sustains the biopolitical aporia). Esposito's attempts to carve out space for an 'affirmative' biopolitics that is able to extricate itself from both the negative and positive articulations of biopolitics in its immunitary key, seems nonetheless to remain entangled in a transcendentalization of Life and in an impractical philosophical commitment to the value of the living – any living – without any available matrix to guide action.

Wolfe's critique, briefly summarized, seems to hit the mark and to reveal Esposito's strategy as fundamentally susceptible to a thanatopolitical return, precisely through the means of a biopolitical racism as operatively applied in relation to nature and, more specifically, to the concept of biodiversity, which, Wolfe suggests, is effectively the operative concept underlying Esposito's flat biocentrism.[140] Biodiversity, here, arguably functions as a transcendental horizon – Life and its components are, in principle, all equally valuable to the extent that each contributes in key – and sometimes unknown or unknowable – ways to the 'health' and 'integrity' of ecosystems. Yet – and this is crucial – biodiversity conservation, while abstractly committed to the equal value of each form of life, operatively discerns with a sharp thanatopolitical gaze.[141] It discerns between services and disservices that ecosystems deliver to humans. It also

135 Thus e.g. Wolfe, *Before the Law*, p. 59.

136 Esposito, *Bíos.*, p. 194.

137 Wolfe, *Before the Law*, p. 60.

138 Ibid.

139 Tim Luke, quoted in Wolfe, *Before the Law*, p. 59.

140 Ibid., p. 60. Wolfe does however recognize how Esposito would likely reject the notion of biodiversity (which he introduces as somewhat equivalent to biocentrism, ibid., p. 60). However, Wolfe also argues that the concept of biodiversity 'would seem in tune with the "qualitative and quantitative expansion" of life that Esposito values', ibid., p. 60.

141 See for example, De Lucia, 'Bare Nature'.

discerns in relation to different species, and in two distinct ways. First, species' value is anchored to their vulnerability and conservation status, so that unique or rare species are more valuable than common species, a point underlined by Wolfe.[142] The second way is an illustration of a far more advanced and intensified operationalization of biopower: the value of species in the Anthropocene (which forcefully demands increasing human intervention in order to *manage* life),[143] is evaluated in relation to their *function*, while their native or wild status becomes increasingly secondary, if not altogether a hindrance.[144] Indeed, Hobbs et al. suggest that '[r]apid, extensive, and ongoing environmental change increasingly demands that humans intervene in ecosystems to maintain or restore ecosystem services and biodiversity'.[145] Consequently, continue Hobbs et al., 'the challenge is to move the focus from preserving existing species and assemblages within particular designated places ... to considering how to conserve systems that are temporally and spatially dynamic'.[146] This is a key shift in focus, from structural to functional integrity, and it may result in the hierarchization of species in relation to their ability to perform valuable functions for the maintenance and enhancement of the productivity of ecosystems – a thorough biopolitical strategy.

From this perspective, ecosystems are not valuable as things-in-themselves. It is rather their 'multi-attribute bundle of various ecosystem goods and services [that they] supply to society that have value (or more appropriately utility) to society'.[147] Consequently, the provision of ecosystem services may very well be 'best achieved through the addition of non-native species',[148] insofar as their functional role is equivalent,[149] or even more effective. This in turn creates an opening for the biopolitical hierarchization of species in relation to the double parameter of functional equivalence and economic efficiency. The biopolitical 'calculated management of life'[150] is here immediately evident.

Returning to Esposito, a further notation is necessary, as his approach may be also understood from another angle. His 'principle of unlimited equivalence for every form of life',[151] rather than expressing a flat biocentrism or a refusal to act

142 Wolfe, *Before the Law*, p. 60.
143 See, e.g. Hobbs et al., 'Intervention Ecology'.
144 Though as we have seen in Chapter 7, there is raging debate within the ranks of conservation biology and ecological sciences on this very set of questions.
145 Ibid., abstract.
146 Ibid., p. 444.
147 Hearnshaw et al., *Ecosystem Health Demystified*, p. 23.
148 Ibid., p.4.
149 This is also the opinion of a number of 'post-natural' ecologists such as D. Cole and L. Yung (eds), *Beyond Naturalness: Rethinking Park and Wilderness Stewardship in an Era of Rapid Change*, Washington, DC: Island Press, 2010. See also G. Steinhoff, 'Ecological Integrity in Protected Areas: Two Interpretations', 3 *Seattle Journal of Environmental Law* 2013, 155.
150 Foucault, *History of Sexuality*, p. 140.
151 Esposito, *Bíos*, p. 186.

altogether, may be interpreted as a *refusal* to establish a priori and fixed hierarchies between different forms of the living. This way of interpreting Esposito's attempt at articulating an affirmative biopolitics (which, by his own admission, is something that he is only able to 'glimpse'),[152] points to something that appears to be close to what Philippopoulos-Mihalopoulos calls, in Deleuzian terms, and from a more immediately critical *legal* perspective, a 'jurisprudence of the middle'.[153]

Finally, thinking law beyond itself, and against biopolitics, also importantly means that one does not need to think against law as such. The latter is the more usual critical strategy. For example, Marxist critiques contemplate, ultimately, the disappearance of law,[154] and critical legal studies look at law most often *only* as an instrument of oppression.[155] Thinking law beyond law, by converse, entails conceiving of law historically and in the plural – as a series of distinct experiences that cannot be reduced to one another, nor to a master version of Law with a capital L. There are diverse modes of existence of law, each historical and contingent, and they can be thought against one another.[156] One can thus think law beyond law, and law against law. Before, however, further exploring what a jurisprudence of the middle looks like in relation to the 'ecosystem approach', it is necessary to understand to what extent and in what manner ecology may help to think law beyond law. I will return to the concept of 'epistemic location' (that emerges as one important element in the productive reading of the ambiguities of the 'ecosystem approach' but remains yet unexplored and unexplained) and transversal ecological thinking.

152 Ibid., p. 194.

153 Philippopoulos-Mihalopoulos, 'Actors or Spectators?'

154 See in this respect especially Pashukanis, *Law and Marxism*.

155 Marella, for example, suggests that, with particular regards to the recent financial crisis, '[i]n the amazingly huge literature on the crisis, the vast majority of the projects that articulate possible radical-leftist exit strategies give little or no significant role to law. In current descriptions of the innumerable forms of resistance and/or civil disobedience against the law in force, law is inescapably depicted as the guardian of the status quo, a neoliberal weapon thrown against the social ethos. The relationship between law and resistance has instead become extremely significant', and she refers most especially to the Italian experience of the 'beni comuni', and law is used 'in many sites of political struggle. The basic idea is not to resist the law but to resist through the law, by means of the law', M. Marella, 'The Constituent Assembly of the Commons (CAC), Open Democracy, 2014a, 28 February https://www.opendemocracy.net/can-europe-make-it/maria-rosaria-marella/constituent-assembly-of-commons-cac, (Accessed 26 September 2018).

156 I am referring here to the theory of the 'juridical experience', developed over the course of several decades by Italian legal scholarship. I cannot explore in any detail this theory here, so I will only refer to some literature: R. Orestano, *Introduzione allo Studio Storico del Diritto Romano*, Bologna: Il Mulino, 1987; P. Grossi, 'Storia di Esperienze Giuridiche e Tradizione Romanistica', 17:1 *Quaderni Fiorentini per la Storia del Pensiero Giuridico Moderno* 1988a, 533.

'Epistemic location' and transversal ecological thinking

Before outlining what, combining the work of Felix Guattari[157] and Lorraine Code,[158] I call *transversal ecological thinking*,[159] it is perhaps useful to rehearse some of the arguments that have been made throughout this book in relation to ecology. As we have seen, ecology is one of the central themes of our contemporaneity, both in relation to the multiple and overlapping ecological crises and in relation to its capacity to fundamentally problematize the categories of (legal) modernity.[160] Ecology, however, raises complex questions. Due its double epistemic role – as a science and as a normative framework – ecology is fundamentally affected by a 'moral ambivalence'.[161] Its unifying premise, moreover, has been linked to the most intense reproduction of the logic of modernity.[162] Finally, ecology is affected by an irreducible genealogical complexity,[163] and lies at the centre of discursive contestations aligned with a variety of political projects, responding to conflicting paradigmatic orientations, and aimed at imposing different hegemonic closures.[164] If ecology is used to derive philosophical implications aimed at challenging the modern construction of nature, it is also deployed as a legitimating framework for the continued enforcement of nature as a modern category and for its exploitation.[165] If the discourse of ecology inspires an apparent radicalization of legal theory,[166] it also enables new biopolitical regimes aimed at the regularization of the provisions of ecosystems goods and services.[167] What of ecology then? Can it provide useful conceptual resources for a critical environmental law, or does it merely reproduce the categories of modernity in an intensified form?

The question, I suggest, needs to be displaced to give a useful answer. I try then to approach ecology not from a substantive point of view (what is ecology), but rather from a methodological one (what does it mean to think ecologically?). In order to re-think ecology in methodological terms, I will draw on

157 F. Guattari. *The Three Ecologies*, London: Continuum, 2008.

158 Code, *Ecological Thinking*.

159 But in a similar direction, and indeed drawing from Deleuze and Guattari, see also Philippopoulos-Mihalopoulos, who speaks of an 'open ecology' where there is 'no distinction along the lines of human/natural/artificial/technological. On the contrary, these categories fold into each other and constantly emerge as epistemological and ontological hybrids', Philippopoulos-Mihalopoulos, 'Actors or Spectators?', p. 857.

160 In at least *some* of its forms, see e.g. Tallacchini, 'Legal Framework'.

161 Worster, *Nature's Economy*, 2nd edn, p. 256.

162 Argyrou, *Logic of Environmentalism*.

163 Golley, *History of the Ecosystem*; Mitman, *State of Nature*; De Lucia, 'Competing Narratives', specifically in relation to how this complex genealogy manifests in international environmental law.

164 Bell, 'Non Human Nature'; De Lucia, 'Competing Narratives'.

165 Darier, *Discourse*; Bell, 'Non Human Nature'.

166 For a comparative discussion of two such legal philosophical approaches see De Lucia, 'Towards an Ecological'

167 Darier, *Discourse*.

existing work that also approaches ecology from a methodological perspective rather than as a multiply orientable field of knowledge, or as the conceptual basis for a multiplicity of ethical frameworks.[168] Ecology, from this perspective, is a *mode of thinking*. Moreover, it is best understood, I suggest, as a *transversal* mode of thinking.

But what does transversal mean, in relation to ecology? Isn't ecology, in and of itself, already transversal, in light of its focus on interconnections between the biotic and abiotic realms? I suggest that transversality goes beyond the already wide gaze of ecology as a scientific discipline. Indeed, transversality exceeds the traditional focus of ecology – that is, nature, however problematic this term is.

In this section,[169] I take on board recent expansions of the scope of ecology[170] for two primary reasons. First, it may help re-orient thinking precisely away from the political epistemology of modernity, premised on conceptual thresholds such as 'nature' and 'culture', 'mind/spirit' and 'matter' etc..[171] Second, it may help to further articulate a critical environmental law. It is French psychoanalyst and philosopher Felix Guattari who offers the first useful re-framing of ecology as a political methodology. Guattari suggests, in fact, that ecology requires us to think 'transversally',[172] that is, it must simultaneously embrace the natural, technical, social and psychological plane, if there is to be hope of addressing the 'increasing deterioration' of each.[173] Lorraine Code similarly suggests that ecological thinking, by subverting the neat separation of the personal, the social, and the natural, allows the formation of transversal links between, say, ecosystems and international capitalism.[174] Such perspectives resonate with Latour's revelation of nature as a methodological device instrumental to the political epistemology of modernity, whose primary function is that of producing a threshold that separates those voices that can be legitimately heard from those who lacks such legitimacy.[175] The displacement of the modern epistemological category of nature, however, does not displace nature as *world*.[176] As Cosi suggested,[177] world is preferable to nature since the latter is a modern category susceptible of logical analysis, while world indicates a complex,

168 deLaplante, 'Environmental Alchemy'.

169 Which draws on De Lucia, 'Re-embodying Law'.

170 deLaplante,'Environmental Alchemy; Guattari, *Three Ecologies*; Code, *Ecological Thinking*.

171 On the political epistemology of modernity in relation to the idea of nature see in particular, B. Latour, *The Politics of Nature*, Harvard University Press, 2004, and Latour, 'An Attempt'.

172 Guattari, *Three Ecologies*, p. 29.

173 Ibid. p. 28.

174 Code, *Ecological Thinking*, p. 19.

175 Latour, 'An Attempt', p. 476.

176 On the preference of the word 'world' (and the semantic and material field it evocates) over that of 'nature' see G. Cosi, 'Tutela del Mondo e Normatività Naturale' in G. Lombardi-Vallauri (ed.), *Il Meritevole di Tutela*, Milano: Giuffrè, 1990.

177 Ibid.

transversal, messy, living mesh which resonates much better with the need to think law beyond law, that is, beyond legal modernity and beyond biopolitics as its crucial horizon of sense. So, if this reading of ecology dismantles nature as a threshold of modernity (at once ideo-ontological and epistemological), it does not dismiss the complex, embodied presence of the living world.[178]

Ecology then, should be understood to entail a plurality of registers engulfing the modern binaries in an irreducible, transversal mesh. This engulfment overwhelms and then displaces the epistemological paradigm of modernity through which independent and autonomous subjects know only isolated objects.[179] Understood in this sense, ecology likewise displaces the ideo-ontological foundation of modernity. It is only on these premises, I argue, that ecology can be mobilized towards the elaboration of a critical legal framework responsive to the demands set out in the introduction: imagining a new (legal) language not reducible to modernity through which one can effect a 'deep re-think' of law. A transversal ecology, then, is neither anthropocentric nor ecocentric; it does not operate from any centre.[180]

Now we can revisit the epistemological perspective of ecology and further elaborate it in relation to the goal of this chapter, namely teasing out the productive ambiguities of the 'ecosystem approach'. A central insight of ecology, as we have seen, relates to the instability of knowledge. Ignorance and uncertainty acquire a crucial epistemic and normative role.[181] This endemic (and perhaps insuperable) uncertainty, in turn, derives from the complexity of non-linear and cross-scalar ecological processes, and is further complicated by 'truth pluralism'.[182] Such pluralism inevitably also aligns differently with different normative, ethical and political commitments arising from both scientific and legal processes,[183] as knowledge and values are entangled on both objective and

178 'Living' is intended here in its widest possible meaning, hence including also, say, rocks and other 'inert' bodies, under the understanding that each and every material body is an embodiment, a *somatic space* 'abundantly filled with meaning' (R. Thomas-Pellicer, What is Kultur?: The Places of God in the Age of Re-embodiments, PhD Thesis, Surrey: University of Surrey, 2012, p. 7, page number refers to chapter 4, single document on file with author), and possessing 'a lively materiality and agential significance of their own', Anna Grear, personal communication.

179 Code criticizes also merely 'contextual' understandings of knowledge, to the extent that, she suggests, in such accounts text and context can still be separated. Contextual accounts then fail to recognize the irreducible entanglement of knower and known in what she calls reciprocal relations of constitution, Code, *Ecological Thinking*, p. 147, but also throughout, as this is a central thesis of Code's book, see e.g. the section called 'Situated knowing, Individualism, and Mastery', ibid. p. 133ff.

180 Philippopoulos-Mihalopoulos, 'Actors or Spectators?', esp. p. 860; Grear, 'Towards a New Horizon'; De Lucia, 'Re-embodying Law'; De Lucia, 'Beyond Anthropocentrism'.

181 Shrader-Frechette, 'Methodological Rules'; Code, *Ecological Thinking*; deLaplante, 'Ecosystem Management'.

182 See e.g. Gutwirth and Naim-Gesbert, 'Science et droit'; see also De Sadeleer, *Environmental Principles*.

183 Tallacchini, 'Legal Framework', p. 1095.

subjective grounds.[184] If knowledge is uncertain, unstable and negotiated – that is, genealogical – it becomes important, following Latour, to embrace a shift in focus from matters of fact to matters of concern.[185] And in the context of this shift, law, no longer defensible as a *sovereign command*, nor as a *biopolitical norm*, must embrace the productive potential of epistemic instability and accept 'the responsibility to solve problems which science cannot decide and that are linked to uncertain outcomes'.[186] What becomes important, as Lorraine Code suggests, is 'epistemic location', that is, a knowledge produced 'down on the ground'.[187] What this means is that 'the nature and conditions of the particular "ground", the situations and circumstances of specific knowers, their interdependence and their negotiations' all become relevant, even crucial, factors.[188] Truth is no longer a regime, but a 'truth to',[189] a form of interpretation (rather than of verification) that is 'textured and responsive',[190] as well as 'responsible [to] local sensitivity'.[191]

Indeed, ultimately truth is a form of life responsive to the embodied world and a form of responsible knowledge, meaning a form of knowledge cognizant of the 'multiply contestable' nature of categories and taxonomies that impose permanent closure on the living world.[192] This is the opposite of a biopolitical epistemology, aimed at subsuming life, without its unbiddable excesses and unpredictable irregularities, within protocols of surveillance and regularization.

Another important effect of the epistemic location of ecological thinking is that formal science, for all its recognized 'force',[193] needs to be *situated* in the

184 Subjective biases may arise from personal biases; from social or cultural preferences; and/or from methodological choices which themselves carry an axiological dimension. Methodological operations such as extrapolation from one context to another 'are never neutral and univocal, but are always influenced by values and goals', ibid at 1096. See also Shrader-Frechette, *Risk and Rationality* and Shrader-Frechette, 'Methodological Rules', pp. 12–39.

185 Latour, 'An Attempt', p. 478.

186 Tallacchini, 'Legal Framework', p. 1095.

187 Code, *Ecological Thinking*, p. 5.

188 Ibid., pp. 5–6.

189 Ibid., p. 7.

190 Ibid.

191 Ibid., p. 8.

192 Ibid., p. 50

193 I use 'force' here rather than truth deliberately, having in mind Hornborg's work (Hornborg, *Power of the Machine*) on the relation between the truth of techno-science and the global capitalist system. Hornborg identifies an intimate, inevitable link between industrial technology and the unequal exchange relations that facilitate the extraction, appropriation and accumulation of ecological and social resources in a world system aligned along a core – periphery continuum. It is capitalist accumulation '*which made industrial technology possible to begin with*', p. 46 (emphasis in the original). 'If specific technologies require and reproduce specific forms of social organization', continues Hornborg – and it is well worth quoting at length – 'it is no less true that industrial technology as a general phenomenon requires and reproduces a specific world order … technological knowledge is "true" (i.e., "works") only within a restricted social space. The social definition of what is technologically feasible is not external to technology but … intrinsic to it', p. 107. In other words, machine

particular place where its regime of truth is to leave traces and marks if it is to be responsive and responsible. Epistemic location recognizes precisely science's character of 'determined abstraction', to use a Marxian concept. If abstraction is a necessary operation of thought, its effects are always produced within a particular social and ecological fabric, in a particular historical place and time. Knowledge, thus, must be *situated*, it must be re-embedded and re-integrated in the richness of life's problematics. Only then will it be able to be responsible and respond to life's demands. This is perhaps a key passage of the transformation Esposito imagines from a politics *on* life, to a politics *of* life. The 'ecosystem approach', read productively, may point exactly to a perspectival shift whereby the epistemic location of knowledges becomes *the model and the condition* for the concretization of abstract regimes of truth on the 'vulnerable living order'.

Knowledge, from the perspective of epistemic location, happens in the middle, rather than being delivered from a centre that reproduces and enforces power through particular regimes of truth. Likewise, I argue following Philippopoulos-Mihalopoulos, law happens *in* the middle, rather than *from* the centre, as it is precisely in the particular 'grounds' that law can be thought beyond law and beyond biopolitics.

Law in the middle of the world

All that we have discussed thus far, including, especially, the shift in critical register from the anthropocentrism/ecocentrism binary to biopower and biopolitics; the immanence of critique; the productive ambiguities we have briefly outlined or even merely alluded to; and the notion of a transversal ecology; all of this points in one direction. The centre is a paradigmatic construction of modernity, the fulcrum of its ideo-ontology and of its epistemological *dispositif*. The idea of middle, by contrast, first introduced as a novel jurisprudential horizon for critical environmental law by Andreas Philippopoulos-Mihalopoulos, offers much critical potential.

Transversal ecology and epistemic location both speak of an inevitable immersion and participation in life, in the *middle* of life. This immersion indeed begins in the middle of that 'plane of immanence'[194] that contains without transcendental excess 'its own origin, causality and teleology'.[195] This middle,

productivity *in itself*, disjointed from global accumulation practices, cannot exist. Considering the intimate relation between science and technology (not only in terms of a theory and its practical applications, but also in terms of the technological nature of science, that is, science requires technology in order to work: data processing being only the most obvious example), this line of reasoning seems to hit the nail right on the head in relation to the argument of this book.

194 Philippopoulos-Mihalopoulos, 'Actors or Spectators?', p. 857, quoting Deleuze and Guattari.

195 Ibid.

suggests Philippopoulos-Mihalopoulos, 'does not allow for a perspective that calls itself an origin'.[196] Thus located in the middle, in the midst of a sensuous and unpredictable life, law is in a position to disarticulate the entire modern logic of presupposition that underlies legal modernity.[197] The perspective of the middle is moreover able to displace the modern obsession with the centre, an obsession represented, within the context of environmental philosophy and law, by the anthropocentrism/ecocentrism binary,[198] and that condemns legal philosophical reflections to oscillate between the sacred and abject views of nature.[199] Indeed, it denies its very possibility.[200] With Cosi, as already suggested, I call this middle of life the 'world'.

If law is epistemically located, if it begins and ends in the middle, where life has no stable, inexorable thresholds and demarcations (and here perhaps one can refer back to Esposito's idea of the suspended equality of all life), then the world is neither sacred nor abject, and law becomes itself a form of life. Yet the world does not offer any guarantee of being *good*. Indeed, as Philippopoulos-Mihalopoulos warns, the world is complex and its space is 'a space of struggle'.[201] This is an important point, as one of the crucial thresholds of modernity is that between order and chaos (in all its declensions: peace and conflict; culture and nature; civilization and savagery, etc.),[202] a threshold that is likewise reproduced in the anthropocentrism/ecocentrism binary. Yet life, being in the middle (and being itself the middle), is necessarily traversed by conflict. We return to genealogy and to the intersection of politics and life. If the intersection happens in the middle, however, there may be room for the imagination of this encounter from the perspective of a different horizon of sense, one where conflict (and this is the premise of an entire philosophical[203] and legal[204] tradition) does not precede order, nor does it follow it (when order unravels). From this perspective, there is no clear demarcation, no Hobbesian threshold between the two that understands conflict as opposite to order, thus requiring neutralization. Through the

196 Ibid., p. 860.

197 Modern subjectivity is instituted through a logic of presupposition, that is, a presupposition of antecedence of the subject with respect to both the world of phenomena, and to itself as a body, Esposito, *Living Thought*.

198 Philippopoulos-Mihalopoulos, 'Actors or Spectators?'.

199 M. Hasley, *Deleuze and Environmental Damage. Violence of the Text*, Farnham: Ashgate, 2006. This oscillation is pervasive, and has been at the root of heated debates in relation to biodiversity conservation for quite some time; see e.g., Kloor, 'Battle for the Soul'.

200 Philippopoulos-Mihalopoulos, 'Actors or Spectators?', p. 860.

201 Ibid.

202 See De Lucia, 'Semantics of Chaos'.

203 Esposito, *Living Thought*; but the productive potential of conflict was already recognized by Heraclitus, in what became a minority position within the context of Greek Philosophy, se e.g. M. Butz, 'Chaos Theory, Philosophically Old, Scientifically New', 39:2 *Counseling & Values* 1995, 84.

204 Bascherini, 'Italian Theories'; Grossi, *L'Ordine Giuridico.*

ambiguous productive potential it generates, conflict is *constitutive* of order, an order however always provisional.[205]

Moreover, the new, mobile ontological commitments that ecology confronts us with, pose problems for modern law, whose grammar and operational mechanics, firmly anchored in the centre, is dependent on clearly delineated subjects and objects and on linear causal relations. As Christopher Stone underlined in his seminal article 'Should trees have standing?'

> there are large problems involved in defining the boundaries of 'natural objects' … from time to time one will wish to speak of that portion of a river that runs through a recognized jurisdiction; at other times one may be concerned with the entire river; or the hydrologic cycle – or the whole of nature. One's ontological choices will have a strong influence on the shape of the legal system.[206]

Indeed, ecosystems inhabit shifting ontologies, depending on the particular perspective, point of view or problem being confronted. How can law, then, accommodate the shifting subjectivities that ecology confronts us with, and give voice to the pervasive and embodied difference of the 'vulnerable living order'?[207] Widening the scope of the subject as a vehicle for giving a legitimate legal voice to natural and other non-human entities may be a practical tactic for the immediate exploitation of the opportunities existing legal systems afford.[208] However, one suspects that (besides the unresolved problems with respect to *shifting* subjectivities) the template of the subject will always tacitly operationalize its exclusionary orientation, by way of inducing ontological and epistemological transformations and simplifications in order for the subsumption of the novel subjectivities to become legally viable.[209]

What can we do then? How can we (re-)imagine law and its categories? A jurisprudence of the middle may offer some help. Thinking law in the middle, as a form of life and as a living practice, denies the perspective of the centre. It calls the bluff of a central gaze that, external to life, can dominate it, control it, regularize it, or from which life can be 'panoptically surveyed'[210] (the perspective of the centre is, in this respect, violent, is a sovereign perspective). Located 'in the middle … of concrete situations',[211] law is able to operate 'without

205 See in general, Esposito, *Living Thought*, especially the parts dedicated to Machiavelli and to Vico.

206 C. D. Stone, 'Should Trees Have Standing? Toward Legal Rights for Natural Objects', 45 *Southern California Law Review* 1972, p. 456.

207 Grear, 'Vulnerable Living Order'.

208 This is for example the approach taken by Cormac Cullinan (personal annotation during the Conference 'Rule of Law for Nature', Oslo, 9–11 May 2012).

209 Tallacchini, *Diritto*, speaks in this respect of a 'field effect', p. 337, my translation.

210 Philippopoulos-Mihalopoulos, 'Actors or Spectators?', p. 860.

211 Ibid., p. 866.

preconceived structures'[212] and without fixed categories of subject and object. No closures. No guarantees either, however. Life (and law) in the middle is fluid.

Towards a new horizon of sense?

I argued earlier that thinking law beyond law implies not so much thinking beyond law in the sense of imagining its disappearance,[213] but rather thinking law *against* law. The 'ecosystem approach', then, can be read against itself – that is, against its biopolitical closure – precisely through a productive reading as that I have briefly proposed in the third section of this chapter. This, however, begs the question: how can this be accomplished? I suggest that the horizon of sense through which the 'ecosystem approach' can be theorized against itself (and against its biopolitical closure), is that of the commons.

The commons has almost already entered the narrative in at least two circumstances in this chapter. First, in relation to Esposito's attempt at articulating an affirmative biopolitics, which, from his perspective, must inevitably pass through the overturning of the immunitary paradigm of modernity. This in turn requires the re-articulation of the intersection of life and politics in terms of *communitas.* Second, when discussing Brunnée and Toope's attempt to imagine ways in which an 'ecosystem approach' in international law might be articulated. They in fact pointed to the concept of common concern as an initial form of limitation of sovereignty exceeding the narrow form of *injuria* (that is, transboundary harm),[214] and extending to areas beyond national jurisdiction, that is global commons.

In the first case, the idea was altogether unexplored and remained perhaps cryptic. However, the complex line of inquiry Esposito follows, a line of inquiry that

212 Ibid.

213 As is common in Marxist critiques of law, see e.g. Pashukanis, *Law and Marxism* and Cava, 'Pashukanis e Negri'. This is not the place for engaging with this question, however it can be underlined that the disappearance of law refers rather more narrowly to the disappearance of the 'juridical form', that is a particular historical instantiation of law as a thoroughly autonomous field of knowledge and practice, disengaged and disembedded from the political, economic and religious forms from which it was almost 'undistinguishable', ibid., p. 10. Understood in this narrower sense, perhaps the distance is not that great between a Marxist critique of law and the critical environmental law articulated here whereby the horizon of critique points towards thinking law against law.

214 The concept of *injuria* operates in two directions. It delimits legal subjects from one another through the territorial boundary, establishing the sphere of freedom of each subject. And it demarcates the legal subject from the objective world, *incapable* of suffering *injuria*; indeed, things cannot have any right, as they lack rationality and, in turn, cannot suffer any wrong. Vitoria would in fact say that animals and in general all irrational creatures '*non possunt pati injuria: ergo non habent jus*', as quoted by L. Baccelli, *Guerra e Diritti. Vitoria, Las Casas e la conquista dell'America*, XXXVII *Quaderni Fiorentini per lo Studio del Pensiero Giuridico Moderno*, 2008, p. 75.

inserts *communitas* within a dialectic with the modern paradigm of immunization, exceeds the scope of this book. In the second case, the horizon opened up by the concept of common concern, while introducing something new and different, remains firmly within the horizon of sense of modern international law, only tinkering at the margin of its structures and categories.[215] The concept of the commons, which is outlined here as a tentative map towards a concept of law in the middle strays significantly from the timid allusions of the concept of common concern, while at the same time maintaining a semantic connection precisely in the reference to something which is common, and not private, nor public.

The commons, once the dominating form of resource governance, were effectively destroyed at the onset of modernity.[216] Some marginal enclaves still endure and resist, but as nothing more, it has been noted, than 'the institutional debris of societal arrangements that somehow fall outside of modernity'.[217] Indeed it has been observed that the commons are 'invisible', or worse, 'unthinkable' to legal modernity.[218] As a dis-homogenous plurality of collective forms of properties,[219] the commons are a 'monstrosity'[220] disruptive of the civilized (legal and economic) order of the State,[221] because they exist outside of the processes of the capitalist circulation of commodities.[222] Yet a number of social practices (which, as we shall see, are *ecological* in the transversal sense

215 And this reflection is also valid, I argue, in relation to the concept of community of interest, even though it moves the bar closer to the horizon of sense the commons, I will show, opens up. Perhaps this movement – of which the community of interest represents the most advanced effect – might be read as an initial rupture, to refer back to Hardt and Negri, through which an exodus can be initially imagined, and theorized. I shall not delve into the history and theory of the community of interest, but it is useful to keep this in mind, as the commons I speak of opens a much more radical horizon, one where the very notion of sovereignty has no legitimate space, however limited, greened or otherwise re-oriented towards sustainability.

216 As Mattei has suggested, the birth of modernity is directly linked to the destruction of the commons, U. Mattei, *Beni Comuni: Un Manifesto*, Bari: Laterza, 2011.

217 Arun Agarwal, quoted in B. Weston and D. Bollier, *Green Governance. Ecological Survival, Human Rights, and the Law of the Commons*, Cambridge: Cambridge University Press, 2013, p. 146.

218 Ibid., p. 173

219 Grossi consistently uses the plural form when referring to collective forms of appropriation, in order to emphasize the conceptual and ideological distance between those and modern private property, utterly singular, uniform and intolerant of diversity and plurality. See for example P. Grossi, 'La Proprietà e le Proprietà nell'Officina dello Storico', 17 *Quaderni Fiorentini per la Storia del Pensiero Giuridico Moderno* 1988b, 359.

220 P. Grossi, 'Assolutismo Giuridico e Proprietà Collettive', 19 *Quaderni Fiorentini per la Storia del Pensioro Giuridico Moderno* 1990, 505, p. 510.

221 Thus Regnoli, an Italian bourgeoisie jurist, as reported in Grossi, 'Assolutismo Giuridico', p. 510.

222 See e.g. ibid. and P. Parajuli., 'Revisiting Gandhi and Zapata: Motion of Global Capital, Geographies of Difference and the Formation of Ecological Ethnicities' in M. Blaser, H. Feit, and G. McRae (eds), *In the Way of Development. Indigenous Peoples, Life Projects and Globalization*, London: Zed Books, 2004.

outlined above) are re-claiming the commons both materially as common goods or resources, and more broadly as a philosophical, political and juridical horizon.[223] Indeed, the genealogy of traditional subsistence commons is currently converging with contemporary political and legal articulations of the new commons, such as 'civic',[224] 'digital'[225] or 'cultural' commons[226] (often collectively called 'immaterial' commons).[227] These are parts of a process of shifting legalities and inter-legalities beginning to occupy spaces in the interstices of modernity, exploiting fractures and moving across porous cultural and conceptual territories,[228] perhaps also possibly coalescing into insurgent trajectories.[229]

A comprehensive account of the commons is beyond the scope of this chapter.[230] I will then only offer some concluding reflections on how the commons may offer a new horizon of sense that could help us to think law against law and the 'ecosystem approach' against the biopolitical closure in such a way that it might be possible to overturn the biopolitical aporia.

First, the practices of the commons are transversal, as they do not respect modern epistemological or ontological boundaries that delineate nature, culture, knowledge, being and having. The commons are practices that traverse all these domains and bring them together.

Second, the practices of the commons happen in the middle, in the world. In this respect, the practices of the commons are epistemically located (inevitably so, one might add). The very embodied social practices of the commons constitute the basis for knowledge as a form of life.[231] Knowledge and truth, then, are neither dependent on 'individual, haphazard experience'[232] nor on

223 Weston and Bollier, *Green Governance*; Mattei, *Beni Comuni*; Hardt and Negri, *Commonwealth*.

224 Civic and/or social commons are those urban commons such as public libraries, schools, parks, theatres etc.

225 Digital commons are, for example, the Internet, various common resources such as Wikis and Open Source Software etc.

226 Such as theatres, a prime example being the 'Teatro Valle' in Rome.

227 Thus e.g. M. Marella, *Oltre il Pubblico e il Privato. Per un Diritto dei Beni Comuni*, Bologna: Ombre Corte, 2012.

228 See e.g. De Sousa Santos, *Towards a New Common Sense*.

229 See e.g. V. De Lucia, 'Law as Insurgent Critique. The Perspective of the Commons in Italy', *Critical Legal thinking*, 2013b, 5 August, http://criticallegalthinking.com/2013/08/05/law-as-insurgent-critique-the-perspective-of-the-commons-in-italy/ (Accessed 26 September 2018); U. Mattei, 'Protecting the Commons: Water, Culture, and Nature: The Commons Movement in the Italian Struggle against Neoliberal Governance', 112:2 *South Atlantic Quarterly* 2013, 366.

230 For different takes on the commons, see Weston and Bollier, *Green Governance*; Hardt and Negri, *Commonwealth*.; Mattei, *Beni Communi*, 2011, op. cit.

231 Hardt and Negri draw here inspiration on Wittgenstein, and his description of linguistic practices: 'to imagine a language mean to imagine a form of life', L. Wittgenstein, *Philosophcial Investigations*, Oxford: Blackwell, 1953, p. 75, as quoted in Hardt and Negri, *Commonwealth*, p. 122.

232 Ibid.

transcendental idea(l)s,[233] but are formed, in their plurality, through a process which historicizes and contextualizes truths as forms of common practice, whose truth-value is an effect of their being attuned to the space/time – social and cultural – created and embodied by such practices. This approach is, in many ways, similar to the notion of 'truth to' articulated by Lorraine Code:[234] ultimately, truth is a form of life responsive to the embodied world. In fact, as Code emphasizes, '[p]hysical, social, or other location … functions neither as backdrop nor context, but as constitutive of the *Lebenswelt*, and the *Leben* within it, shaping possibilities of knowing, and demanding, themselves, to be known'.[235] In the same way, such physical, social or other locations are constitutive of the practices of the commons. Inhabiting a place – and a place can be a mountain, a forest, a city or a particular occupied/liberated building – the commons is a form of 'responsible knowledge',[236] a social practice and a juridical institution.[237] The legal-modern form of law, constructed as an autonomous field of knowledge and practice is disarticulated.

Third, the commons are conflictual. The commons in fact reject any essentialism and cannot be framed as something already there. This is true even in relation to natural resources, often framed *objectively* as commons.[238] The commons is,

233 In this way Hardt and Negri, *Commonwealth*, mean to overcome both reformist hopes nurtured within the discourses of 'hypermodernist' scholars such as Ulrik Beck and Jürgen Habermas ('There is no faith here that the core principles of modernity can be reformed and perfected', p. 114) and post-modern discouragement as regards the impossibility of making any claim to truth ('[I]n contrast to most propositions of postmodernity [the perspective of the commons] provides a strong notion of new values, new knowledges, and new practices; in short [it] constitutes a new *dispositive* for the production of subjectivity', p. 115, emphasis in the original).

234 Code, *Ecological Thinking*, p. 7.

235 Ibid., p. 8.

236 Ibid., p. 50.

237 In the sense given to institutions as the springboard of any juridical phenomenon by Italian legal theorist Santi Romano. Romano describes an institution very widely as 'every social entity or body' that has a significant enough measure of stable pattern, form and/or organization, S. Romano, *L'Ordinamento Giuridico*, 2nd edn, Florence: Sansoni, 1946. Romano thus radically identifies the social and the legal (so that neither has causal, logical or temporal priority). A legal order ('ordinamento giuridico'), from this perspective, 'is the concretisation of a social fact … the effectiveness of its structure', F. Fontanelli, 'Santi Romano and L'ordinamento giuridico: The Relevance of a Forgotten Masterpiece for Contemporary International, Transnational and Global Legal Relations', 2:1 *Transnational Legal Theory* 2001, 67, p. 79. This is not the place for elaborating on Romano's theory. For an English account of his theory see Fontanelli, 'Santi Romano' and M. La Torre, *Law as Institution*, London and New York: Springer, 2010, especially pp. 98 – 115, where La Torre discusses Romano's theory and then compares it with MacCormick and Weinberger's (new) institutionalism.

238 Importantly however, certain resources are immanently more susceptible to be claimed as commons, to the extent that they are necessarily implicated in the sustenance of life. There is no commons though without commoners, even if, as Rodotà underlines, the legal regimes proceeds from the resource, and not vice versa. Both Weston and Bollier, *Green Governance*

instead, an open category. As Mattei emphasizes, the commons are discovered, affirmed and re/produced through social practices and/or struggles, and their continued existence is contingent upon the continued practices which affirm and (re-)claim it:[239] indeed, the very social conflict that leads to the recognition of something as a commons is considered an integral part of that commons.[240]

Crucially, the commons responds to a language of *doing* and *becoming*, rather than to a language of *having*.[241] The commons, accordingly, is not primarily an object of (proprietary or sovereign) rights (which in turn requires a subject), but is rather a *relation* and a *practice*.

An ecosystem, understood from the perspective of the commons then is no longer a system that must be monitored, optimized and regularized; or a problem of resource flows, of services and market transactions, or species censuses or of genetic resources. It is rather a world to inhabit. An 'ecosystem *approach*', in turn, becomes a form of life that can be organized in flexible, even adaptive institutional form through a law that *lives* in the practices of the commons. Life and politics continue, inevitably, to intersect, but in the middle there is no centre, and no top, no immunitary *dispositif* that transforms the protection of life into its hindrance and subjugation.

Conclusions

This chapter has endeavoured to outline some of the productive ambiguities that are available to a reading of the 'ecosystem approach' that moves from the perspective of critical environmental law, that is, a reading that is simultaneously *inside* and *against* biopower. Through this kind of productive reading, it has been possible, as this chapter has sought to show, to traverse and perhaps puncture the biopolitical closure that wraps and folds the 'ecosystem approach' in both its anthropocentric and ecocentric articulations. Through this productive reading, a path is open towards thinking law beyond law and biopolitics beyond biopolitics, yet not in a dialectical sense, but rather in the sense of a conceptual displacement that *unthinks* the categories of legal modernity, especially in its biopolitical

and Rodotà in this respect identify some resources that are commons by necessity (Rodotà in particular considers commons by necessity all those that are capable of providing utilities that are functional to the realization of the fundamental rights inscribed in the (Italian) Constitution, however thus reducing the capacity of the commons to present a new horizon of sense, see Rodotà Commission Bill, Delegated Legislation to Reform the Civil Code Articles Concerning Public Property, Act of the Senate of the Republic n. 2031, XVI Legislature).

239 Mattei, *Beni Comuni*, p. 54.

240 M. Marella, 'La Parzialità dei Beni Comuni contro l'Universalismo del Bene Comune', 2014b, 6 May, http://www.euronomade.info/?p=2282, (Accessed 26 September 2018).

241 Mattei speaks of a distinction between a language of 'being' and a 'language' of having. I rather speak of a language of 'doing and becoming' to avoid static representations of being susceptible of essentialization; the political intention of this differentiation however is similar, and it relates to fundamentally delineating the commons against the modern horizon of property (i.e. having) see Mattei, *Beni Comuni*.

declension. This direction has been further illustrated through a series of conceptual frames, or modes of thinking, that, I argued, offer a map for future research and critical thinking that may disarticulate the biopolitical closure, and resist the biopolitical aporia of modernity: these are the notion of transversal ecological thinking; epistemic location; and a jurisprudence of the middle. All these notions, or dimensions, finally, have been linked to the horizon of the commons, which I argue, operate precisely at the point of convergence of these notions, and thus offer a critical path for thinking inside and against the 'ecosystem approach'. Perhaps the term 'ecosystem approach' ought to be entirely discarded, as it retains conceptual legacies and commitments – such as the idea of the system embedded in the term eco*system* – that are perhaps not well suited to be carried over towards a new horizon of sense such as that that the commons promise to offer. This chapter concludes thus by pointing ahead, and without claiming to offer any stable, conclusive construction, or theory; it invites open thinking and the pursuit of the trajectories of research I have endeavoured to sketch.

Conclusions

It is now time to revisit the research question. In the first chapter I presented the research question, and a series of secondary questions that the book would grapple with in a preliminary manner. The importance of the preliminary questions lies in setting out the parameters and the necessary reference points without which the overall research question that has driven the research and the writing could not be addressed and answered. At its most basic level, the research question that drove this book was *whether and to what extent the 'ecosystem approach' represents a paradigm shift*. However, a set of preliminary questions were built at the onset into the main research question, and steered the development of the argument. For example: what is the 'ecosystem approach'? Is there a uniform, universal concept of the 'ecosystem approach', or are there variations on a theme, perhaps due to the novelty of the concept, which inevitably leads to instabilities? Having now reached the concluding chapter of the book, what have we learned about the 'ecosystem approach'? Is it a paradigm shift or does it stand in a relation of continuity with legal modernity? Before finally offering any answer, it is expedient to sum up the journey undertaken thus far.

In the course of this book we have seen that the 'ecosystem approach' is situated at a particularly complex crossroad where shifting critical registers upend and unhinge every narrative through perspectival complexity and seemingly circular re-calibrations. Caught in this complex situation, the 'ecosystem approach' appears to simultaneously signal the potential for a decisive shift in the philosophical commitments of law, a shift enacting a postmodern ecological worldview, and the operationalization of an intensified biopolitical framework of control. Its epistemic and legal orientation, we have seen, depends on the perspective, or point of view, from which the question is asked, evaluated and answered, in line with the genealogical methodology that has driven the book. The question then (of whether the 'ecosystem approach' represents a paradigm shift or is in conceptual and operative continuity with legal modernity) has been accordingly explored through multiple registers. In particular, the analysis has undergone a key shift from the usual modern frame hinging on the anthropocentrism/ecocentrism binary, to a critical theoretical one deploying biopolitics as the primary framework of analysis.

After locating the 'ecosystem approach' within the broad field of international law, one part of the book (Part II) carried out a broad conceptual analysis, interrogating the 'ecosystem approach' in order to understand what it is, in both scientific and legal terms. What this part revealed is that it is impossible to speak of a singular, well defined and delineated concept, for a number of interconnected reasons, some being general to ecology, some being local to the 'ecosystem approach' as a particular conceptual frame. In order to carry out this analysis I have deployed a genealogical method. Genealogy aims at the reconstruction of the contingencies and contestations constituting the complex history of (legal) concepts. Through this methodological approach, the 'ecosystem approach' has been revealed to be the result of complex 'genealogies', and has been found to be stretched between diverging and sometimes irreconcilable ideological projects. As international environmental law, as we have already mentioned, is a 'propitious breeding ground' for broad 'indefinite' (and contested) principles and concepts,[1] approaching the analysis of the 'ecosystem approach' through this conceptual and methodological angle has arguably been very fruitful, and has allowed for a rich critical analysis premised on the central role of narratives in the construction of legal meaning. The two broad narratives through which the 'ecosystem approach' is articulated, narratives responding to two broadly competing logics, have been identified with ecocentrism and anthropocentrism.

The key finding of Part II of the book is that there is no *true* or *essential* 'ecosystem approach', but only the contingent and contested results that emerge, in the different regimes, from the 'play of forces'. This is different from asserting that there is no single, or at least converging, trajectory of the 'ecosystem approach', a point borne out by the biopolitical analysis of the two main case studies carried out in Part III of the book.

The starting point of Part III is that the anthropocentrism/ecocentrism binary is unsatisfying as a framework of critique, for a number of reasons that are carefully explored and explained, but ultimately hinging on the fact that the (a) binary isn't able to capture the complexities involved in the discourse and practice of the 'ecosystem approach'; that (b) the binary fosters a linear narrative that from a genealogical perspective is indefensible; and, finally that (c) each of the terms of the binary is affected by important problems. I then enact a shift in the critical register, and to this purpose deploy an analytics of biopolitics as the method and framework of analysis. Chapter 8 explores and explains the concepts of biopower and, especially, biopolitics, through the work of Foucault, and the subsequent elaborations and corrections offered by Agamben, Hardt and Negri and Esposito. The biopolitical frame is articulated specifically in relation to ecology (as opposed to the more common application in relation to human populations) in order to further problematize ecology, and to articulate a biopolitical understanding of law as it relates to the environment. Subsequently, this

1 De Sadeleer, *Environmental Principles*, p. 1.

'analytics of biopolitics' is applied to a case study, the 'ecosystem approach' as it is articulated in the context of the CBD. I chose the CBD given its centrality in the articulation of the 'ecosystem approach' in international law, and the complexities traversing the 'ecosystem approach' therein.

What Part III reveals is that the 'ecosystem approach' is fully inserted in the modern biopolitical *dispositif*, which encompasses both anthropocentric and ecocentric articulations, and provides an overarching horizon of sense. This biopolitical horizon of sense operates so as to continuously transform protection into services, optimization into predictability, enhancement into productivity, conservation into subjugation.

Yet, ultimately the analysis suggests that the 'ecosystem approach' is *neither* a paradigm shift *nor* can it be situated in perfect continuity with legal modernity. As a genealogical concept, 'the ecosystem approach' cannot escape the conflictual confrontations of competing narratives trying to impose particular closures on it. Rather, 'the ecosystem approach' is caught precisely in the middle of the 'play of forces', simultaneously signalling the potential for a decisive shift in the philosophical commitments of law *and* the operationalization of a biopolitical framework of control that is in continuity with, and even intensifies, the paradigm of legal modernity. Yet, the latter case – the continuation and intensification of the paradigm of legal modernity – emerges as the dominant tonality of the 'ecosystem approach' in the case analysed in this book.

However, it is exactly the fact that 'the ecosystem approach' can be activated in multiple, contradictory ways, that opens up a productive space which critical legal scholarship can exploit in order to engage with a concept which may open a potentially important novel horizon, a productive space which I attempt to tease out, albeit in preliminary form, through the discussion of some of the productive ambiguities available to an immanent critique of the 'ecosystem approach'.

Any productive reading requires though an 'authentic commitment',[2] a responsible engagement which, recognizing the immanent genealogical character of life, and of law, uses critique to *open up* the conversation, to disarticulate the 'givens' and the 'objectivity' that grips (environmental) legal discourse.[3] In line with the perspective outlined in the introduction, this book is an attempt to respond to the increasingly urgent calls to deeply re-think law in the Anthropocene, rather than embrace its biopolitical declension, entailing *more* control, *more* management, *more* subsumption of life under the care of biopower; a subsumption, moreover, which always transforms into subjugation, in light of the complicities that link biopower, law and capitalism, complicities that become concretized through the biopolitical *dispositif*, and further downstream, through particular conceptual and legal trajectories such as the 'ecosystem approach'. Hence, in teasing out the productive ambiguities that can be read *inside* and *against* both

2 Koskenniemi, *From Apology to Utopia*, p. 546.

3 Koskenniemi makes a similar argument in more specifically in relation to international law, see ibid., pp. 545–6.

biopolitics and the 'ecosystem approach', I attempt to offer reflections aimed at thinking law beyond law, and biopolitics beyond biopolitics. This, suggests Esposito, can be done by way of overturning the logic of biopolitics (and its continuous thanatopolitical transformation). Considering the very concrete, material effects of legal concepts on the bodies of the 'vulnerable living world', it is arguably an important task of the critical legal scholar to operate precisely inside and against biopower, and to take sides.

I could only offer some very preliminary reflections, given the scope of this book and the space available – a space already overstretched due to the complexity of the endeavour. I could therefore only briefly allude at some of those productive ambiguities, cognizant of the fact that there may be others, or that these very ones could be further explored in a variety of directions. Yet I endeavour to explore and draw out some of their implications for thinking law beyond law and against the biopolitical closure, knowing that at best it is a necessarily limited, provisional discussion, and a tentative map of future trajectories of research.

So I sketch the ways in which ecology can be mobilized – despite its moral and epistemological ambiguity – towards the critical project attempted here. And I briefly outline how both epistemology and law, while inevitably inside biopolitics, can be thought *against* and can overturn the biopolitical closure at work in the categories of legal modernity through a radical lateral shift, perhaps not unlike the 'exodus' Hardt and Negri imagine. Law is no longer to be thought in the centre, but in the middle: in the middle of life, in the middle of the world understood as the plane of immanence where life streams and flows, unpredictable. Law then, I suggest, following Philippopoulos-Mihalopoulos, is suddenly revealed to be in the middle, a living law that is no longer thinkable as sovereign command or technical norm, but as a living practice that is ontologically embodied, and epistemologically located. This living law, I finally suggest, points in the direction of a new horizon of sense that, while still (and intensely) at the intersection of life and politics, is no longer a biopolitical, immunitary horizon.

Bibliography

Scholarly literature (Books, chapter in books, articles)

Agamben, G., *Homo Sacer. Sovereign Power and Bare Life*, Stanford: Stanford University Press, 1998.

Agamben, *State of Exception*, Chicago: University of Chicago Press, 2005, p. 3.

Agrawal, A., and Redford, K., 'Conservation and Displacement: An Overview', *Conservation and Society*, 7:1, 2009, 1.

Ahteensuu, M., 'Defending the Precautionary Principle Against Three Criticisms', *TRAMES A Journal of the Humanities and Social Sciences*, 11:4, 2007, 366.

Alberts, P., 'Responsibility Towards Life in the Early Anthropocene', *Angelaki: Journal of the Theoretical Humanities*, 2011, 16:4, 5.

Allen, T., Zellmer, C. and Wuennenberg, C., 'The Loss of Narrative' in K. Cuddington and B. Beisner (eds) *Ecological Paradigms Lost: Routes of Theory Change*, Burlington, MA: Elsevier Academic Press, 2005.

Anghie, A., *Imperialism, Sovereignty and the Making of International Law*, Cambridge: Cambridge University Press, 2004.

Aqorau, T., 'Obligations to protect marine ecosystems under international conventions and other instruments' in M. Sinclair and G. Valdimarsson (eds) *Responsible Fisheries in the Marine Ecosystem*, Rome, Italy and Wallingford: FAO and CAB International, 2003, 25.

Archer-Hind, R. (ed.) *The Timaeus of Plato*, London: Macmillan, 1888.

Argyrou, V. *The Logic of Environmentalism. Anthropology, Ecology and Postcoloniality*, New York and Oxford: Berghahn Books, 2005.

Armstrong, K. and Ball, S., 'DNA Barcodes for Biosecurity: Invasive Species Identification', *Philosophical Transactions of the Royal Society*, 360:1462, 2005, 1813.

Arthurs, H., *Law and Learning: Report to the Social Sciences and Humanities Research Council of Canada by the Consultative Group on Research and Education in Law*, Information Division, Social Sciences and Humanities Research Council of Canada, Ottawa, 1983, 63.

Baccelli, L., 'Guerra e Diritti. Vitoria, Las Casas e la conquista dell'America', *XXXVII Quaderni Fiorentini per lo Studio del Pensiero Giuridico Moderno*, 2008.

Bacon, F., *Selected Philosophical Works*, in Rose-Mary Sargent (ed.) Indiapolis and Cambridge: Hackett Publishing Company Inc., 1999.

Barabanova, Y., Emerging Climate Justice Discourse: Perspectives of Grassroots Networks in the UK, PhD Thesis, Budapest: Central European University, 2013.

Barcellona, P., *Diritto senza Società: dal Disincanto all'Indifferenza*, Bari: Edizioni Dedalo, 2003.

Bascherini, G., 'Italian Theories. Spunti attorno all'esperienza giuridica a partire da un recente saggio di Roberto Esposito', *Costituzionalismo*, 2013, 1, HTTP: http://www.costituzionalismo.it/articoli/438/(Accessed 26 September 2018).

Baumann, Z., *Liquid Modernity*, Cambridge: Polity Press, 2000.

Bax, N., Williamson, A., Aguero, M., Gonzalez, E., and Geeves, W., 'Marine Invasive Alien Species: a Threat to Global Biodiversity', *Marine Policy*, 27:4, 2003, 313.

Beck, A., 'Foucault and Law: the Collapse of Law's Empire', *Oxford Journal of Legal Studies*, 16:3, 1996, 489.

Beckett, J., *Critical International Legal Theory Oxford Bibliographies*, HTTP: http://www.oxfordbibliographies.com/view/document/obo-9780199796953/obo-9780199796953-0007.xml (Accessed 26 September 2018).

Bell, A., 'Frog Reflections: The Ecosystem Approach to Conservation' in G. Beck and B. Littlejohn (eds) *Voices from the Watershed: Environmental Issues in the Great Lakes-St Lawrence Drainage Basin*, Montreal: McGill-Queen's University Press, 2000.

Bell, A., 'Non Human Nature and the Ecosystem Approach. The Limits of Anthropocentrism in Great Lakes Management', *Alternatives Journal*, 20:3, 2004, 5.

Bell, M., 'Thomas Berry and an Earth Jurisprudence: An Exploratory Essay', *The Trumpeter: Journal of Ecosophy*, 19:1, 2003, 69.

Belsky M., 'Using Legal Principles to Promote the "Health" of an Ecosystem', *Tulsa Journal of Comparative and International Law*, 3, 1995, 183.

Berkes, F., 'Implementing ecosystem-based management: evolution or revolution?', 13:4 *Fish and Fisheries*, 2011, 465, p. 466 and p. 473 respectively.

Berkes, F., Colding J., and Folke, C., 'Rediscovery of Traditional Ecological Knowledge as Adaptive Management', *Ecological Applications*, 10:5, 2000, 125.

Bernini, L., 'Sovereignty and Biopolitics in Hobbes', *Leviathan*, working paper, 2011, p. 2, HTTP: https://www.academia.edu/3721237/Sovereignty_and_Biopolitics_in_Hobbes_Leviathan_May_2011_ (Accessed 26 September 2018), originally published in Italian as L. Bernini,'Il Cristallo Biopolitico di Hobbes' in L. Bernini (ed.) *Apocalissi Queer. Elementi di Teoria Antisociale*, Pisa: Edizioni ETS, 2013,146.

Bernstein, S., *The Compromise of Liberal Environmentalism*, New York: Columbia University Press, 2001.

Berry, T., *The Great Work: Our Way into the Future*, New York: Bell Tower, 1999.

Biermann, C. and Mansfield, B., 'Biodiversity, Purity, and Death: Conservation Biology as Biopolitics', *Environment and Planning D: Society and Space*, 32:2, 2014, 257.

Birnie, P., Boyle, A. and Redgwell, C., *International Law and the Environment*, Oxford: Oxford University Press, 2009.

Bocking, S., 'Visions of Nature and Society. A History of the Ecosystem Concept', *Alternatives*, 20:3, 1994, 12.

Boehmer Christiansen, S., 'The Precautionary Principle in Germany: Enabling Government' in T. O'Riordan and J. Cameron (eds) *Interpreting the Precautionary Principle*, London: Earthscan Publications Ltd, 1994.

Bosselmann, K., *The Principle of Sustainability: Transforming Law and Governance*, Aldershot: Ashgate 2008.

Bosselmann, K., 'Losing the Forest for the Trees: Environmental Reductionism in the Law', *Sustainability*, 2:8, 2010, 2424.

Bosselmann, K. and Montini, M., 'The Oslo Manifesto: From Environmental Law to Ecological Law: A Call for Re-Framing Law and Governance', IUCN WCEL Ethics Specialist Group Workshop, IUCN Academy of Environmental Law Colloquium, University of Oslo, 21 June 2016.

Botkin, D., *Discordant Harmonies: A New Ecology for the Twenty-first Century*, Oxford: Oxford University Press, 1st edition, 1990.

Botkin, D., 'Adjusting Law to Nature's Discordant Harmonies', *Duke Environmental Law and Policy Forum*, 7:1, 1996, 25.

Boyle, A., 'The Convention on Biological Diversity' in L. Campiglio, L. Pineschi, D. Siniscalco and T. Treves (eds), *The Environment after Rio: International Law and Economics*, London/Leiden: Graham & Trotman/Martinus Nijhoff, 1994, p. 115.

Boyle, R., 'The Implications of a Paradigm Shift in Ocean Resource Management for Fisheries Stock Assessment' in R. Beamish and B. Rothschild (eds) *The Future of Fisheries Science in North America*, Heidelberg: Springer, 2009.

Brand, U. and Vadrot, A., 'Epistemic selectivities towards the valorization of nature in the Nagoya Protocol and the making of the Intergovernmental Science-Policy Platform on Biodiversity and Ecosystem Services (IPBES)', *LEAD – Law, Environment and Development Journal*, 9:2, 2013, 202.

Brännström, L., 'Law, Objectives of Government, and Regimes of Truth: Foucault's Understanding of Law and the Transformation of the Law of the EU Internal Market', *Foucault Studies*, 18, 2014, 173.

Brels, S., Coates, D., and Loures, F., *Transboundary Water Resources Management: The Role of International Watercourse Agreements in Implementation of the CBD*, CBD Technical Series No. 40, Montréal: Secretariat of the Convention on Biological Diversity, 2008.

Brooks, R. O., Jones, R., and Virginia, R. A., *Law and Ecology: The Rise of the Ecosystem Regime*, Burlington: Ashgate, 2002.

Brownlie, I., *Principles of Public International Law*, Oxford: Oxford University Press, 7th edition, 2008.

Brunnée, J., 'Environmental Security in the Twenty-First Century: New Momentum for the Development of International Environmental Law?', *Fordham Journal of International Law*, 18, 1995, 1742.

Brunnée, J., 'COPing with Consent: Law Making Under Multilateral Environmental Agreements', *Leiden Journal of International Law*, 15:1, 2002, 1.

Brunnée, J. and Toope, S., 'Environmental Security and Freshwater Resources: A Case for International Ecosystem Law', *Yearbook of International Environmental Law*, 5:1, 1994, 41.

Brunnée, J. and Toope, S., 'Environmental Security and Freshwater Resources: Ecosystem Regime Building', *The American Journal of International Law*, 91:1, 1997, 26.

Burdon, P., *Exploring Wild Law: The Philosophy of Earth Jurisprudence*, Kent Town, S. A.: Wakefield Press, 2011a.

Burdon, P., Earth Jurisprudence: Private Property and Earth Community, PhD Thesis, Adelaide Law School the University of Adelaide, 2011b.

Burdon, P., 'Environmental Protection and The Limits of Rights Talk', *Right Now*, 10 August 2012.

Burdon, P., *Earth Jurisprudence: Private Property and the Environment*, New York: Glasshouse/Routledge, 2014.

Burhenne-Guilmin, F. and Casey-Lefkowitz, S., 'The Convention on Biological Diversity: A Hard Won Global Achievement', *Yearbook of International Environmental Law*, 3:1, 1992, 43.

Büscher, B., Sullivan, S., Neves, K., Igoe, J., and Brockington, D., 'Towards a Synthesized Critique of Neoliberal Biodiversity Conservation', *Capitalism Nature Socialism*, 23:2, 2012, 4.

Butchart, S., Walpole M., et al., 'Global Biodiversity: Indicators of Recent Declines', *Science*, 328:5982, 2010, 1164.

Butz, M., 'Chaos Theory, Philosophically Old, Scientifically New', *Counseling & Values*, 39:2, 1995, 84.

Caldwell, L. 1970, 'The Ecosystem as a Criterion for Public Land Policy', *Natural Resources Journal* 10:2, 1970, 203.

Campbell, T., 'Interview: Roberto Esposito', *Diacritics*, 36:2, 2006a, 49.

Campbell, T., 'Bios, Immunity, Life: The Thought of Roberto Esposito', *Diacritics*, 36:2, 2006b, 2.

Campbell, T., 'Bios, Immunity, Life: The Thought of Roberto Esposito' in R. Esposito (ed.) *Bios: Biopolitics and Philosophy*, Minneapolis: Minnesota University Press, 2008.

Campbell, T. and Sitze, A. (eds) *Biopolitics. A Reader*, London: Duke University Press, 2013.

Capra, F., *The Turning Point: Science, Society, and the Rising Culture*, Toronto: Bantam Books, 1984.

Cassese, A., *States: Rise and Decline of the Primary Subject of International Law* in B. Fassbender, A. Peters, A.S. Peter and D. Högger (eds) *The Oxford Handbook of the History of International Law*, Oxford: Oxford University Press, 2013.

Cava, B., 'Pashukanis e Negri: do antidireito ao direito do comum', *Revista Direito e Praxis*, 4, 2013, 2.

Cecchetti, M., 'La Disciplina Giuridica della Tutela Ambientale come Diritto dell'ambiente', *federalismi.it, Rivista di Diritto Pubblico Italiano, Comunitario e Comparato*, HTTP: https://federalismi.it/nv14/articolo-documento.cfm?artid=6610 (Accessed 26 September 2018).

Chandler, M., 'The Biodiversity Convention: Selected Issues of Interest to the International Lawyer', *Colorado Journal of International Environmental Law*, 4, 1993, 141.

Chaplin, S., 'Fictions of Origin: Law, Abjection, Difference', *Law and Critique*, 16:2, 2005, 161.

Christensen, N. L., et al., 'The Report of the Ecological Society of America Committee on the Scientific Basis for Ecosystem Management', *Ecological Applications*, 6:3, 1996, 665.

Christie, W. J., Becker, M., Cowden, J. W. and Vallentyne, J. R., 'Managing the Great Lakes Basin as a Home', *Journal of Great Lakes Research*, 12:1, 1986, 2.

Chynoweth, P., 'Legal Research' in L. Ruddock and Knight, A. (eds) *Advanced Research Methods in the Built Environment*, Oxford: Wiley-Blackwell, 2008, 28.

Clark, T. and Zaunbrecher, D., 'The Greater Yellowstone Ecosystem: The Ecosystem Concept in Natural Resource Policy and Management', *Renewable Resources Journal*, 5:3, 1987, 8.

Clements, F., *Plant Succession: An Analysis of the Development of Vegetation*, Washington DC: Carnegie Institution of Washington, 1916.

Clowney, D. and Mosto, P. (eds), *Earthcare: An Anthology in Environmental Ethics*, Lanham, MD: Rowman & Littlefield, 2009, p. 161.

Code, L., *Ecological Thinking: The Politics of Epistemic Location*, New York and Oxford: Oxford University Press, 2006.

Code, L., 'Doubt and Denial: Epistemic Responsibility Meets Climate Change Scepticism', *Oñati Socio-Legal Series*, 3:5, 2013, 838.

Cohen, J. L., *Sovereignty in the Context of Globalization: a Constitutional Pluralist Perspective* in S. Besson and J. Tasioulas (eds) *The Philosophy of International Law*, Oxford: Oxford University Press, 2010.

Cole, D. and Yung, L. (eds) *Beyond Naturalness: Rethinking Park And Wilderness Stewardship in an Era of Rapid Change*, Washington, DC: Island Press, 2010.

Coleman, M., and Grove, K., 'Biopolitics, Biopower, and the Return of Sovereignty', *Environment and Planning D: Society and Space*, 27:3, 2009, 489.

Commoner, B., *The Closing Circle: Nature, Man and Technology*, New York: Alfred Knopf, 1971, 1971, p. 16.

Cooney, R., *Sustainable Use: Concepts, Ambiguities, Challenges*, 2007, paper prepared as background for meeting of the IUCN Species Survival Commission's Sustainable Use Specialist Group Strategic Planning Meeting 10–13 July 2007 White Oak Plantation, Florida.

Cosi, G., 'Tutela del Mondo e Normatività Naturale' in G. Lombardi-Vallauri (ed.) *Il Meritevole di Tutela*, Milano: Giuffrè, 1990.

Costanza, R., d'Arge, R., de Groot, R., et al., 'The Value of the World's Ecosystem Services and Natural Capital', *Nature*, 387, 1997, 253.

Costanza, R., Norgaard, R., Daly, H., et al., *An Introduction to Ecological Economics*, (e-book) 2010, HTTP: http://library.uniteddiversity.coop/Measuring_Progress_and_Eco_Footprinting/An_Introduction_to_Ecological_Economics.pdf (Accessed 26 September 2018)

Costanza, R., Groot, R. S. de, Sutton, P., Ploeg, S. van der, Anderson S. J., Kubiszewski, I., Farber, S. and Turner, R. K., 2014, 'Changes in the Global Value of Ecosystem Services', *Global Environmental Change*, 26, 2014, 152.

Cotterrel, R., 'Law in Culture', *Ratio Juris*, 17:1, 2004, 1.

Cover, R., 'The Supreme Court, 1982 Term — Foreword: Nomos and Narrative', *Harvard Law Review*, 97, 1983, 4.

Cover, R., 'Violence and the Word', *Yale Law Journal*, 95, 1986, 1601.

Crawford, J., *Brownlie's Principles of Public International Law*, 8th edition, Oxford: Oxford University Press, 2012a.

Crawford, J., 'Sovereignty as Legal Value' in J. Crawford and M. Koskenniemi (eds) *The Cambridge Companion to International Law*, Cambridge: Cambridge University Press, 2012b.

Crober, A., The Ecosystem Approach to Ecosystem Management, Senior Honours Thesis, Department of Geography, Faculty of Environmental Studies, University of Waterloo. Unpublished, 1999.

Cullinan, C., *Wild Law: A Manifesto for Earth Justice*, South Africa: Siber Ink, 2002.

Cullinan, C. *Voices of Earth Jurisprudence An Introduction To Earth Jurisprudence: Guiding Principles And Wild Law Possibilities*, Center for Earth Jurisprudence, 2011.

Cunningham, W., Cunningham, M. and Woodworth Saigo, B., *Environmental Science: A Global Concern*, 12th edn, New York: McGraw-Hill, 2012,

Currie, D., *Ecosystem-Based Management in Multilateral Environmental Agreements: Progress towards Adopting the Ecosystem Approach in the International Management of Living Marine Resources*, WWF, 2007.

Curry, P., 'Re-Thinking Nature: Towards an Eco-Pluralism', *Environmental Values*, 12:3, 2003, 337.

Curry, P., 'Nature Post-Nature', *New Formations*, 26, 2008, 51.

Curry, P., *Ecological Ethics: An Introduction*, 2nd edition, Cambridge: Polity, 2011.

Daly, H., *Economics, Ecology, Ethics: Essays Toward a Steady-State Economy*, New York: W.H. Freeman and Company, 1980.

Daly, H., 1997, *Beyond Growth. The Economics of Sustainable Development*, Boston, MA: Beacon Press, 1997, esp. chapter 2, p. 45ff.

Daily, G., 'Introduction: What are ecosystem services?' in G. Daily (ed.) *Nature's Services: Societal Dependence on Natural Ecosystems*, Washington, DC: Island Press, 1997.

Darier, E. (ed.) *Discourse of the Environment*, Oxford: Blackwell Publishers, 1999a.

Darier, E., 'Foucault and the Environment: An Introduction' in E. Darier (ed.) *Discourses of the Environment*, Oxford: Blackwell Publishers, 1999b.

Davies, M., *Asking the Law Question*, 3rd edition, Lawbooks, 2008.

De Almeida, O. E., *Environment and Development: The Founex Report on Development and Environment*, New York: Carnegie Endowment for International Peace, 1972.

De Anna, G., 'Modernità e Immanenza: l'Azione Umana in Tommaso D'Aquino e Thomas Hobbes' in L. Parisoli (ed.) *Il Soggetto e la Sua Identità. Mente e Norma, Medioevo e Modernità*, Palermo: Officina di Studi Medievali, 2010.

De Klemm, C., 'Migratory Species: A review of Existing International Instruments', *Environmental Policy and Law* 15, 1985, 81.

Delaney, D., 'Making nature/Marking Humans: Law as a site of (Cultural) Production', *Annals of the Association of American Geographers*, 9:3, 2004, 487.

deLaplante, K., 'Environmental Alchemy: How to Turn Ecological Science into Ecological Philosophy', *Environmental Ethics*, 26:4, 2004, 361.

deLaplante, K., 'Is Ecosystem Management a Postmodern Science?' in K. E. Cuddington, and B. E. Beisner (eds) *Ecological Paradigms Lost: Routes of Theory Change*, Burlington, MA: Elsevier Academic Press, 2005.

deLaplante, K. and Odenbaugh, J.,'What Isn't Wrong with Ecosystem Ecology' in R. A. Skipper, C. Allen, R. Ankeny, C. F. Craver, L. Darden, G. M. Mikkelson, and R. C. Richardson (eds) *Philosophy Across the Life Sciences*, Cambridge, MA: MIT Press, 2006.

deLaplante, K. and Picasso, V., 'The Biodiversity-Ecosystem Function Debate in Ecology' in K. deLaplante, B. Brown and A. Peacock (eds), 2010, *Handbook of the Philosophy of Science. Volume 11: Philosophy of Ecology*, North Holland: Elsevier, 2010.

De Lucia, V., 'La Conferenza di Copenaghen tra Fallimento e Nuovi Modelli di Governance Globale', NENS Nuova Economia Nuova Società, Occasional Paper, Marzo 2010, HTTP: https://www.nens.it/sites/default/files/cop15-analisi-nuovi-modelli-governance-vdl-25032010.pdf (Accessed 26 September 2018).

De Lucia, V., 'Towards an Ecological Philosophy of law: a comparative discussion', *Journal of Human Rights and the Environment*, 4:2, 2013a, 167.

De Lucia, V., 'Law as Insurgent Critique. The Perspective of the Commons in Italy' Critical Legal thinking, August 5 2013b, HTTP: http://criticallegalthinking.com/2013/08/05/law-as-insurgent-critique-the-perspective-of-the-commons-in-italy/ (Accessed 26 September 2018).

De Lucia, V., 'Competing Narratives and Complex Genealogies: The Ecosystem Approach in International Environmental Law', *Journal of Environmental Law*, 27:1, 2015a, 91.

De Lucia, V., 'Chaos, Modernity and the Commons', Paper presented at the 4th Workshop of Europe's Common, 26–27 March 2015b, Centre for Critical Legal Thought, Kent Law School.

De Lucia, V. 'Re-embodying Law: Transversal Ecology and the Commons' in R. Thomas-Pellicer, De Lucia V. and Sullivan S. (eds) *Law, Philosophy, Ecology: Exploring Re-embodiments*, London: Routledge, 2016.

De Lucia, V., 'Beyond Anthropocentrism and Ecocentrism: A Biopolitical Reading of Environmental Law', 8:2 *Journal of Human Rights and Environment* 2017a, 181.

De Lucia, V., 'Critical Environmental Law and the Double Register of the Anthropocene: A Biopolitical Reading' in L. Kotzé (ed.) *Environmental Law and Governance in the Anthropocene*, Hart Publishing, 2017b.

De Lucia, V., 'A Critical Interrogation of the Relation between the Ecosystem Approach and Ecosystem Services', *RECIEL*, 27:2, 2018a.

De Lucia, V., 'Semantics of Chaos. Law, Modernity and the Commons', *Pólemos Journal of Law*, 12:2, Literature and Culture, 2018b.

De Lucia, V., 'Bare Nature. The Biopolitical Logic of the International Regulation of Invasive Alien Species', *Journal of Environmental Law*, 2018c, Advance Online.

De Sadeleer, N., *Environmental Principles: From Political Slogans to Legal Rules*, Oxford: Oxford University Press, 2002.

De Sadeleer, N., 'The Precautionary Principle in EU Law', *Aansprakelijkheid Verzekering En Schade*, 5, 2010, 173.

De Sousa Santos, B., *Toward a New Common Sense. Law, Science and Politics in a Paradigmatic Transition*, New York, London: Routledge, 1995.

Debord, G., *The Society of the Spectacle*, London: Black and Red, 1997.

Dillon, M. and J. Reid, J., *The Liberal Way of War: Killing to Make Life Live*, London and New York: Routledge, 2009.

Di Plinio, G., *Diritto Pubblico dell'Ambiente e Aree Naturali Protette*, Torino: UTET, 1994

Doremus, H., 'The Rhetoric and Reality of Nature Protection: Toward a New Discourse', *Washington and Lee Law Review*, 57.1, 2000, 11.

Douzinas, C., *The End of Human Rights*, Oxford: Hart Publishing, 2000.

Douzinas, C. and Geary, A., *Critical Jurisprudence. The Political Philosophy of Justice*, Oxford and Portland, OR: Hart Publishing, 2005.

Dunn, R., 'Global Mapping of Ecosystem Disservices: The Unspoken Reality that Nature Sometimes Kills us', *Biotropica*, 42:5, 2010, 555.

Dupuy, P., 'Soft Law and the International Law of the Environment', *Michigan Journal of International Law*, 12:2, 1991, 420.

Ebbesson, J., *Piercing the State Veil in Pursuit of Environmental Justice* in J. Ebbesson and P. Okowa (eds) *Environmental Law and Justice in Context*, Cambridge: Cambridge University Press, 2009.

Eliot, C., 'The Legend of Order and Chaos: Communities and Early Community Ecology' in K. deLaplante, B. Brown and A. Peacock (eds), *Handbook of the Philosophy of Science. Volume 11: Philosophy of Ecology*, North Holland: Elsevier, 2010.

Emmenegger, S., and Tschentscher, A., 'Taking Nature's Rights Seriously: The Long Way to Biocentrism in Environmental Law', *Georgetown International Environmental Law Review*, 6, 1994, 545, 547.

Endicott, T., *The Logic of Freedom of Power* in S. Besson, and J. Tasioulas (eds) *The Philosophy of International Law*, Oxford: Oxford University Press, 2010.

Erasga, D., 'Biopolitics: Biodiversity as Discourse of Claims' in D. Ersaga (ed.) *Sociological Landscape – Theories, Realities and Trends*, Rijeka: Intech, 2012.

Esposito, R., *Bíos: Biopolitics and Philosophy*, Minneapolis: University of Minnesota Press, 2008.

Esposito, R., *Pensiero Vivente. Origine e Attualità della Filosofia Italiana*, Turin: Einaudi, 2010.

Esposito, R., *Immunitas. The Protection and Negation of Life*, Cambridge: Polity, 2011.

Esposito, R., *Living Thought. The Origins and Actuality of Italian Philosophy*, Stanford, CA: Stanford University Press, 2012.

Esposito, R., *Community, Immunity, Biopolitics*, New York: Fordham University Press, 2013a.

Esposito, R., 'Community, Immunity, Biopolitics', *Angelaki*, 18:3, 2013b, 83.

Evernden, N., *The Social Creation of Nature*, Baltimore, MD: The Johns Hopkins University Press, 1992.

Ewald, F., 'Norms, Discipline, and the Law', *Representations Special Issue: Law and the Order of Culture*, 30, 1990, 138.

Fabra, A. and Gascón, V., 'The Convention on the Conservation of Antarctic Marine Living Resources (CCAMLR) and the Ecosystem Approach', *International Journal of Marine and Coastal Life*, 23, 2008, 567.

Ferré, F., 'Personalistic Organicism: Paradox or Paradigm?' in R. Attfield and A. Belsey (eds) *Philosophy and the Natural Environment*, London: Royal Institute of Philosophy Supplement, 36, Cambridge: Cambridge University Press, 1994, 59.

Finlayson, C., Davidson, N., Pritchard, D., Randy Milton, G. and MacKay, H., 'The Ramsar Convention and Ecosystem-Based Approaches to the Wise Use and Sustainable Development of Wetlands', *Journal of International Wildlife Law & Policy*, 14:3/4, 2011, 176–198.

Fisher, D., 'Legal and Paralegal Rules for Biodiversity Conservation: a Sequence of Conceptual, Linguistic and Legal Problems' in M. Jeffery, J. Firestone and K. Bubna-Litic (eds) *Biodiversity Conservation, Law and Livelihoods: Bridging the North-South Divide*, Cambridge: Cambridge University Press, 2008.

Fisher, E., Lange, B. E., Scotford, A. K. and Carlarne, C. P., 'Maturity and Methodology: Starting a Debate about Environmental Law Scholarship', *Journal of Environmental Law*, 21:2, 2009, 213–250.

Fisher, E., 'Environmental Law as "Hot" Law', *Journal of Environmental Law*, 25:3, 2013, 347.

Fitz-Henry, E., 'Between Rights and Services', *Earth Jurisprudence: Building Theory and Practice From*, Third Wild Law Conference, Griffith University, 16–18 September 2011.

Fitzpatrick, P., *Mythology of Modern Law*, London, New York: Routledge, 1992.

Fitzsimmons, A., 'Ecosystem Management: an Illusion?', *PERC Reports*, 17:5, 1999a, 1.

Fitzsimmons, A., *Defending Illusions. Federal Protection of Ecosystems*, Lanham, MD: Rowman & Littlefield, 1999b.

Fontanelli, F., 'Santi Romano and L'ordinamento giuridico: The Relevance of a Forgotten Masterpiece for Contemporary International, Transnational and Global Legal Relations', *Transnational Legal Theory*, 2:1, 2001, 67.

Forst, M., 'The Convergence of Integrated Coastal Zone Management and the Ecosystems Approach', *Ocean and Coastal Management*, 52:6, 2009, 294.

Foucault, M., 'Nietzsche, Genealogy, History' in D. F. Bouchard (ed.) *Language, Counter-Memory, Practice: Selected Essays and Interviews*, Ithaca, NY: Cornell University Press, 1977.

Foucault, M., *The History of Sexuality. Volume I: an Introduction*, New York: Pantheon Books, 1978.

Foucault, M., *Discipline and Punish. The Birth of the Prison*, 2nd edition, New York, Toronto: Vintage Books, 1995.

Foucault, M., 'What is Enlightenment?' in P. Rabinow (ed.) *Ethics: Subjectivity and Truth (Essential Works of Foucault 1954–1984, volume 1)*, London: Penguin Books, 2000a.

Foucault, M., 'The Birth of Social Medicine' in M. Foucault (ed.) *Essential Works III: Power*, New York: The New Press, 2000b.

Foucault, M., *Society Must be Defended.* Lectures at the Collège de France 1975–1976, London: Penguin Books, 2004.

Foucault, M., *Security, Territory, Population.* Lectures at the Collège de France 1977 – 1978, New York: Palgrave Macmillan, 2009.

Foucault, M., *The Birth of Biopolitics.* Lectures at the Collège de France 1978–1979, New York: Palgrave Macmillan, 2010.

Frankenberg, G. 'Critical Comparison: Re-thinking Comparative Law', 26 *Harvard International Law Journal* 1985.

Frost, M. 'The Convergence of Integrated Coastal Zone Management and the Ecosystems Approach', 6 *Ocean and Coastal Management*, 52:6, 2009, 294.

Funtowicz, S. O. and Ravetz, J. R., 'Three types of risk assessment and the emergence of postnormal science' in D. Golding and S. Krimsky (eds) *Social Theories of Risk*, Westport, CT: Praeger, 1992.

Funtowicz, S. O. and Ravetz, J. R., 'Science for the Postnormal Age', *Futures*, 25:7, 1993, 739.

Galindo-Leal, C. and Bunnell, F., 'Ecosystem Management: Implications and Opportunities of a New Paradigm', *The Forestry Chronicle*, 71:5, 1995, 601.

Garcia, S., Rice, J., and Charles, A., 'A tale of two streams: Synthesizing governance of marine fisheries and biodiversity conservation' in S. Garcia, J. Rice and A. Charles (eds) *Governance of Marine Fisheries and Biodiversity Conservation: Interaction and Coevolution*, Oxford: Wiley, 2014.

Garcia, S., Zerbi, A., Aliaume, C., Do Chi, C. T. and Lasserre, G. *The Ecosystem Approach to Fisheries: Issues, Terminologies, Principles, Institutional Foundations, Implementation and Outlook*, FAO Fisheries Technical Paper 443, Food and Agriculture Organization of The United Nations, Rome, 2003, 21.

Geertz, C., *Local Knowledge: Further Essays in Interpretive Anthropology*, London: Fontana Press, 1983.

Gentili, D., *Italian Theory: Dall'Operaismo alla Biopolitica*, Blogna: Il Mulino, 2012.

Gentili, D. and Stimilli, E., *Differenze Italiane. Politica e Filosofia: Mappe e Sconfinamenti*, Rome: DeriveApprodi, 2015.

Gentili, D., Stimilli, E. and Garelli, G., *Italian Critical Thought: Genealogies and Categories*, Rowman & Littlefield, 2018.

Gillespie, A., *International Environmental Law, Policy and Ethics*, Oxford: Clarendon Press 1997.

Gillespie, A., *The Illusion of Progress: Unsustainable Development in International Law and Policy*, London: Earthscan, 2001.

Glowka, L., Burhenne-Guilmin, F., Synge H., McNeely, J. A. and Gündling, L., *A Guide to the Convention on Biological Diversity*, IUCN Environmental Law Centre (ELC), IUCN, Biodiversity Programme, 1994.

Godden, L., Nature as Other: The Legal Ordering of the Natural World, PhD Thesis, Faculty of Law, Griffith University, 2000.

Goldman, M., Nadasdy P. and Turner M. (eds) *Knowing Nature: Conversations at the Intersection of Political Ecology and Science Studies*, Chicago: University of Chicago Press, 2011.

Goldstein, B., 'Can Ecosystem Management Turn an Administrative Patchwork into a Greater Yellowstone Ecosystem?', *Northwest Environmental Journal*, 8, 1992, 285.

Golley, F., *A History of the Ecosystem Concept in Ecology: More than the Sum of its Parts*, New Haven, CT and London: Yale University Press, 1993.

Gordon, C., 'Expelled Questions: Foucault, the Left and the Law' in B. Golder (ed.) *Re-reading Foucault on Law, Power and Rights*, New York: Routledge, 2014.

Gramsci, A., *Selections from the Prison Notebooks*, Q. Hoare and G. Nowell-Smith (eds), London: Lawrence &and Wishart, 2007.

Grassi, S., 'Tutela dell'Ambiente (diritto amministrativo)', in A. Falzea, P. Grossi and E. Cheli (eds) *Enciclopedia del Diritto: Annali I [Accertamento – Tutela]*, Milan: Giuffrè, 2007, 1114.

Grear, A., 'Law's Entities: Complexity, Plasticity and Justice', *Jurisprudence*, 4:1, 2013, 76.

Grear, A., *Rethinking Human Rights. Facing the Challenge of Corporate Legal Humanity*, London: Palgrave Macmillan, 2009.

Grear, A., 'The Vulnerable Living Order: Human Rights and the Environment in a Critical and Philosophical Perspective', *Journal of Human Rights and the Environment*, 2:1, 2011, 23–44.

Grear, A., 'Human Rights, Property and the Search for "Worlds Other"', 3:2 *Journal of Human Rights and Environment* 2012a, 173.

Grear, A. (ed.) *Should Trees Have Standing? 40 Years on*, Cheltenham: Edward Elgar, 2012b.

Grear, A., 'Towards a New Horizon: in Search of a Renewing Socio-Juridical Imaginary', *Oñati Socio-Legal Series*, 3:5, 2013a, 966–990.

Grear, A. 'Law's Entities: Complexity, Plasticity and Justice', 4:1 *Jurisprudence* 2013b, 76.

Grear, A., 'Deconstructing Anthropos: A Critical Legal Reflection on "Anthropocentric" Law and Anthropocene "Humanity"', *Law and Critique*, 26:3, 2015, 225.

Grear, A., and Grant, E. (eds) *Thought, Law, Action and Rights in the Age of Environmental Crisis*, Cheltenham: Edward Elgar, 2015.

Grear, A. and Kotzé, L. (eds) *Research Handbook on Human Rights and the Environment*, Cheltenham, Glos. and Northampton, MA: Edgar Elgar, 2015.

Greenberg, J. and Park, T., 'Political Ecology', *Journal of Political Ecology*, 1, 1994, 11.

Griffin, P., *The Ramsar Convention: A New Window for Environmental Diplomacy?* Institute for Environmental Diplomacy & Security, University of Vermont, 2012.

Grossi, P., *Storia di Esperienze Giuridiche e Tradizione Romanistica*, Quaderni Fiorentini per la Storia del Pensiero Giuridico Moderno 17, 1988a, p. 533.

Grossi, P., *La Proprietà e le Proprietà nell'Officina dello Storico*, Quaderni Fiorentini per la Storia del Pensiero Giuridico Moderno, 17, 1988b, p. 359.

Grossi, P. 'Assolutismo Giuridico e Proprietà Collettive', Quaderni Fiorentini per la Storia del Pensioro Giuridico Moderno, 19, 1990, 505.

Grossi, P. *L'Ordine Giuridico Medievale*, Laterza, 2006.

Grumbine, E. R., 'What is Ecosystem Management?', *Conservation Biology*, 8:1, 1994, 27.

Grumbine, E. R., 'Reflections on "What Is Ecosystem Management?"', *Conservation Biology*, 11:1, 1997, 41.

Grzegorczyk, C., 'Le concept de bien juridique: l'impossible définition?', *Archives de philosophie du droit*, 24, 1979, Les biens et les choses, 259.

Guattari, F., *The Three Ecologies*, London: Continuum, 2008.

Gurevich, A. J., *Categories of Medieval Culture*, London: Routledge & Kegan Paul, 1985.

Gutwirth, S. and Naim-Gesbert, E., 'Science et droit de l'environnement: réflexions pour le cadre conceptual du pluralism de vérités', *Revue interdisciplinaire d'études juridiques*, 34, 1995, 33.

Habermas, J., 'Modernity versus Postmodernity', *New German Critique*, 22, 1981, 3.

Hagen, J. B., *An Entangled Bank: The Origins of Ecosystem Ecology*, New Brunswick, NJ: Rutgers University Press, 1992,

Handl, G., 'Trail Smelter in Contemporary International Environmental Law', in R. Bratspies, and A. Miller (eds) *Transboundary Harm in International Law: Lessons From The Trail Smelter Arbitration*, New York: Cambridge University Press, 2006.

Hansen, J., *Storms of My Grandchildren: The Truth about the Coming Climate Catastrophe and Our Last Chance to Save Humanity*, New York: Bloomsberg Publishing, 2009.

Hardin, G., 'The Tragedy of the Commons', *Science*, 162:3859, 1968, 1243.

Hardt, M., and Negri, A., *Empire. The New World Order*, Cambridge, MA: Harvard University Press, 2000.

Hardt, M., and Negri, A., *Commonwealth*, Cambridge: Belknap Press of Harvard University Press, 2011.

Hartje, V., Klaphake, A. and Schliep, R., 'The International Debate on the Ecosystem Approach – Critical Review – International Actors – Obstacles and Challenges', *BfN – Skripten*, 2003, 80.

Harvey, D., *The New Imperialism*, Oxford: Oxford University Press, 2003.

Hasley, M., *Deleuze and Environmental Damage. Violence of the Text*, Aldershot: Ashgate, 2006.

Hasley, M., 'Majesty and Monstrosity: Deleuze and the Defence of Nature' in A. Philippopoulos-Mihalopoulos (ed.) *Law and Ecology. New Environmental Foundations*, London: Routledge, a Glasshouse Book, 2011, 218–219.

Hatcher, B. and Bradbury, R. 'Marine Ecosystem Management. Is the Whole Greater than the Sum of the Parts?' in D. Rothwell and D. L. VanderZwaag (eds) *Towards Principled Ocean Governance: Australian and Canadian Approaches and Challenges*, New York: Routledge, 2013.

Hayward, T., *Ecological Thought. An Introduction*, Cambridge, UK: Polity Press, 1995.

Hayward, T., 'Anthropocentrism: A Misunderstood Problem', *Environmental Values*, 6:1, 1997, 49.

Hawken, P., Lovins, A. and Lovins, H., *Natural Capitalism: Creating the Next Industrial Revolution*, Back Bay Books; 1st edition, 2008.

Hearnshaw, E. J. S., Cullen, R. and Hughey, K. F. D., *Ecosystem Health Demystified: an Ecological Concept Determined by Economic Means*, Conference paper presented at the Economics and Environment Network Workshop, held at the Australian National University, Canberra, 4–6 May 2005.

Henriksen, T., 'Conservation and Sustainable Use of Arctic Marine Biodiversity: Challenges and Opportunities', *Arctic Review on Law and Politics*, 1:2, 2010, 249.

Hey, E., 'The Persistence of a Concept: Maximum Sustainable Yield', *The International Journal of Marine and Coastal Law*, 27:4, 2012, 763.

Hicks, B., 'Treaty Congestion in International Environmental Law: The Need for Greater International Coordination', 32 *University of Richmond Law Review*, 1998–99, 32.

Hincapíe, G. and Restrepo, R., 'La Constitución Encriptada. Nuevas formas de emancipación del poder global', *Revista de Derechos Humanos y Estudios Sociales*, 4:8, 2012, 97.

Hobbes, T., *The Elements of Law Natural and Politic*, XXVIII, 1640, (1), http://www.constitution.org/th/elements.htm (Accessed 26 September 2018).

Hobbes, T., *Leviathan*, eBooks@Adelaide The University of Adelaide Library, 1651.

Hobbes, T., *De Cive: Philosophical Rudiments Concerning Government and Society*, 1651, XIII (4), http://www.constitution.org/th/decive.htm (Accessed 26 September 2018).

Hobbs, R., Hallett, L., Ehrlich, P., and Mooney, H., 'Intervention Ecology: Applying Ecological Science in the Twenty-first Century', *BioScience*, 61:6, 2011, 442.

Hoegh-Guldberg, O., et al.,. *Reviving the Ocean Economy: the case for action - 2015*. Geneva: WWF International, 2015.

Holder, J., 'New Age: Rediscovering Natural Law', *Current Legal Problems*, 53:1, 2000, 151.

Holling. C., and Meffe. G., 'Command and Control and the Pathology of Natural Resource Management', *Conservation Biology*, 10:2, 1996, 328.

Holling, C., 'Engineering Resilience versus Ecological Resilience', in P. Schulze (ed.) *Engineering within Ecological Constraints*, Washington, DC, The National Academy Press, 1996.

Holt, S. and Talbot, L., 'New Principles for the Conservation of Wild Living Resources', *Wildlife Monographs, Supplement to the Journal of Wildlife Management*, 59, 1978.

Hopkins, C., 'The Concept of Ecosystem Health and Association with the Ecosystem Approach to Management and Related Initiatives', *ICES BSRP/HELCOM/UNEP Regional Sea Workshop on Baltic Sea Ecosystem Health Indicators*, 30 March–1 April 2005, Sopot, Poland.

Horkheimer, M. and Adorno, T., *Dialectic of Enlightenment*, London: Verso Classics, 1997.

Hornborg, A., *The Power of the Machine. Global Inequalities of Economy, Technology, and Environment*, Walnut Creek, CA: Altamira Press, 2001.

Howarth, W., 'The Progression towards Ecological Quality Standards', *Journal of Environmental Law*, 18:3, 2006, 3.

Humble, L. and Allen, E., 'Forest Biosecurity: Alien Invasive Species and Vectored Organisms', *Canadian Journal of Plant Pathology*, 28:1, 2006, 256.

Hunt, A., 'Foucault's Expulsion of Law: Toward a Retrieval', *Law and Social Inquiry*, 17:1, 1992, 1.

Hunt, A., and Wickham G., *Foucault and Law: Towards a Sociology of Law as Governance*, London: Pluto Press, 1994.

Jakobsen, I., 'Marine Protected Areas in International Law: A Norwegian Perspective', PhD Dissertation, University of Tromsø, 2009.

Jax, K., 'On the Different Perceptions of the Ecosystem Concept within the Ecosystem Approach: Problems and Potentials' in H. Korn, R. Schliep and J. Stadler (eds) *Report of the International Workshop on the "Further Development of the Ecosystem Approach"*, Bonn, Germany: BfN - Skripten 78, 2003.

Jeffrey, M., 'Biodiversity Conservation in the Context of Sustainable Human Development: A Call to Action' in M. Jeffery, J. Firestone and K. Bubna-Litic (eds) *Biodiversity Conservation, Law and Livelihoods: Bridging the North-South Divide*, Cambridge: Cambridge University Press, 2008.

Jennings, S., Smith, A., Fulton, E. and Smith, D., 'The Ecosystem Approach to Fisheries: Management at the Dynamic Interface between Biodiversity Conservation and Sustainable Use', *Annals of the New York Academy of Sciences*, 1322, 2014, 48.

Jing, L., *Preservation of Ecosystems of International Watercourses and the Integration of Relevant Rules*, Leiden: Brill/Nijhoff, 2014.

Johannsdottir, A., 'The Convention on Biological Diversity. Supporting Ecological Sustainability or Prolonging Denial?', *Nordic Environmental Law Journal*, 1, 2010, 81.

Johnston, S. 'The Convention on Biological Diversity: The Next Phase', *Review of European, Comparative and International Environmental Law*, 6:3, 1997, 219.

Jonas, H., 'The Scientific and Technological Revolutions', *Philosophy Today*, 15:2, 1971, 76.

Kaime, T., 'Symposium Foreword: Framing the Law and Policy for Ecosystem Services', 2 *Transnational Environmental Law* 2013, 211.

Kareiva, P., 'New Conservation: Setting the Record Straight and Finding Common Ground', *Conservation Biology*, 28:3, 2014, 634.

Kareiva, P., Marvier, M. and Lalasz, R. 'Conservation in the Anthropocene: Beyond Solitude and Fragility', *Breakthrough Journal*, Winter 2011.

Karkkainen, B., 'Towards Ecologically Sustainable Democracy?' in A. Fung and E. Wright (eds) *Deepening Democracy Institutional Innovations in Empowered Participatory Governance*, London, New York: Verso, 2003.

Karkkainen, B., 'Post-Sovereign Environmental Governance', *Global Environmental Politics*, 4:1, 2004, 72.

Karkkainen, B. 'Marine Ecosystem Management and a "Post-Sovereign" Transboundary Governance', *San Diego International Law Journal*, 6, 2004–5, 113.

Katz, E. and Oechsli, L. 'Moving beyond Anthropocentrism: Environmental Ethics, Development, and the Amazon', *Environmental Ethics*, 15:1, 1993, 49, p. 58

Kay, J., Regier, H., Boyle, M. and Francis, G. 'An Ecosystem Approach for Sustainability: Addressing the Challenge of Complexity', *Futures* 31:7, 1999, 721, p. 722.

Keiter, R., and Boyce, M., *The Greater Yellowstone Ecosystem: Redefining America's Wilderness Heritage*, New Haven, CT: Yale University Press, 1994.

Keller, D. (ed.) *Environmental Ethics. The Big Questions*, Malden, MA: Wiley-Blackwell, 2010.

Keller, D. and Golley, F. (eds) *The Philosophy of Ecology: From Science to Synthesis*, Athens, GA: University of Georgia Press, 2000.

Kennedy, D., 'Form and Substance in Private Law Adjudication', *Harvard Law Review*, 89:8, 1979a, 1685.

Kennedy, D., 'The Structure of Blackstone's Commentaries', *Buffalo Law Review*, 28, 1979b, 211.

Kibaroglu, A., Kirschner, A., Mehring, S. and Wolfrum, R., *Water Law and Cooperation in the Euphrates-Tigris Region*, Leiden: Martinus Nijhoff, 2013.

Kidd, S., Plater A., and Frid, C., *The Ecosystem Approach to Marine Planning and Management*, New York: Routledge, 2011.

Kim, R. and Bosselmann, K., 'International Environmental Law in the Anthropocene: Towards a Purposive System of Multilateral Environmental Agreements', *Transnational Environmental Law*, 2:2, 2013, 285.

Kim, R., and Bosselmann, K., 'Operationalizing Sustainable Development: Ecological Integrity as a Grundnorm of International Law', *Review of European, Comparative & International Environmental Law*, 24:2, 2015, 194.

Kimball, L., *International Ocean Governance. Using International Law and Organizations to Manage Marine Resources Sustainably*, Gland, Switzerland and Cambridge, UK: The World Conservation Union, 2001.

Kingsland, S. E., 'Essay review: the history of ecology', *Journal of the History of Biology*, 27:2, 1994, 349.

Kirk, E., 'The Ecosystem Approach and the Search for an Objective and Content for the Concept of Holistic Ocean Governance', *Ocean Development and International Law*, 46:1, 2015, 33.

Kloor, K., 'The Battle for the Soul of Conservation Science', *Issues in Science and Technology*, 31:2, 2015, 73.

Knop, K., *Statehood: Territory, People, Government* in J. Crawford and M. Koskenniemi (eds) *The Cambridge Companion to International Law*, Cambridge: Cambridge University Press, 2012.

Koester, V., 'The Biodiversity Convention Negotiation Process and Some Comments on the Outcome', *Environmental Policy and Law*, 27:3, 1997, 175.

Korn, H., Schliep, R., and Stadler J. (eds) 'Report of the International Workshop on the "Further Development of the Ecosystem Approach"', Bonn, Germany: BfN - Skripten 78, 2003.

Koskenniemi, M., *From Apology to Utopia. The Structure of International Legal Argument.* Reissue with a new epilogue, Cambridge: Cambridge University Press, 2005.

Kotsakis, A., The Biological Diversity Complex: A History of Environmental Government, PhD Thesis, London School of Economics, 2011.

Kotzé, L. and Marauhn, T. (eds) *Transboundary Governance of Biodiversity*, The Hague: Martinus Nijhoff, 2014.

Kuhn, T., *The Strucutre of Scientific Revolutions*, Chicago: University of Chicago Press, 2nd edition, enlarged, 1970, reprinted in International Encyclopedia of Unified Science, Volume 2, Number 2.

Kuokkanen, T., *International Law and the Environment: Variations on a Theme*, Leiden, Boston: Martinus Nijhoff, 2002.

Lackey, R., 'Seven Pillars of Ecosystem Management', *Landscape and Urban Planning*, 40, 1998a, 40, 21.

Lackey, R., 'Ecosystem Management: In Search of the Elusive Paradigm', *Human Ecology Review*, 4:2, 1998b, 107.

Lackey, R., 'Appropriate use of ecosystem health and normative science in ecological policy' in D. J. Rapport, W. L. Lasley, D. E. Rolston, N. O. Nielsen, C. O. Qualset and A. B. Damania, *Managing for Healthy Ecosystems*, Boca Raton, FL: CRC Press, 2002.

Laclau, E. and Mouffe, C., *Hegemony and Socialist Strategy: Towards a Radical Democratic Politics*, 2nd edition, London: Verso, 2001.

La Torre, M., *Law as Institution*, London and New York: Springer, 2010.

Lange, B., 'Focuauldian-inspired Discourse Analysis: A Contribution to Critical Environmental Scholarship?' in A. Philippopoulos-Mihalopoulos (ed.) *Law and Ecology: New Environmental Foundations*, London: Routledge Glasshouse Book, 2011.

Larsen, T., 'Progress in Polar Bear Research and Conservation in the Arctic Nations', *Boston College Environmental Affairs Law Review*, 4:2, 1975, 295.

Lascoumes, P., *L'éco-pouvoir. Environnements et Politiques*, Paris: La Découverte, 1994.

Latour, B., *The Politics of Nature*, London: Harvard University Press, 2004.

Latour, B., 'An Attempt at a "Compositionist Manifesto"', *New Literary History*, 41, 2010, 471.

Leary, D. and Pisupati, B., *The Future and International Environmental Law*, Tokyo: United Nations University, 2010.

Lechner, S., *Sovereignty and Territoriality: An Essay in Medieval Political Theory*, Paper prepared for the 6 SGIR Pan-European Conference on International Relations, Turin, 12 – 15 Sept. 2007.

Leiss, W., *The Domination of Nature*, Montreal and Quebec: McGill-Queens University Press, 1994.

Lemke, L., *Biopolitics. An Advanced Introduction*, New York: New York University Press, 2011.

Lemke, T., 'The Birth of Bio-politics: Michael Foucault's Lectures at the Collège de France on Neo-liberal Governmentality', *Economy and Society*, 30:2, 2001, 190–207.

Leonardi, E., Biopolitics of Climate Change: Carbon Commodities, Environmental Profanations, and the Lost Innocence of Use-Value, PhD Thesis, The School of Graduate and Postdoctoral Studies, The University of Western Ontario, 2012.

Lévêque, C., *Ecology from Ecosystem to Biosphere*, Enfield, NH: Science Publishers, 2003

Liftin, K. (ed.) *The Greening of Sovereignty in World Politics, Cambridge*, Cambridge, MA and London: MIT Press, 1998.

Light, A., 'Contemporary Environmental Ethics from Metaethics to Public Philosophy', *Metaphilosophy* 33:4, 2002, 426, p. 439.

Litowitz, D., 'Gramsci, Hegemony and the Law', *Brigham Young University Law Review*, 2, 2000, 515.

Locke, H. and Dearden, P., 'Rethinking Protected Area Categories and the New Paradigm', *Environmental Conservation* 32:1, 2010, 1.

Lombardi Vallauri, L., *Corso di Filosofia del Diritto*, Padova: CEDAM, 1978.

Long, R., 'Legal Aspects of Ecosystem-Based Marine Management in Europe' in: A. Chircop, M. L. McConnell and S. Coffen-Smou (eds) *Ocean Yearbook*, The Hague: Martinus Nijhoff, 2012.

Long, R., Charles, A. and Stephenson, R., 'Key principles of marine ecosystem-based management', *Marine Policy*, 57, 2015, 53.

Loreau, M., *From Populations to Ecosystems Theoretical Foundations for a New Ecological Synthesis*, Princeton, NJ: Princeton University Press, 2010.

Loveland, T. R. and Merchant, J. M., 'Ecoregions and Ecoregionalization: Geographical and Ecological Perspectives', *Environmental Management*, 34, Supplement 1, 2004, 1.

Lowe, V., 'Sustainable Development, Unsustainable Argument' in A. Boyle and C. Freestone (eds) *International Law and Sustainable Development. Past Achievements and Future Challenges*, Oxford: Oxford University Press, 1999.

Lowe, V., 'The Politics of Law-Making: Are the Method and Character of Norm Creation Changing?' in M. Byers (ed.) *The Role of Law in International Politics. Essays in International Relations and International Law*, Oxford: Oxford University Press, 2000.

Maffei, M. C., 'New Trends in the Protection of Whales' in F. Francioni and T. Scovazzi (eds) *International Law for Antarctica*, 2nd edition, The Hague: Kluwer Law, 1996, 171.

Malm A., and Hornborg, A., 'The Geology of Mankind? A Critique of the Anthropocene Narrative', *The Anthropocene Review*, 1:1, 2014, 62.

Maltby, E., 'Some European Perspectives on the Ecosystem Approach' in H. Korn, J. Stadler, E. Maltby and A. J. Kerr (eds) *Report on the Scientific Workshop on "The Ecosystem Approach - What Does it Mean for European Ecosystems?*, Bonn: Bundesamt für Naturschutz, 1999.

Maltby, E., 'Ecosystem Approach: from Principle to Practice', Ecosystem Service and Sustainable Watershed Management in North China International Conference, Beijing, P.R. China, August 23–25, 2000, 205, HTTP: http://www.biotechnology.uni-koeln.de/inco2-dev/common/contribs/18_maltb.pdf, (Accessed 26 September 2018).

Mangel, M., et al., 'Principles for the Conservation of Wild Living Resources', *Ecological Applications*, 6:2, 1996, 338.

Marella, M., *Oltre il Pubblico e il Privato. Per un Diritto dei Beni Comuni*, Bologna: Ombre Corte, 2012.

Marella, M., 'The Constituent Assembly of the Commons (CAC)', *Open Democracy*, 28 February 2014a, HTTP: https://www.opendemocracy.net/can-europe-make-it/maria-rosaria-marella/constituent-assembly-of-commons-cac (Accessed 26 September 2018).

Marella, M., 'La Parzialità dei Beni Comuni contro l'Universalismo del Bene Comune', 6 May, 2014b, HTTP http://www.euronomade.info/?p=2282, (Accessed 26 September 2018).

Marella, M., 'The Constituent Assembly of the Commons (CAC), Open Democracy, 28 February 2014a, HTTP: https://www.opendemocracy.net/can-europe-make-it/maria-rosaria-marella/constituent-assembly-of-commons-cac (Accessed 26 September 2018).

Mariscalco, D. and Maltese, P. (eds), *Introduzione a Vita, Politica, Rappresentazione. A Partire dall'Italian Theory*, Verona: Ombre Corte.

Martinez-Alier, J., *Ecological Economics: Energy, Environment and Society*, Oxford: Basil Blackwell, 1990.

Marvier, M., 'New Conservation is True Conservation', *Conservation Biology*, 28:1, 2013, 1.

Marvier, M. and Kareiva, P., 'The Evidence and Values Underlying "New Conservation"', *Trends in Ecology and Evolution*, 29:3, 2014, 131.

Mattei, U. and Nader, L., *Plunder: When the Rule of Law is Illegal*, Malden, MA: Wiley-Blackwell, 2008.

Mattei, U., *Beni Comuni. Un Manifesto*, Bari: Laterza, 2011.

Mattei, U., 'Protecting the Commons: Water, Culture, and Nature: The Commons Movement in the Italian Struggle against Neoliberal Governance', *South Atlantic Quarterly*, 112:2, 2013, 366–3676.

Mazzocchi, F., 'Western Science and Traditional Knowledge: Despite Their Variations, Different Forms of Knowledge Can Learn from Each Other', *EMBO Reports* 7:5, 2006, 463.

McCaffrey, *The Law of International Watercourses: Non-Navigational Uses*, Oxford: Oxford University Press, 2003.

McConnell, F., *The Biodiversity Convention. A Negotiating History*, London: Kluwer Law International, 1996.

McGrath, D. and Greenwalt, T., 'Valuation and Payment for Ecosystem Services as Tools to Improve Ecosystem Management' in K. Robbins (ed.) *The Laws of Nature*, Akron, OH: The University of Akron Press, 2014.

McIntyre-Mills, J., *Systemic Ethics and Non-Anthropocentric Stewardship. Implications for Transdisciplinarity and Cosmopolitan Politics*, Switzerland: Springer International Publishing, 2014.

McIntyre, O., 'The Emergence of an "Ecosystem Approach" to the Protection of International Watercourses Under International Law', *Review for European, Comparative and International Environmental Law*, 13:1, 2004, 1–14.

McIntyre, O., *The Ramsar Convention on Wetlands and General International Water Law: Mutually Supportive Regimes*, Report prepared for the Ramsar Convention Secretariat, 2014a.

McIntyre, O., 'The Protection of Freshwater Ecosystems Revisited: Towards a Common Understanding of the "Ecosystems Approach" to the Protection of Transboundary Water Resources', *RECIEL (Review of European, Comparative & International Environmental Law)* 23:1, 2014b, 88.

McLeod, K., Lubchenco, J., Palumbi, S., and Rosenberg, A., *Scientific Consensus Statement on Marine Ecosystem-Based Management*, 2005. The statement was signed by 219 scientists and policy experts, and published by the Communication Partnership for Science and the Sea, HTTP: http://marineplanning.org/wp-content/uploads/2015/07/Consensusstatement.pdf (Accessed 26 September 2018).

Meffe, G., Nielsen, L., Knight, R. L. and Schenborn, D., *Ecosystem Management: Adaptive, Community-Based Conservation*, 1st edition, Washington, DC: Island Press, 2002.

Merchant, C., *The Death of Nature. Women, Ecology and the Scientific Revolution*, San Francisco: Harper One, 1990.

Metzer, E., *The Quest for Sustainable International Fisheries. Regional Efforts to Implement the 1995 United Nations Fish Stock Agreement*, Ottawa: NRC Research Press, 2009.

M'Gonigle, M., and Takeda, L., 'The Liberal Limits of Environmental Law: A Green Legal Critique', *Pace Environmental Law Review*, 30:3, 2013, 1005.

Miéville, C., *Between Equal Rights. A Marxist Theory of International Law*, Chicago, IL: Haymarket Books, 2006.

Millennium Ecosystem Assessment, *Ecosystems and Human Well-being: Synthesis*, Washington, DC: Island Press, 2005.

Miller, F., et al., 'Resilience and Vulnerability: Complementary or Conflicting Concepts?' *Ecology and Society*, 15:3, 2010, 11.

Minter, B., *Nature in Common? Environmental Ethics and the Contested Foundations of Environmental Policy*, Philadelphia: Temple University Press, 2009.

Minter, B., 'Geoengineering and Ecological Ethics in the Anthropocene', *BioScience*, 62:10, 2012, 857.

Mitman, G., *The State of Nature: Ecology, Community, and American Social Thought*, 1900–1950, Chicago: University of Chicago Press, 1992.

Monateri, P. G., 'Legal Formants and Competitive Models: Understanding Comparative Law from Legal Process to Critique in Cross-System Legal Analysis', 2008 http://ssrn.com/abstract=1317302 (Accessed 26 September 2018).

Morgera, E. and Tsioumani, E., 'Yesterday, Today, and Tomorrow: Looking Afresh at the Convention on Biological Diversity', *Yearbook of International Environmental Law*, 21:1, 2011, 3.

Morishita, J., 'What is the Ecosystem Approach for Fisheries Management?', *Marine Policy*, 32, 2008, 19.

Munro, R. and Lammers, J., *Environmental Protection and Sustainable Development. Legal Principle and Recommendations*, London: Graham & TrotMan/Martinus Nijhoff, 1987.

Munro, V., 'Legal Feminism and Foucault – A Critique of the Expulsion of Law', *Journal of Law and Society*, 28:4, 2001, 546.

Murphy, T., 'BritCrits: Subversion and Submission, Past, Present and Future', *Law and Critique*, 10:3, 1999, 237.

Naffine, N., 'Who are Law's Persons? From Cheshire Cats to Responsible Subjects', *Modern Law Review*, 66:3, 2003, 346.

Nanda, V. and Pring, G., *International Environmental Law and Policy for the 21st Century*, Leiden: Martinus Nijhoff, 2012.

Nayar, R. and Ong, D., 'Developing Countries, "Development" and the Conservation of Biological Diversity' in M. Bowman and C. Redgewell (eds) *International Law and the Conservation of Biological Diversity*, London: Kluwer Law International, 1996.

Næss, A., 'The Shallow and the Deep, Long-range Ecology Movement. A Summary', *Inquiry*, 16:1,1973, 95.

Negri, A., *Cinque Lezioni di Metodo su Moltitudine e Impero*, Soveria Mannelli: Rubbettino Editore, 2003.

Negri, A., *Dentro/contro il Diritto Sovrano. Dallo Stato dei Partiti ai Movimenti della Governance*, Verona: Ombre Corte, 2009.

Nollkaemper, A., Sovereignty and Environmental Justice in International Law' in J. Ebbesson. and P. Okowa (eds) *Environmental Law and Justice in Context*, Cambridge: Cambridge University Press, 2009.

Norton, B., 'Environmental Ethics and Weak Anthropocentrism', *Environmental Ethics* 6:2, 1984, 131.

Noss, S. F. and Cooperrider, A. *Saving Nature's Legacy: Protecting and Restoring Biodiversity*, Washington, DC: Island Press, 1994.

Ntona, M. and Morgera, E., 'Connecting the Dots between SDG 14 and the Other SDGs: The Value Added of the Ecosystem Services Concept and the Integration of Equity Through Marine Spatial Planning', SCELG Working Paper No 6 (2017) 7.

O'Donoghue, A., 'Sovereign Exception: Notes on the Thought of Giorgio Agamben, Critical Legal Thinking', HTTP: http://criticallegalthinking.com/2015/07/02/sovereign-exception-notes-on-the-thought-of-giorgio-agamben/(Accessed 26 September 2018).

Odum, E., *Fundamentals of Ecology*, Philadelphia: W. B. Saunders Company, 1953.

Ong, D., 'International Environmental Law Governing Threats to Biological Diversity' in M. Fitzmaurice, D. Ong and P. Merkouris (eds) *Research Handbook on International Environmental Law*, Cheltenham: Edward Elgar, 2010.

Orestano, R., *Introduzione allo Studio del Diritto Romano*, Milan: Giuffrè, 1987.

Osborne, P., 'Modernity is a Qualitative, Not a Chronological, Category', in F. Barker, P. Hulme and M. Iverson (eds) *Postmodernism and the Rereading of Modernity*, Manchester: Manchester University Press, 1992.

Ost, F., *La Nature hors la loi; L'écologie à l'épreuve du droit*, Paris: La Découverte, 2003.

Palmer, C., 'Stewardship: A Case Study in Environmental Ethics' in R. Berry (ed.) *Environmental Stewardship: Critical Perspectives, Past and Present*, London: T&T Clark, 2006.

Parajuli., P., 'Revisiting Gandhi and Zapata: Motion of Global Capital, Geographies of Difference and the Formation of Ecological Ethnicities', in M. Blaser, H. Feit, and G. McRae (eds) *In the Way of Development. Indigenous Peoples, Life Projects and Globalization*, London: Zed Books, 2004.

Pardy, B., 'Changing Nature: The Myth of the Inevitability of Ecosystem Management', *Pace Environmental Law Review*, 20:2, 2003, 675.

Pardy, B., 'Ecosystem Management in Question: A Reply to Ruhl', *Pace Environmental Law Review*, 23:1, 2006, 212.

Pardy, B., 'The Pardy-Ruhl Dialogue on Ecosystem Management, Part V: Discretion, Complex-Adaptive Problem Solving and the Rule of Law', *Pace Environmental Law Review*, 25:2, 2008, 341.

Pashukanis, E., *Law and Marxism: A General Theory*, London: Pluto Press, 1989.

Passmore, J., *Man's Responsibility for Nature: Ecological Problems and Western Traditions*, London: Scribner, 1974.

Paterson, M., 'Legitimation and Accumulation in Climate Change Governance', *New Political Economy*, 15:3, 2010, 345.

Pauly, D. and Froese, R. 'Comments on FAO's State of Fisheries and Aquaculture, or "SOFIA 2010"', *Marine Policy*, 36:3, 2012, 746.

Peel, J., 'Precaution – a Matter of Principle, Approach or Process?', *Melbourne Journal of International Law*, 5, 2004.

Percival, R., 'Global Law and the Environment', *Washington Law Review*, 86, 2001, 579.

Petersen, H., 'Gender and Nature in Comparative Legal Cultures' in D. Nelken (ed.) *Comparing Legal Cultures*, Aldershot: Dartmouth, 1997.

Peterson, G. D., 'Ecological Management: Control, Uncertainty, and Understanding', in K. E. Cuddington and B. E. Beisner (eds) *Ecological Paradigms Lost: Routes of Theory Change*, Burlington, MA: Elsevier Academic Press, 2005.

Pezzano, G., 'Comunità, Immunità, Alterità: una Biopolitica Affermativa e Oltre-umana ?' *Tropos*, 2, 2011, 167.

Philippopoulos-Mihalopoulos, A., 'Towards a Critical Environmental Law' in A Philippopoulos-Mihalopoulos (ed.) *Law and Ecology. New Environmental Foundations*, New York: Routledge, 2011a.

Philippopoulos-Mihalopoulos, A., 'Looking for the Space between Law and Ecology' in A. Philippopoulos-Mihalopoulos (ed.) *Law and Ecology. New Environmental Foundations*, New York: Routledge, 2011b.

Philippopoulos-Mihalopoulos, A., 'Actors or Spectators? Vulnerability and Critical Environmental Law', *Oñati Socio-Legal Series*, 3:5, 2013, 854.

Philippopoulos-Mihalopoulos, A., 'Epistemologies of Doubt' in A. Grear and L. Kotzé (eds) *Research Handbook in Human Rights and the Environment*, Cheltenham, UK: Edward Elgar, 2015.

Pickett, B., *On the Use and Abuse of Foucault for Politics*, Lanham, MD: Lexington Books, 2005.

Pieraccini, M., 'Reflections on the Relationship between Environmental Regulation, Human Rights and beyond – with Heidegger' in A. Grear and E. Grant (eds) *Thought, Law, Action and Rights in the Age of Environmental Crisis*, Cheltenham: Edward Elgar, 2015.

Pihälä, P., 'HELCOM Baltic Sea Action Plan: An Ecosystem Approach to the Management of Human Activities' in M. Reckermann, K. Brander, B. MacKenzie and A. Omstedt (eds) *Climate Impacts on the Baltic Sea: From Science to Policy*, Heidelberg: Springer Verlag, 2012.

Pinchot, G., *The Training of a Forester*, London: JB Lippincott Co., 1914.

Platjouw, F., *Environmental Law and the Ecosystem Approach: Maintaining Ecological Integrity through Consistency in Law*, London: Routledge, 2016, p. 79.

Princen, T., *The Logic of Sufficiency*, Cambridge, MA: MIT Press, 2005.

Rabinow, P. and Rose, N., 'Biopower Today', *BioSocieties*, 1:2, 2006, 195.

Redford, K. and Adams, W., 'Payment for Ecosystem Services and the Challenge of Saving Nature', *Conservation Biology*, 23:4, 2009, 785.

Reid, K., Croxall, J. P., Briggs, D. R. and Murphy, E. J., 'Antarctic ecosystem monitoring: quantifying the response of ecosystem indicators to variability in Antarctic krill', *ICES Journal of Marine Science*, 62:3, 2005, 366.

Rieu-Clarke, A. and Spray, C., 'Ecosystem Services and International Water Law: Towards a More Effective Determination and Implementation of Equity?', *Potchefstroom Electronic Law Journal*, 16:2, 2013, 12.

Rolston III, H. *A New Environmental Ethics: The Next Millennium for Life on Earth*, London: Routledge, 2012.

Romano, S., *L'Ordinamento Giuridico*, 2nd edition, Florence: Sansoni, 1946.

Rosendal, K., 'The Convention on Biological Diversity: A Viable Instrument for Conservation and Sustainable Use' in H. Bergesen, G. Parmann, and Ø. Thommessen (eds) *Green Globe Yearbook of International Cooperation on Environment and Development*, Oxford: Oxford University Press, 1995.

Rosen, L., *Law as Culture: An Invitation*, Princeton, NJ: Princeton University Press, 2008.

Rota, R., 'Sussidiarietà e Ambiente: la Centralità dell'Uomo', Paper presented at the Conference 'Il governo del territorio e dell'ambiente per l'uomo', held on 22 June 2007 at Università of Rome 'Tor Vergata'.

Routley, R., and Routley, V., 'Against the Inevitability of Human Chauvinism' in K. E. Goodpaster and K. M. Sayre (eds) *Ethics and Problems of the 21st Century*, Indiana: University of Notre Dame Press, 1979.

Rowe, S., 'Review of Discordant Harmonies: A New Ecology for the 21st Century', *The Trumpeter Journal of Ecophilosophy*, 12:4, 1995, HTTP: http://trumpeter.athabascau.ca/index.php/trumpet/article/view/292/431 (26 September 2018).

Ruhl, J., 'The Myth of What Is Inevitable under Ecosystem Management: A Response to Pardy', *Pace Environmental Law Review*, 21:2, 2004, 315.

Ruhl, J., 'The Pardy-Ruhl Dialogue on Ecosystem Management, Part IV: Narrowing and Sharpening the Questions', *Pace Environmental Law Review*, 24:1, 2007, 25.

Ruhl, J., 'Ecosystem Services and Ecosystem Management – How Good a Fit?' in K. Robbins (ed.) *The Laws of Nature: Reflections on the Evolution of Ecosystem Management Law and Policy*, Akron, OH: University of Akron Press, 2013.

Ruhl, J. and Salzman, J., 'The Law and Policy Beginnings of Ecosystem Services', 22 *Journal of Land Use and Environmental Law* 2007, 157.

Rutherford, P., 'The Entry of Life into History' in E. Darier (ed.) *Discourses of the Environment*, Oxford: Blackwell Publishers, 1999, 37.

Rutherford, P., The Problem of Nature in Contemporary Social Theory, PhD Thesis, The Australian National University, 2000.

Sacco, R., 'Legal Formants: A Dynamic Approach to Comparative Law (Installment I of II)', *The American Journal of Comparative Law*, 39:1, 1991a, 1.

Sacco, R., 'Legal Formants: A Dynamic Approach to Comparative Law (Installment II of II)', *The American Journal of Comparative Law*, 39:2, 1991b, 343.

Sachs, W., *Global Ecology. A New Arena of Political Conflict*, London: ZED Books, 1993.

Sage, B., 'Precautionary Coastal States' Jurisdiction', *Ocean Development and International Law*, 37:3/4, 2006, 359.

Sagoff, M., 'The Plaza and the Pendulum: Two Concepts of Ecological Science', *Biology and Philosophy*, 18:4, 2003, 529.

Sagoff, M., 'The Rise and Fall of Ecological Economics a Cautionary Tale', *Breakthrough Journal*, 2012, HTTP: http://thebreakthrough.org/index.php/journal/past-issues/issue-2/the-rise-and-fall-of-ecological-economics (Accessed 26 September 2018).

Sainsbury, K., Punt, A., and Smith, A., 'Design of Operational Management Strategies for Achieving Fishery Ecosystem Objectives', *ICES Journal of Marine Science*, 57:3, 2000, 731.

Sasaki, N. and Putz, F., 'Critical need for new definitions of "forest" and "forest degradation" in global climate change agreements', *Conservation Letters*, 2:5, 2009, 226.

Sassen, S., *Territory Authority and Rights. From Medieval to Global* Assemblages, Princeton, NJ: Princeton University Press, 2006.

Scarlett, L. and Boyd, J., 'Landscape-wide Conservation and Ecosystem Services. Leveraging Federal Policies' in K. Robbins (ed.) *The Laws of Nature: Reflections on the Evolution of Ecosystem Management Law and Policy*, Akron, OH: University of Akron Press, 2013.

Scheiber, H. N., 'From Science to Law to Politics: An Historical View of the Ecosystem Idea and Its Effect on Resource Management', *Ecology Law Quarterly*, 24, 1997, 631.

Schillmoller, A. and Pelizzon, A., 'Mapping the Terrain of Earth Jurisprudence: Landscape, Thresholds and Horizons', *Environmental and Earth Law Journal*, 3:1, 2013, 1.

Schillmoller, A. and Ricketts, A., 'Recognising Rights for Nature: A Negotiation of Principle and Pragmatism', *Transformations*, 21, 2012, 1.

Schlaepfer, R., *Ecosystem-Based Management of Natural Resources: A Step Towards Sustainable Development* (IUFRO Occasional Paper No. 6, International Union of Forest Research Organizations), Washington, DC: IUFRO Secretariat, 1997.

Schlegel, J., 'Critical Legal Studies' in A. Hadden and A. Brophy (eds) *A Companion to American Legal History*, Chichester: Wiley-Blackwell, 2013.

Schmitt, C., *The Concept of the Political*, Chicago: University of Chicago Press, 1996.

Schneider, J., 'Endangered Species Markets. A Focus for Criminology?' in M. Natarajan (ed.) *International Crime and Justice*, Cambridge: Cambridge University Press, 2011.

Schrift, A., *Nietzsche's French Legacy. A Genealogy of Poststructuralism*, New York, London: Routledge, 1995.

Schrijver, N., *Sovereignty over Natural Resources. Balancing Rights and Duties*, Cambridge: Cambridge University Press, 1997.

Schröter, M., van der Zanden, E., van Oudenhoven, A. P. E., Remme, R. P., Serna-Chavez, H. M., De Groot, R. S. and Opdam, P., 'Ecosystem Services as a Contested Concept: A Synthesis of Critique and Counter-Arguments', 7 *Conservation Letters* 7, 2014, 27, 514.

Sharpe, S., *Redeeming the Time: A Political Theology of the Environment*, London, UK: Bloomsbury Publishing, 1998.

Shrader-Frechette, K. S., *Risk and Rationality. Philosophical Foundations for Populist Reforms*, Berkeley: University of California Press, 1991.

Shrader-Frechette, K. S., 'Methodological Rules for Four Classes of Scientific Uncertainty' in J. Lemons (ed.) *Scientific Uncertainty and Environmental Problem Solving*, Oxford: Blackwell Science, 1996, 12.

Siemen, S. P., 'Earth Jurisprudence: Toward Law in Nature's Balance', *Barry Law Review*, 21:1, 2008, 1.

Singh, S., 'The Critic (-al) Subject', in W. Werner, A. de Hoon and M. Galán (eds), *The Law of International Lawyers Reading Martti Koskenniemi*, Cambridge: Cambridge University Press 2017.

Siron, R., Sherman, K., Skjoldal, H. and Hiltz, E, 'Ecosystem-Based Management in the Arctic Ocean: A Multi-Level Spatial Approach', *Arctic*, 61, 2008, 86.

Smith, J. C., and Weisstub, D. N., *The Western Idea of Law*, London: Butterworth-Heinemann, 1984.

Smith, R., and Maltby, E., *Using the Ecosystem Approach to implement the CBD: A global synthesis report drawing lessons from three regional pathfinder workshops*, 2003, Hunterdale, UK: RHIER.

Soirila, U., 'The European Court of Human Rights, Islam and Foucauldian Biopower', *Helsinki Law Review*, 2, 2011, 365.

Soulé, M., 'The Social Siege of Nature', in M. Soulé and G. Lease (eds) *Reinventing Nature? Responses to Postmodern Deconstruction*, Washington, DC: Island Press, 1995.

Soulé, M., 'The New Conservation', *Conservation Biology*, 27:5, 2013, 895–897.

Soulé, M. and Wilcox, B. (eds), *Conservation Biology: An Evolutionary-Ecological Perspective*, Sunderland, MA: Sinauer, 1980.

Speth, J. and Haas, P., *Global Environmental Governance*, Washington, DC: Island Press, 2006.

Stanley, T., 'Ecosystem Management and the Arrogance of Humanism', *Conservation Biology*, 9:2, 1995, 255.

Steinhoff, G., 'Ecological Integrity in Protected Areas: Two Interpretations', *Seattle Journal of Environmental Law*, 3, 2013, 155.

Stevens, T., 'Rigor and Representativeness in Marine Protected Area Design', *Coastal Management*, 30:2, 2002, 237.

Stone, C. D., 'Should Trees have Standing? Toward Legal Rights for Natural Objects', *Southern California Law Review*, 45, 1972, 450.

Stone, C. D., 'Should Trees Have Standing? Revisited: How Far Will Law and Morals Reach? A Pluralist Perspective', *California Law Review*, 59, 1985, 1.

Stone, C. D., *Should Trees Have Standing? Law, Morality, and the Environment*, 3rd edition, Oxford: Oxford University Press, 2010.

Stone, C. D., Should Trees Have Standing? Law, Morality and the Environment, Oxford: Oxford University Press, 2004.

Sullivan, S., 'Green Capitalism, and the Cultural Poverty of Constructing Nature as Service Provider', *Radical Anthropology* 3, 2009, 18.

Sullivan, S., 'Banking Nature? The Spectacular Financialisation of Environmental Conservation', *Antipode* 45, 2013, 198. S. Sullivan, 2009, 'Green Capitalism, and the Cultural Poverty of Constructing Nature as Service Provider', *Radical Anthropology* 3, 18).

Tadros, V., 'Between Governance and Discipline: The Law and Michel Foucault', *Oxford Journal of Legal Studies*, 18:1, 1998, 75.

Tallacchini, M., *Diritto per la Natura. Ecologia e Filosofia del Diritto*, Turin: Giappichelli Editore, 1996.

Tallacchini, M., 'A Legal Framework from Ecology', *Biodiversity and Conservation*, 9:8, 2000, 1085.

Tallis, H. and Lubchenco, J., 'A Call for Inclusive Conservation', *Nature*, 515, 2014, 27–28.

Tanaka, Y., *A Dual Approach to Ocean Management. The Cases of Zonal and Integrated Management in International Law*, Alderhot: Ashgate, 2008.

Tansley, A. G., 'The Use and Abuse of Vegetational Concepts and Terms', *Ecology*, 16:3, 1935, 284.

Tarlock, D., 'The Nonequilibrium Paradigm in Ecology and the Partial Unraveling of Environmental Law', *Loyola L.A. Law Review*, 27, 1994, 1009.

Tarlock, D., 'International Water Law and the Protection of River System Ecosystem Integrity', *Brigham Young University Journal of Public Law*, 10:2, 1996, 181.

Tarlock, D.,'The Influence of International Environmental Law on U.S. Pollution Control Law', 21 *Vermont Law Review* 1997.

Teubner, G., 'Legal Irritants: Good Faith in British Law or How Unifying Law Ends Up in New Differences', *Modern Law Review*, 61, 1998, 11.

Thomas-Pellicer, R., What is Kultur?: The Places of God in the Age of Re-embodiments, PhD Thesis, University of Surrey, 2012

Tigar, M. and Levy, M., *Law and the Rise of Capitalism*, Delhi: Aakar Books, 2005.

Trouwborst, A., 'The Precautionary Principle and the Ecosystem Approach in International Law: Differences, Similarities and Linkages', *Review of Comparative, European and International Environmental Law*, 2009, 18:1; 26.

Tomasello, F., Le stagioni dell'Operaismo Italiano', *COSMOPOLIS, IV:1*, 2011, available HTTP: http://www.cosmopolisonline.it/20110715/tomasello.php (Accessed 26 September 2018).

Tronti, M., *Operai e Capitale*, Rome: DeriveApprodi, 2006.

Tudela, S. and Short, K., 'Paradigm Shifts, Gaps, Inertia, and Political Agendas in Ecosystem-Based Fisheries Management', *Marine Ecology Progress Series*, 300, 2005, 282.

Tuori, K., *Critical Legal Positivism*, Aldershot: Ashgate, Applied Legal Philosophy series, 2002.

Turkel, G., 'Michel Foucault: Law, Power, and Knowledge', *Journal of Law and Society*, 17:2, 1990, 170.

Turrell, W. R., *The Policy Basis of the "Ecosystem Approach" to Fisheries Management*, EuroGOOS Publication 21, 2004.

Tushnet, M., 'Critical Legal Studies: A Political History' *The Yale Law Journal*, 100:5, 1991, 1515.

Unger, R., *Law in Modern Society: Toward a Criticism of Social Theory*, New York: Free Press, 1976.

Uy, N. and Shaw, R. (eds), *Ecosystem-based Management*, Bradford, UK: Emerald Group Publishing Limited 2012.

Vadrot, A., *The Politics of Knowledge and International Biodiversity*, London: Routledge, 2014.

Vallentyne, J. and Beeton, A., 'The "Ecosystem" Approach to Managing Human Uses and Abuses of Natural Resources in the Great Lakes Basin', *Environmental Conservation*, 15, 1988, 58.

van den Belt, M. et al., 'Scientific Understanding of Ecosystem Services', in United Nations, *The First Global Integrated Marine Assessment: World Ocean Assessment 1*, Cambridge: Cambridge University Press, 2016, p. 6.

Van Heijnsbergen, P., *International Protection of Wild Flora and Fauna*, Amsterdam: IOS Press, 1997.

Vierros, M., 'The Ecosystem Approach of the Convention on Biological Diversity' in G. Bianchi and H. Skjoldal (eds) *The Ecosystem Approach to Fisheries*, Rome: CABI Publishing and FAO, 2008.

Villey, M., 'Dikaion-Torah' in M. Villey, *Seize essais de philosophie du droit*, Paris: Dallzo, 1969.

Vogel, S., *Against Nature: The Concept of Nature in Critical Theory*, New York: SUNY Press, 1996.

Voigt, C., 'The Principle of Sustainable Development. Integration and Ecological Integrity' in C. Voigt (ed.) *Rule of Law for Nature: New Dimensions and Ideas in Environmental Law*, Cambridge: Cambridge University Press, 2013.

Wallace, M. G., Cortner, H. J., Moote, M. A. and Burke S.,'Moving Toward Ecosystem Management: Examining a Change in Philosophy for Resource Management', *Journal of Political Ecology*, 3, 1996, 1.

Wang, H., 'Ecosystem Management and Its Application to Large Marine Ecosystems: Science, Law, and Politics', *Ocean Development & International Law*, 35:1, 2004, 41.

Weber, M., *The Protestant Ethics and the Spirit of Capitalism*, Translated by Talcott Parsons. With an Introduction by Anthony Giddens, London & New York: Routledge Classics, 2001.

Welchman, J., 'A Defence of Environmental Stewardship', *Environmental Values*, 21, 2012, 297.

Werksman, J., 'The Conference of Parties to Environmental Treaties' in J. Werksman (ed.) *Greening International Institutions*, London: Earthscan 1996.

West, M., 'The Future of the International Law of Capture Fisheries', *Journal of Northwestern Atlantic Fisheries Sciences*, 23, 1998, 19.

Westman, W., 'How Much Are Nature's Services Worth', *Science* 197, 1977, 960.

Weston, B. and Bollier, D., *Green Governance. Ecological Survival, Human Rights, and the Law of the Commons*, Cambridge: Cambridge University Press, 2013.

Whiteside, K., *Divided Natures: French Contributions to Political Ecology*, Cambridge, MA: MIT Press 2002.

Wiber, M. G., 'The Spatial and Temporal Role of Law in Natural Resource Management: The Impact of State Regulation of Fishing Space' in F. Von Benda-Beckmann and A. Griffiths (eds) *Spatializing Law: An Anthropological Geography of Law and Society*, 2009, Farnhan: Ashgate, 75.

Wickham, G., 'Foucault, Law, and Power: A Reassessment', *Journal of Law and Society*, 33:4, 2006, 596.

Wiener, J. 'Something Borrowed for Something Blue: Legal Transplants and the Evolution of Global Environmental Law', 27 *Ecology Law Quarterly* 2001, 1295.

Wilkinson, D.,'Using Environmental Ethics to Create Ecological Law' in J. Holder and D. McGillivray (eds) *Locality and Identity: Environmental Issues in Law and Society*, Ashgate and Dartmouth: Dartmouth Pub Co, 1999.

Wilson, D., 'The Paradoxes of Transparency Science and the Ecosystem Approach to Fisheries Management in Europe', 5 MARE Publication Series 5, Amsterdam: Amsterdam University Press, 2009, p. 167

Wittgenstein, L., *Philosophcial Investigations*, Translated by G. E. M. Anscombe, Oxford: Blackwell, 1953.

Wolfe, C., *Before the Law: Humans and Other Animals in a Biopolitical Frame*, Chicago, IL: University of Chicago Press, 2013.

Wood, M. C., 'Advancing the Sovereign Trust of Government to Safeguard the Environment for Present and Future Generations (Part I): Ecological Realism and the Need for a Paradigm Shift', *Environmental Law Review*, 39:1, 2009, 43.

Woodley, S., 'Ecological Integrity: A Framework for Ecosystem-Based Management' in D. N. Cole and L. Yung (eds) *Beyond Naturalness: Rethinking Park And Wilderness Stewardship In An Era Of Rapid Change*, Washington, DC: Island Press, 2010.

Worrell, R. and Appleby, M., 'Stewardship Of Natural Resources: Definition, Ethical And Practical Aspects', *Journal of Agricultural and Environmental Ethics*, 12, 2000, 263.

Worster, D., *Nature's Economy: The Roots of Ecology*, San Francisco: Sierra Club Books, 1977.

Worster, D., 'The Vulnerable Earth: Towards a Planetary History', *Environmental Review*, 11:2, 1987, 87.

Worster, D., 'The Ecology of Order and Chaos', *Environmental History Review*, 14:1/2, 1990, 1.

Worster, D., *Nature's Economy: The Roots of Ecology*, 2nd edition, Cambridge: Cambridge University Press, 1994.

Yaffee, S. L., 'Three Faces of Ecosystem Management', *Conservation Biology*, 13:4, 1999, 713.

Youatt, R., 'Counting Species: Biopower and the Global Biodiversity Census', *Environmental Values*, 17, 2008, 393.

European Parliament and Council Regulation (EU) 1380/2013 of 11 December 2013 on the Common Fisheries Policy [2013] OJ L354/22.

Treaties, conventions, protocols and other international agreements

1902 Convention for the Protection of Birds Useful to Agriculture IPE, Vol. IV, 1615 3, 177.

1933 Montevideo Convention on the Rights and Duties of States (signed 26 December 1933, entered into force 26 December 1934), League of Nations Treaty Series, vol. 165, pp. 20 – 43.

1945 Constitution of the United Nations Food and Agriculture Organization, 40 AJIL Supp. 76 CTS 1945/32.

1945 Convention for the Protection of Birds Useful to Agriculture, IPF, Vol. IV, 16153, 177.

1946 International Convention for the Regulation of Whaling, 161 UNTS 72 (1946).

1958 Convention on Fishing and Conservation of the Living Resources of the High Seas, adopted 29 April 1958, entered into force 20 March 1966, United Nations Treaty Series, vol. 559, p. 285.

1958 Geneva Convention on Fishing and the Conservation of the Living Resources of the High Seas 559 UNTS 285.

1963 Agreement for the Establishment of the General Fisheries Commission for the Mediterranean as amended by the General Fisheries Council for the Mediterranean at its First Special Session (May 1963), at its Thirteenth Session (July 1976), at its Twenty-second Session (October 1997) and by the General Fisheries Commission for the Mediterranean at its Thirty-eighth Session (May 2014) and approved by the FAO Conference at its Twelfth Session (December 1963) and by the FAO Council at its Seventieth Session (December 1976), at its Hundred and Thirteenth Session (November 1997) and at its One Hundred and Fiftieth Session (December 2014).

1970 FAO Technical Conference on Marine Pollution and its Effects on Living Resources and Fishing, Rome,

1971 Convention on Wetlands of International Importance especially as Waterfowl Habitat, 996 UNTS 245.

1972 FAO Technical Conference on Fishery Management and Development, Vancouver,

1973 Convention on International Trade in Endangered Species of Wild Fauna and Flora, 993 UNTS 243.

1973 Agreement on Conservation of Polar Bears, 13 ILM 13.

1975 Convention Concerning the Protection of the World Cultural and Natural Heritage (adopted 16 November 1972, entered into force 17 December 1975) 1037 UNTS 151.

1978 Agreement between Canada and the United States of America on Great Lakes Water Quality (30 UST 1383).

1979 Convention on Conservation of Migratory Species of Wild Animals, 1651 UNTS 333.

1980 Convention on the Conservation of Antarctic Marine Living Resources (CCAMLR), 1329 UNTS 48; 19 ILM 841

1980 Vienna Convention of the Law of the Treaties, adopted 23 May 1969, entered into force 27 January 1980, UNTS 1155, p. 331.

1982 United Nations Convention on the Law of the Sea, Montego Bay (UNCLOS), 10 December 1982, entered into force on 16 November 1994, 1833 UNTS 3; 21 ILM 1261.

1992 Nagoya Protocol on Access to Genetic Resources and the Fair and Equitable Sharing of Benefits Arising from their Utilization to the Convention on Biological Diversity, 31 ILM 818.

1992 United Nations Framework Convention on Climate Change, 1771 UNTS 107; 31 ILM 849.

1992 Convention on Biological Diversity, Rio de Janeiro, 5 June 1992, 1760 UNTS 79; 31 ILM 818.

1995 Convention on Future Multilateral Cooperation in the Northwest Atlantic Fisheries, 1135 UNTS 369; 34 ILM 1452.

1995 Agreement for the Implementation of the Provisions of the United Nations Convention on the Law of the Sea of 10 December 1982 Relating To The Conservation And Management Of Straddling Fish Stocks And Highly Migratory Fish Stocks, 1995, 34 ILM 1542.

1997 Convention on the Law of the Non-Navigational Uses of International Watercourses, UN Doc A/RES/51/229, 36 ILM 700.

2000 Cartagena Protocol on Biosafety to the Convention on Biological Diversity, 2226 UNTS 208; 39 ILM 1027.

2001 Convention on the Conservation and Management of Fishery Resources in the South East Atlantic Ocean, 41 ILM 257.

2001 International Treaty on Plant Genetic Resources for Food and Agriculture, 2400 UNTS 303.

2012 Great Lakes Water Quality Agreement of 2012, (Protocol Amending the Agreement Between Canada and the United States of America on Great Lakes Water Quality, 1978 (30 UST 1383), as Amended on October 16, 1983 (35 UST 2370), and on November 18, 1987 (2185 UNTS 504), signed September 7, 2012, entered into force February 12, 2013), 2012.

Court decisions

1997 ICJ, *Gabčíkovo-Nagymaros Project* (Hungary/Slovakia), Judgment, ICJ Reports 1997, p. 7

2015 ITLOS, *Request for an Advisory Opinion Submitted by the Sub-Regional Fisheries Commission* (SRFC Advisory Opinion), Advisory Opinion of Apr. 2, 2015.

CBD documents

Preparatory works

1987 Decision 14/26 'Rationalization of international conventions on biological diversity', UNEP, in the Report of the Governing Council on the Work of its Fourteenth Session, 8–19 June 1987 General Assembly Official Records: Second Session Supplement 25 (A/42/25).

1990 Report of the Ad Hoc Working Group on the Work of Its Second Session in Preparation for a Legal Instrument on Biological Diversity of the Planet, 23 February 1990 (UNEP/Bio.Div.2/3, 1990).

1991 Report of the Ad Hoc Working Group of Legal and Technical Experts on Biological Diversity on the Work of its Second Session, 7 March 1991 (UNEP/Bio.Div/WG.2/2/5, 1991).

1991 Report of the Intergovernmental Negotiating Committee for a Convention on Biological Diversity on the Work of its Third Session/ Fifth Negotiating Session, 4 December 1991 (UNEP/Bio.Div/N5-INC. 3/4, 1991).

COP decisions

1994 Decision I/7, 'Subsidiary Body on Scientific, Technical and Technological Advice', in Report of the First Ordinary Meeting of the Conference of the Parties to the Convention on Biological Diversity; 28 November–9 December 1994 – Nassau, Bahamas; UNEP/CBD/COP/1/17 (UNEP/CBD/COP/DEC/I/7).

1995 Decision II/8, 'Preliminary Consideration of Components of Biological Diversity Particularly Under Threat and Action Which Could Be Taken under the Convention', in the Report of the Second Meeting of the Conference of the Parties to the Convention of Biological Diversity at its Second meeting, Jakarta, 6–17 November 1995, UNEP/CBD/COP/2/19 (UNEP/CBD/COP/DEC/II/8).

1995 Decision II/10, 'Conservation and Sustainable Use of Marine and Coastal Biological Diversity', 6–17 November 1995, Jakarta, Indonesia, UNEP/CBD/COP/2/19 (UNEP/CBD/COP/DEC/II/10).

1995 Decision II/16, 'Statement to the international technical conference on the conservation and utilization of plant genetic resources for food and agriculture' in Report of the Second Ordinary Meeting of the Conference of the Parties to the Convention on Biological Diversity, 6–17 November 1995, Jakarta, Indonesia, UNEP/CBD/COP/2/19 (UNEP/CBD/COP/DEC/II/16).

1996 Decision III/3, 'Use of languages in the meetings of the Subsidiary Body on Scientific, Technical and Technological Advice', in Report of the Third Ordinary Meeting of the Conference of the Parties to the Convention on Biological Diversity, 4 – 15 November 1996 - Buenos Aires, Argentina, UNEP/CBD/COP/3/3 (UNEP/CBD/COP/3/III/3).

1996 Decision III/11, 'Conservation and sustainable use of agricultural biological diversity', in the Third Ordinary Meeting of the Conference of the Parties to the Convention on Biological Diversity, 4 – 15 November 1996, Buenos Aires, Argentina, UNEP/CBD/COP/3/38 (UNEP/CBD/COP/3/III/11).

1998 Decision IV/1, 'Report and recommendations', of the third meeting of the Subsidiary Body on Scientific, Technical and Technological Advice, and instructions by the Conference of the Parties to the Subsidiary Body on Scientific, Technical and Technological Advice, in the Report of the Fourth Ordinary Meeting of the Conference of the Parties to the Convention on Biological Diversity, 4–15 May 1998 – Bratislava, Slovakia, UNEP/CBD/COP/4/27 (UNEP/CBD/COP/IV/1).

1998 Decision IV/4, 'Status and trends of the biological diversity of inland water ecosystems and options for conservation and sustainable use' in Report of the Fourth Ordinary Meeting of the Conference of the Parties to the Convention on Biological Diversity; 4 – 15 May 1998 – Bratislava, Slovakia, UNEP/CBD/COP/4/27 (UNEP/CBD/COP/IV/4).

1998 Decision IV/5, 'Conservation and sustainable use of marine and coastal biological diversity, including a programme of work' in Report of the Fourth Meeting of the Conference of the Parties to the Convention on Biological Diversity; 4–15 May 1998 – Bratislava, Slovakia, UNEP/CBD/COP/4/27 (UNEP/CBD/COP/IV/5).

2000 Decision V/6, 'Ecosystem Approach', in the Report of the Fifth Meeting of the Conference of the Parties to the Conference of the Parties to the Convention of Biological Diversity, 15–26 May 2000, Nairobi, UNEP/COP/5/23 (UNEP/CBD/COP/DEC/V/6).

2002 Decision VI/8, 'Global Taxonomic Initiative', adopted by the Conference of the Parties to the Convention on Biological Diversity at its Sixth Meeting, The Hague, 7 – 19 April 2002, UNEP/CBD/COP/6/20 (UNEP/CBD/COP/6/VI/8).

2002 Decision VI/12, 'Ecosystem approach', adopted at the Sixth Ordinary Meeting of the Conference of the Parties to the Convention on Biological Diversity, 7–19 April 2002, The Hague, Netherlands, UNEP/CBD/COP/D (UNEP/CBD/COP/6/VI/12).

2002 Decision VI/26, 'Strategic Plan for the Convention on Biological Diversity', in the Report of the Sixth Ordinary Meeting of the Conference of the Parties to the Convention on Biological Diversity, 7–19 April 2002 – The Hague, Netherlands, UN Doc. UNEP/CBD/COP/6/20 (UNEP/CBD/COP/DEC/VI/26).

2003 Recommendation IX/6, 'Ecosystem approach: further elaboration, guidelines for implementation and relationship with sustainable forest management' SBSTTA 9, Montreal, 10 – 14 November 2003, Ninth Meeting of the Subsidiary Body on Scientific, Technical and Technological Advice, 10–14 November 2003 – Montreal, Canada

2004 Decision VII/11, 'Ecosystem Approach', in the Report of the Seventh Ordinary Meeting of the Conference of the Parties to the Convention on Biological Diversity, 9–20 February 2004 – Kuala Lumpur, Malaysia; UNEP/CBD/COP/7/21/PART1 (UNEP/CBD/COP/DEC/VII/11).

2004 Decision VII/12, 'Sustainable Use (Article 10) ', in the Report of the Seventh Ordinary Meeting of the Conference of the Parties to the Convention on Biological Diversity, 9–20 February 2004 – Kuala Lumpur, Malaysia; UNEP/CBD/COP/7/21/PART1 (UNEP/CBD/COP/DEC/VII/12).

2004 Decision VII/28, 'Protected areas (Articles 8 (a) to (e))', in the Report of the Seventh Ordinary Meeting of the Conference of the Parties to the Convention on Biological Diversity, 9–20 February 2004 – Kuala Lumpur, Malaysia; UNEP/CBD/COP/7/21/PART1 (UNEP/CBD/COP/DEC/VII/28).

2006 Decision VIII/9, 'Implications of the findings of the Millennium Ecosystem Assessment', in the Report of the Eighth Ordinary Meeting of the Conference of the Parties to the Convention on Biological Diversity, 20–-31 March 2006 – Curitiba, Brazil, UNEP/CBD/COP/8/31 (UNEP/CBD/COP/DEC/VIII/9).

2006 Decision VIII/10, 'Operations of the Convention', in the Report of the Eighth Ordinary Meeting of the Conference of the Parties to the Convention on Biological Diversity, 20–31 March 2006 – Curitiba, Brazil, UNEP/CBD/COP/8/31, (UNEP/CBD/COP/DEC/VIII/10).

2006 Decision VIII/24 on 'Protected Areas', Eighth Ordinary Meeting of the Conference of the Parties to the Convention on Biological Diversity, 20–31 March 2006, Curitiba, Brazil, UNEP/CBD/COP/8/31 (UNEP/CBD/COP/DEC/VIII/24).

2010 Decision X/2, 'Strategic Plan for Biodiversity 2011 – 2020 and the Aichi Biodiversity Targets', in the Report of the Tenth meeting of the Conference of the Parties to the Convention on Biological Diversity, 18–29 October 2010 – Nagoya, Aichi Prefecture, Japan, UNEP/CBD/COP/10/27 (UNEP/CBD/COP/X/2).

2010 Decision X/4, 'The Third Edition of the Global Biodiversity Outlook: Implications for the Future Implementation of the Convention', in the Report of the Tenth meeting of the Conference of the Parties to the Convention on Biological Diversity, 18–29 October 2010 – Nagoya, Aichi Prefecture, Japan, UNEP/CBD/COP/10/27 (UNEP/CBD/COP/X/4).

2010 Decision X/39, 'Global Taxonomy Initiative', adopted by the Conference of the Parties to the Convention on Biological Diversity at its Tenth Meeting, Nagoya, Japan, 18-29 October 2010, UNEP/CBD/COP/DEC/X/39 (UNEP/CBD/COP/X/39)

2012 Decision XI/2, 'Review of progress in implementation of national biodiversity strategies and action plans and related capacity building support to Parties', in the Report of the Eleventh meeting of the Conference of the Parties to the Convention on Biological Diversity, 8–19 October 2012 – Hyderabad, India, UNEP/CBD/COP/11/35 (UNEP/CBD/COP/DEC/XI/2).

2014 Decision XII/25, 'Intergovernmental Science-Policy Platform on Biodiversity and Ecosystem Services', in the Report of the Twelfth meeting of the Conference of the Parties to the Convention on Biological Diversity, 6–-17 October 2014 – Pyeongchang, Republic of Korea, UNEP/CBD/COP/12/29 (UNEP/CBD/COP/DEC/XII/25).

Technical documentation (reports, recommendations of technical bodies etc.)

1995 Recommendation I/3 on 'Alternative ways and means in which the Conference of the Parties could start the process of considering the components of biological diversity particularly those under threat and the identification of action which could be taken under the Convention', in the Report of the of the First Meeting of the Subsidiary Body on Scientific, Technical and Technological Advice, 4–8 September 1995 – Paris, France, (UNEP/CBD/COP/2/5).

1995 Recommendation I/8 on 'Scientific, technical and technological aspects of the conservation and sustainable use of coastal and marine biological diversity', in the Report of the First Meeting of the Subsidiary Body on Scientific, Technical and Technological Advice, 4–8 September 1995 – Paris, France, (UNEP/CBD/COP/2/5).

1996 Recommendation II/2 on 'Agenda Item 3.4: Practical approaches for capacity-building for taxonomy', in the report of the Second Meeting of the Subsidiary Body on Scientific, Technical and Technological Advice 2–6 September 1996 – Montreal, Canada (UNEP/CBD/COP/3/3).

1998 Report of the Workshop on the Ecosystem Approach, Lilongwe, Malawi, 26–28 January 1998 (UNEP/CBD/COP/4/Inf.9, 1998).

2000 Malawi Principles, in Recommendation V/10 on 'Ecosystem approach: further conceptual elaboration', in the Report of the Fifth Meeting of the

Subsidiary Body on Scientific, Technical and Technological Advice Montreal, 31 January – 4 February 2000, Canada (UNEP/CBD/COP/5/3).

2003 Expert Meeting on The Ecosystem Approach, *Review of the Principles of the Ecosystem Approach and Suggestions for Refinement: A Framework for Discussion*, Montreal 7–11 July 2003 (UNEP/CBD/EM-EA/1/3, 2003).

2003 Recommendation IX/6 on 'Ecosystem approach: further elaboration, guidelines for implementation and relationship with sustainable forest management' in the Report of the Ninth Meeting of the Subsidiary Body on Scientific, Technical and Technological Advice, 10–14 November 2003, Montreal, Canada, (UNEP/CBD/SBSTTA/9/INF/4).

2014 Summary Report on Progress in Implementing the Work Programme of the Intergovernmental Science-Policy Platform on Biodiversity and Ecosystem Services, Twelfth Meeting, 6–17 October 2014, Pyeongchang, Republic of Korea (UNEP/CBD/COP/12/INF/28).

Other CBD documents

1996 'Modus Operandi of the Subsidiary Body on Scientific, Technical and Technological Advice' (2-6 September 1996) (SBSTTA 2nd Meeting) UN Doc UNEP/CBD/SBSTTA/2/16', pp. 1 – 2.

2003 Review of the Principles of the Ecosystem Approach and Suggestions for Refinement: A Framework for Discussion, Note by the Executive Secretary, Expert Meeting on the Ecosystem Approach, Montreal 7–11 July 2003 (UNEP/CBD/EM-EA/1/3).

2004 Secretariat of the CBD, *Addis Ababa Principles and Guidelines for the Sustainable Use of Biodiversity* (CBD Guidelines), Montreal: Secretariat of the Convention on Biological Diversity, 2004.

2004 Secretariat of the CBD, *The Ecosystem Approach*, HTTP: https://www.cbd.int/doc/publications/ea-text-en.pdf (Accessed 26 September 2018).

2007 Potsdam Initiative – Biological Diversity 2010, adopted at Environment Ministers Meeting, Potsdam, 15–17 March 2007, HTTP: http://www.g-8.de/Content/EN/__Anlagen/2007-03-18-potsdamer-erklaerung-en,property=publicationFile.pdf (Accessed 26 September 2018).

2007 Message from the Executive Secretary, Ahmed Djoghlaf, on the Occasion of the International Day of the World's Indigenous People, 2007.

2012 Statement by Mr. Braulio, F. De Souza Dias, Executive Secretary Convention on Biological Diversity to the International Seminar 'Towards Linking Ecosystems and Ecosystem Services to Economic and Human Activity', New York, United States of America 27 – 29 November 2012.

2010 Secretariat of the Convention on Biological Diversity, *Global Biodiversity Outlook 3*, Montréal: Secretariat of the Convention on Biological Diversity, 2010.

2014 Secretariat of the Convention on Biological Diversity, *Global Biodiversity Outlook 4. A Mid-term Assessment of Progress towards the Implementation of the Strategic Plan for Biodiversity 2011 – 2020*, Montréal: Secretariat of the Convention on Biological Diversity, 2014.

Convention on Biological Diversity, *Global Taxonomy Initiative*, HTTP: https://www.cbd.int/gti/default.shtml (Accessed 26 September 2018).

United Nations documents (General Assembly Resolutions, Declarations etc.)

1973 Declaration of the United Nations Conference on the Human Environment, UN Doc. A/Conf.48/14/Rev. 1.

1987 WCED, *Our Common Future: Report of the World Commission on Environment and Development*, 4 August 1987 (A/42/427, 1987).

1992 Agenda 21, A/CONF.151/26 (Vol. I and Vol. II).

1992 Rio Declaration on Environment and Development, U.N. Doc. A/Conf. 151/26 (Vol. I).

2002 Plan of Implementation of the World Summit on Sustainable Development, 4 September 2002 (A/CONF.199/20).

2002 Report of the World Summit on Sustainable Development Johannesburg, South Africa, 26 August–4 September 2002 (UN Doc. A/CONF.199/20).

2004 Oceans and the law of the sea, Report of the Secretary-General, 18 August 2004, Addendum (UN Doc A/59/62/Add.1).

2006 Report on the work of the United Nations Open-ended Informal Consultative Process on Oceans and the Law of the Sea at its seventh meeting, New York, 12–16 June 2006 (UNDOC/A/61/156, 2006).

2008 UNGA, Sustainable fisheries, including through the 1995 Agreement for the Implementation of the Provisions of the United Nations Convention on the Law of the Sea of 10 December 1982 relating to the Conservation and Management of Straddling Fish Stocks and Highly Migratory Fish Stocks, and related instruments, 28 February 2008, UN Doc. A/RES/ 62/177.

2010 UNGA, 'Implementation of Agenda 21, the Programme for the Further Implementation of Agenda 21 and the outcomes of the World Summit on Sustainable Development', 31 March 2010, UN Doc. A/ RES/64/236.

2010 UNGA, 'Progress to Date and Remaining Gaps in the Implementation of the Outcomes of the Major Summits in the Area of Sustainable Development, as well as an Analysis of the Themes of the Conference Report of the Secretary-General', 1 April 2010, UN Doc. A/CONF.216/PC/2.

2012 Report of the United Nations Conference on Sustainable Development Rio de Janeiro, Brazil 20–22 June 2012 (UN Doc. A/CONF.216/16).

2012 UNGA, 'The Future We Want', 27 July 2012, UN Doc. A/RES/66/288.

2013 UNGA, 'Sustainable fisheries, including through the 1995 Agreement for the Implementation of the Provisions of the United Nations Convention on the Law of the Sea of 10 December 1982 relating to the Conservation and Management of Straddling Fish Stocks and Highly Migratory Fish Stocks, and related instruments', 30 April 2013, UN Doc. A/RES/67/79.

2014 UNGA, 'Sustainable fisheries, including through the 1995 Agreement for the Implementation of the Provisions of the United Nations Convention on the Law of the Sea of 10 December 1982 relating to the Conservation and Management of Straddling Fish Stocks and Highly Migratory Fish Stocks, and related instruments', 25 February 2014, UN Doc. A/RES/ 68/71.

Documents of International Governmental Organizations and other Intergovernmental Arrangements (resolutions, recommendations, as well as reports and technical guidelines etc.)

Arctic Council

2013 Arctic Council, *Ecosystem-Based Management in the Arctic Report submitted to Senior Arctic Officials by the Expert Group on Ecosystem-Based Management* (J05-ARCTIC-007, 2013).

2014 The Protection of the Arctic Marine Environment Working Group (PAME), *The Ecosystem Approach to Management*, Concept Paper, 2014.

Food and Agricultural Organization (FAO)

2001 Statement by D. Jacques, FAO Director-General, in *Report of the Reykjavik Conference On Responsible Fisheries in The Marine Ecosystem*, Reykjavik, Iceland, 1–4 October 2001, FAO Fisheries Report No. 658.

2003 FAO, *Fisheries Management. The Ecosystem Approach to Fisheries.* FAO Technical Guidelines for Responsible Fisheries. (No 4, Suppl. 2.).

2008 FAO Fisheries and Aquaculture Report No. 890 (FIEL/R890 (Bi)).

2011 FAO, *State of the World's Forests*, Rome: FAO.

2014 FAO, *The State of World Fisheries and Aquaculture. Opportunities and Challenges*, Rome: FAO.

International Council for the Exploration of the Seas (ICES)

2005 ICES, *Guidance on the Application of the Ecosystem Approach to Management of Human Activities in the European Marine Environment*, ICES Cooperative Research Report 273, 2005.

2014 ICES, Action Areas: Integrated Ecosystem Assessments; HTTP: http://ices.dk/explore-us/Action%20Areas/Pages/Integrated-ecosystem-assessments.aspx (Accessed 26 September 2018).

International Law Commission

1994 ILC, Draft articles on the law of the non-navigational uses of international watercourses and commentaries thereto and resolution on transboundary confined groundwater, in Yearbook of the International Law Commission, 1994, vol. II, Part Two.

1994 ILC, Report of the International Law Commission on the Work of its Forty-Ninth Session, UN Doc. GAOR A/49/10/1994 (1994).

IUCN

1972 IUCN, *Polar Bears: Proceedings of the Third Working Meeting of the Polar Bear Specialist Group*, organized by the Survival Service Commission of IUCN, 7–10 February 1972, Morges, Switzerland, IUCN publications new series, Supplementary Paper 35, Gland: Switzerland, 1972.

1980 IUCN, *World Conservation Strategy: Living Resource Conservation for Sustainable Development*, 1980.

1999 Minister of International Development and Human Rights Hilde F. Johnson, Norway/UN Conference on the Ecosystem Approach for Sustainable Use of Biological Diversity Trondheim, 6–10 September 1999, HTTP: https://www.regjeringen.no/nb/aktuelt/norway-un_conference_on_the_ecosystem/id263361/(Accessed 26 September 2018).

2012 IUCN, Statutes, Including Rules of Procedure of the World Conservation Congress, and Regulations, International Union for Conservation of Nature and Natural Resources, Gland: Switzerland, 2012.

United Nations Environment Programme

2003 UNEP, *Ecosystems and Human Well-being, A Framework for Assessment*, Washington, DC: Island Press.

2005 UNEP, *Ecosystems and Human Well-being, Current State and Trends*, Volume 1, Washington, DC: Island Press.

2010 UNEP, Busan Outcome in Report of the Third ad hoc intergovernmental and multi-stakeholder meeting on an intergovernmental science-policy platform on biodiversity and ecosystem services, 11 June 2010, Annex, UNEP/IPBES/3/3.

2010 UNEP, Six Priority Areas Factsheets - Ecosystem management.

2011 Report of the Governing Council of the United Nations Environment Programme on its Eleventh special session, UNGA Resolution 65/162, 11 March 2011, (A/RES/65/162).

2011a UNEP, *Securing a Green Economy through Ecosystem Management*, International Ecosystem Management Partnership, UNEP-IEMP.

2011b UNEP, *Restoring the natural foundation to sustain a Green Economy. A century-long journey for Ecosystem Management*, International ecosystem management programme (IEMP) UNEP-IEMP, Policy Series paper, 2011a, available at HTTP http://mahenvis.nic.in/Pdf/Report/report_eco_Restoring%20the%20natural%20foundation.pdf (Accessed 26 September 2018).

2011c UNEP, *Towards a Green Economy. Pathways to Sustainable Development and Poverty Eradication, A Synthesis for Policy Makers.*

2011d UNEP Decision 26/4, 'Intergovernmental science-policy platform on biodiversity and ecosystem services', adopted by the United Nations Environment Programme Governing Council/Global Ministerial Environment Forum at its twenty-fifth session – Nairobi, Kenya, 2011, UNEP/GC.26/4.

2011e UNEP, *Taking Steps Toward Marine and Coastal Ecosystem-Based Management. An Introductory Guide.*

2012 UNEP, *The Transition to a Green Economy: Benefits, Challenges and Risks from a Sustainable Development Perspective*, Report by a Panel of Experts to Second Preparatory Committee Meeting for United Nations Conference on Sustainable Development, 2012b (Prepared under the direction of: Division for Sustainable Development, UN-DESA) available at HTTP: https://plagiarism.repec.org/trica-papuc/trica-papuc2.pdf (Accessed 26 September 2018).

2012 UNEP, *Global Environment Outlook 5: Environment for the Future We Want*, UN Doc. DEW/1417/NA, 2012a.

Ramsar Convention

2005 Ramsar, COP9, Resolution IX/1, Annex A, 'A Conceptual Framework for the wise use of wetlands and the maintenance of their ecological character', adopted at 9th Meeting of the Conference of the Parties to the Convention on Wetlands, Kampala, Uganda, 8 – 15 November 2005.

2011 Ramsar Secretariat, *Wise Use of Wetlands: Concepts and Approaches for the Wise Use of Wetlands*, Ramsar handbooks for the wise use of wetlands, 4th edition Ramsar Convention Secretariat, 2010.

2011 Ramsar Convention Secretariat, *Ramsar's Liquid Assets. 40 years of the Convention on Wetlands*, Gland, Switzerland: Ramsar Secretariat, 2011.

Others

1970 UNESCO, *Use and Conservation of the Biosphere*, Proceedings of the Intergovernmental Conference of Experts on the Scientific Basis for Rational Use and Conservation of the Resources of the Biosphere Paris, 4–13 September 1968.

1978 Great Lakes Research Advisory Board, *The Ecosystem Approach. Scope and Implications of an Ecosystem Approach to Transboundary Problems in the Great Lakes Basin*, Special Report to the International Joint Commission, Windsor: Ontario, Canada, July 1978.

1992 OSPAR Convention for the Protection of the Marine Environment of the North-East Atlantic), adopted 22 September 1992, entered into force 25 March 1998, 2354 UNTS 67.

1993 UNECE, *Guidelines on the Ecosystem Approach in Water Management*, 1993.

1994 CITES Resolution Conf. 9.24 (Rev. CoP16), 'Criteria for amendment of Appendices I and II', Ninth meeting of the Conference of the Parties Fort Lauderdale (United States of America), 7–18 November 1994.

2002 CMS, Cooperation with Other Bodies CBD/CMS, Joint Work Programme (2002–2005), Seventh Meeting of the Conference of the Parties Bonn, 18–24 September 2002 (UNEP/CMS/Inf.7, 2002).

2003 OSPAR Convention, BDC, Dublin, 20 – 24 January 2003, Summary Record, Annex 13.

2009 OECD, The Bioeconomy to 2030: Designing a Policy Agenda, Findings and Policy Conclusions, 2009.

2010 OSPAR Commission, *Quality Status Report*, 2010

2010 OSPAR Commission, *The North-East Atlantic Environment Strategy of the OSPAR Commission for the Protection of the Marine Environment of the North-East Atlantic 2010–2020*, OSPAR Agreement 2010 – 2013.

2003 HELCOM/OSPAR Statement on the Ecosystem Approach to the Management of Human Activities 'Towards an Ecosystem Approach to the Management of Human Activities', First Joint Ministerial Meeting of the Helsinki and OSPAR Commissions (JMM) (Bremen: 25–26 June 2003, Agenda item 6) JMM 2003/4-Rev.1-E.

EU legislation and other legal documents

2000 Communication from the Commission on the Precautionary Principle, Brussels, 2.2.2000, COM (2000) 1 final

2008 Directive 2008/56/EC of the European Parliament and of the Council of 17 June 2008 establishing a framework for community action in the field of marine environmental policy (OJ L 164/19 25 June 2008).

2008 European Commission Communication. *The role of the CFP in implementing an ecosystem approach to marine management.* COM (2008)187, Brussels 11 April 2008, p. 3.

2010 Commission Decision 2010/477/EU of 1 September 2010 on criteria and methodological standards on good environmental status of marine waters (OJ L 232/16 2 September 2010).

2011 European Parliament and Council Regulation 1255/2011 of 30 November 2011 establishing a Programme to support the further development of an Integrated Maritime Policy, (OJ L321/1, 5/12/2011).

2013 European Parliament and Council Regulation (EU) 1380/2013 of 11 December 2013 on the Common Fisheries Policy [2013] OJ L354/22.

Domestic legislation

1948 Federal Water Pollution Control Act (FWPCA) (P.L. 80-845, 62 Stat. 1155).

1972 Clean Water Act, 33 U.S.C. §1251 et seq.

2007 Commissione Rodotà – per la modifica delle norme del codice civile in materia di beni pubblici (14 giugno 2007) HTTP: https://www.giustizia.it/giustizia/it/mg_1_12_1.wp?facetNode_1=0_10&facetNode_2=0_10_21&previsiousPage=mg_1_12&contentId=SPS47617 (Accessed 26 September 2018).

2010 Rodotà Commission Bill, Delegated Legislation to Reform the Civil Code Articles Concerning Public Property, Act of the Senate of the Republic n. 2031, XVI Legislature.

Other documents

Intergovernmental Panel on Climate Change

2007 IPCC, 'Summary for Policymakers' in S. Solomon, D. Qin, M. Manning, Z. Chen, M. Marquis, K. B. Averyt, M. Tignor and H. L. Miller (eds) *Climate Change 2007: The Physical Science Basis, Contribution of Working Group I to the Fourth Assessment Report of the Intergovernmental Panel on Climate Change*, Cambridge, United Kingdom and New York: Cambridge University Press, 2007.

2013 IPCC, 'Summary for Policymakers' in T. F. Stocker, D. Qin, G.-K. Plattner, M. Tignor, S. K. Allen, J. Boschung, A. Nauels, Y. Xia, V. Bex and P. M. Midgley (eds) *Climate Change 2013: The Physical Science Basis. Contribution of Working Group I to the Fifth Assessment Report of the*

Intergovernmental Panel on Climate Change, Cambridge, United Kingdom and New York: Cambridge University Press, 2013.

2014 IPCC, 'Summary for policymakers' in C. B. Field, V. R. Barros, D. J. Dokken, K. J. Mach, M. D. Mastrandrea, T. E. Bilir, M. Chatterjee, K. L. Ebi, Y. O. Estrada, R. C. Genova, B. Girma, E. S. Kissel, A. N. Levy, S. MacCracken, P. R. Mastrandrea and L. L.White (eds) *Climate Change 2014: Impacts, Adaptation and Vulnerability. Part A: Global and Sectoral Aspects. Contribution of Working Group II to the Fifth Assessment Report of the Intergovernmental Panel on Climate Change*, Cambridge, United Kingdom and New York: Cambridge University Press, 2014.

Intergovernmental Platform on Biodiversity and Ecosystem Services (IPBES)

2012 IPBES, Resolution 'Intergovernmental Science-Policy Platform on Biodiversity and Ecosystem Services' in Report of the second session of the plenary meeting to determine modalities and institutional arrangements for an intergovernmental science-policy platform on biodiversity and ecosystem services, 18 May 2012, UNEP/IPBES.MI/2/9.

2012 IPBES, Functions, operating principles and institutional arrangements of the Intergovernmental Science-Policy Platform on Biodiversity and Ecosystem Services, adopted by the second session of the plenary meeting to determine the modalities and institutional arrangements for IPBES, held from 16–21 April 2012 in Panama City, Panama.

2013 IPBES, Decision IPBES-2/5: Work programme for the period 2014–2018, adopted at its second meeting of the Platform's Plenary (IPBES-2) held in Antalya, Turkey, from 9 to 14 December 2013.

The Economics of Ecosystems and Biodiversity

2008 TEEB, The Economics of Ecosystems and Biodiversity: An Interim Report, European Communities, 2008.

2010 TEEB, The Economics of Ecosystems and Biodiversity Ecological and Economic Foundations, P. Kumar (ed.) London and Washington: Earthscan, 2010.

2015 The Economics of Ecosystems & Biodiversity, available at HTTP http://www.teebweb.org (Accessed 26 September 2018).

Other miscellaneous documents

World Ocean Review, *The Future of Fish – The Fisheries of the Future*, Maribus, 2013.

Novogratz, A. and Velings, M. 'The end of fish', *Washington Post*, 3 June 2014, available at HTTP: http://www.washingtonpost.com/posteverything/wp/2014/06/03/the-end-of-fish/(Accessed 26 September 2018).

Call for Papers of the Conference 'The Changing Nature of International Environmental Law: Evolving Approaches of the United States and the European Union', held in Geneva on 22 – 23 November2013, and organized jointly by the European and American Societies of International Law's International Environmental Law

Interest Groups, HTTP: http://asilesilchangingnature.wordpress.com/conference-programme/(Accessed 26 September 2018).

Office of Marine Programs (University of Rhode Island), Census of Marine Life – About the Census, 2010, HTTP: http://www.coml.org/about-census (Accessed 26 September 2018).

UN News Center, Doha meeting must take decisive action to tackle growing crisis of climate change, 4 December 2012, HTTP: http://www.un.org/apps/news/story.asp?NewsID=43671&C#.UMDNsoN2PwA (Accessed 26 September 2018).

Index

For Product Safety Concerns and Information please contact our EU representative GPSR@taylorandfrancis.com
Taylor & Francis Verlag GmbH, Kaufingerstraße 24, 80331 München, Germany

www.ingramcontent.com/pod-product-compliance
Lightning Source LLC
LaVergne TN
LVHW010630300926
849162LV00001B/107

* 9 7 8 1 1 3 8 5 5 7 2 6 0 *